THE
LIVING
CELL

EIGHT

THE LIVING CELL

Readings from **SCIENTIFIC AMERICAN**

With Introductions by

DONALD KENNEDY
STANFORD UNIVERSITY

W. H. FREEMAN AND COMPANY
SAN FRANCISCO AND LONDON

COPY 1

Each of the SCIENTIFIC AMERICAN articles in "The Living Cell" is available as a separate Offprint at twenty cents each. For a complete listing of approximately 500 articles now available as Offprints, write to W. H. Freeman and Company, 660 Market Street, San Francisco, California 94104.

Copyright, © 1958, 1960, 1961, 1962, 1963, 1964, 1965,
by SCIENTIFIC AMERICAN, INC.

The publisher reserves all rights to reproduce this book in whole or in part, with the exception of the right to use short quotations for review of the book.

Printed in the United States of America. (B4)

Library of Congress Catalogue Card Number: 65-21160

Preface

When *Scientific American* began, almost two decades ago, to publish articles by leading scientific specialists for the intelligent layman, teachers quickly recognized that this target audience nicely encompassed their own capable but not-yet-specialized students. (Many teachers have, in fact, found a less general but equally effective use for the magazine in the preparation of course lectures on unfamiliar subjects!) The magazine's wide use as supplementary reading material led eventually to the reprinting of certain articles, and these reprints have in the past few years become a staple in the academic diet—not only in biology, but in the physical sciences, social sciences, and psychology as well. Indeed, the ready availability of *Scientific American* Offprints and other reprints and original paperback books has led to a real revolution in the use of textual materials in teaching: a revolution based on an increased reliance upon the separate contributions of "expert testimony."

Since *Scientific American* articles first became available as offprints, more than 12,000,000 of them have been used in specific courses in colleges and universities. As one of a number of teachers who have experimented with them in various ways, I have appreciated the flexibility that allows one to combine different sets of offprints to fit various course requirements. I have also found, however, that certain combinations covering basic and well-defined segments of courses are persistently effective.

This volume, which is intended for courses that deal, at least in part, with cell biology, "packages" one such combination. The twenty-four articles range in publication date from 1958 to 1964; several (including those that begin Sections I, III, IV, V and VI) are from the special *Scientific American* issue of September, 1961, "The Living Cell." The articles from this issue are of special value to the student, because they are deliberately designed to cover a broader reach of material than the usual treatment of a specific research topic. But each section in this book also contains articles with the second, more specialized approach. Hopefully, the blend of the two will provide a useful overview of cell biology plus an authoritative and recent coverage of special areas of interest.

Cross-references within the articles are of three kinds. A reference to an article included in this book consists of the title followed by the number of the page on which it begins; to an article not included, but available as an offprint, of the title and offprint number; to an article not available as an offprint, of the title and the month and year of its publication in *Scientific American*.

DONALD KENNEDY

Contents

I. LEVELS OF COMPLEXITY

Introduction · 2
1. BRACHET The Living Cell · 5
2. JACOB AND WOLLMAN Viruses and Genes · 17
3. MOROWITZ AND TOURTELLOTTE The Smallest Living Cells · 31

II. ORGANELLES

Introduction · 42
4. ROBERTSON The Membrane of the Living Cell · 45
5. SATIR Cilia · 53
6. GREEN The Mitochondrion · 63
7. DE DUVE The Lysosome · 72

III. ENERGETICS

Introduction · 82
8. LEHNINGER How Cells Transform Energy · 85
9. ARNON The Role of Light in Photosynthesis · 97
10. BASSHAM The Path of Carbon in Photosynthesis · 109
11. MCELROY AND SELIGER Biological Luminescence · 122

IV. SYNTHESIS

Introduction · 136
12. ALLFREY AND MIRSKY How Cells Make Molecules · 138
13. NIRENBERG The Genetic Code: II · 148
14. RICH Polyribosomes · 160
15. KENDREW The Three-dimensional Structure of a Protein Molecule · 170
16. GROSS Collagen · 187

V. DIVISION AND DIFFERENTIATION

Introduction · 196
17. MAZIA How Cells Divide · 199
18. MOSCONA How Cells Associate · 213
19. FISCHBERG AND BLACKLER How Cells Specialize · 222
20. BEERMANN AND CLEVER Chromosome Puffs · 232

VI. SPECIAL ACTIVITIES

Introduction · 242
21. KATZ How Cells Communicate · 245
22. MILLER, RATLIFF, AND HARTLINE How Cells Receive Stimuli · 257
23. HAYASHI How Cells Move · 268
24. HUXLEY The Contraction of Muscle · 279

Bibliographies and Biographies · 290

Part I

LEVELS OF COMPLEXITY

I Levels of Complexity

INTRODUCTION

One of the insights that has been most helpful to biologists is the so-called Cell Theory. It is difficult to say precisely what that phrase encompasses. In part, this is because the cellular nature of living systems was revealed not in a dramatic single resolution but rather through a series of clarifications that were developed throughout most of the mid-nineteenth century. The basic postulates of the Cell Theory are, first, that all organisms are composed of subunits resembling one another in the possession of a certain set of organelles and a boundary; second, that these entities arise only through the division of pre-existing cells. Certain organisms, observed a century and a half earlier by Leeuwenhoek and the other early lensmakers, are themselves unicellular; a significant new finding indicated that the complex structure of large multicellular plants and animals appeared to be composed of microscopic units resembling these solitary microbes. The equally remarkable corollary adds that the cell is the basic unit of reproduction; indeed, we know of no example (though it would not seem to be theoretically unreasonable) in which an element of less-than-cellular complexity serves as the reproductive unit in any organism that has fully cellular organization itself.

The establishment of the Cell Theory, though attended by far less ceremony than the nearly contemporaneous Darwinian Revolution, has made it possible for biologists to deal coherently with what would otherwise be a hopeless welter of diversity. The doctrine of evolution offered explanations for the enormous breadth of the living spectrum; the cell theory offered hopeful assurances that these variations, despite their extent, had a theme, that the theme was the cellular organization of living systems, and that one might hope to comprehend some of the basic whys of life without inspecting an infinite series of special cases. That expectation has been amply fulfilled; the articles that follow document the recent progress that has resulted from the discovery that cells everywhere show a remarkable similarity of mechanisms for obtaining and using energy and for replicating themselves and their parts.

This section deals with the general properties of cells and, to some extent, with the question of *levels of complexity* in cells. The first selection, Jean Brachet's "The Living Cell," was originally written as the introductory article for the September, 1961, issue of *Scientific American,* also titled "The Living Cell." It is a general account of the properties of cells, dealing in particular with the many new revelations made by electron microscopy and correlating these structural insights with the functional roles assigned by biochemical techniques. Brachet discusses those cells characteristic of multicellular organisms or of the higher protists like protozoa and algae—cells characterized by well-developed organelles and a discrete nucleus with visible chromosomes. Such *eucaryotic* cells thus display much more internal complexity and order than the *procaryotic* cells characteristic of the bacteria and blue-green algae. In the latter, defined nuclei with patent chromosomes are missing, and the functions normally associated with mitochondria appear to be associated with the membrane instead.

If such simplicity obscures a direct structural analysis, more inferential techniques are useful. Even though the linear structure of the bacterial chromosome cannot be seen, experiments that demonstrate it have been performed; these are described in "Viruses and Genes" by François Jacob and Elie Wollman. Such ingenious genetic manipulations not only contribute to our knowledge of the nature of the bacterial system; they also lead to speculation about the origin and nature of viruses. These particles, consisting of a protein coat and nucleic acid core, are the simplest living systems known—so simple, in fact, that many biologists would contest the adjective "living." They must intercept the metabolic mechanism of a cellular host in order to reproduce; this property arouses speculation about whether viruses are minimal, possible "primitive" organisms or, as Jacob and Wollman suggest, whether some may be nonintegrated strays from the genetic material of a cell.

Interest in the levels of cellular complexity and in the history of cellular organization prompts one to try to discover what constitutes minimum cellular equipment. The account of pleuropneumonia-like organisms, "The Smallest Living Cells" by Harold J. Morowitz and Mark E. Tourtellotte, illustrates an attempt to arrive at some answers by inspecting presumably minimal cells. The biochemical machinery of these cells suggests that they may indeed be approaching a theoretical lower size limit, about which the authors provide an illuminating discussion.

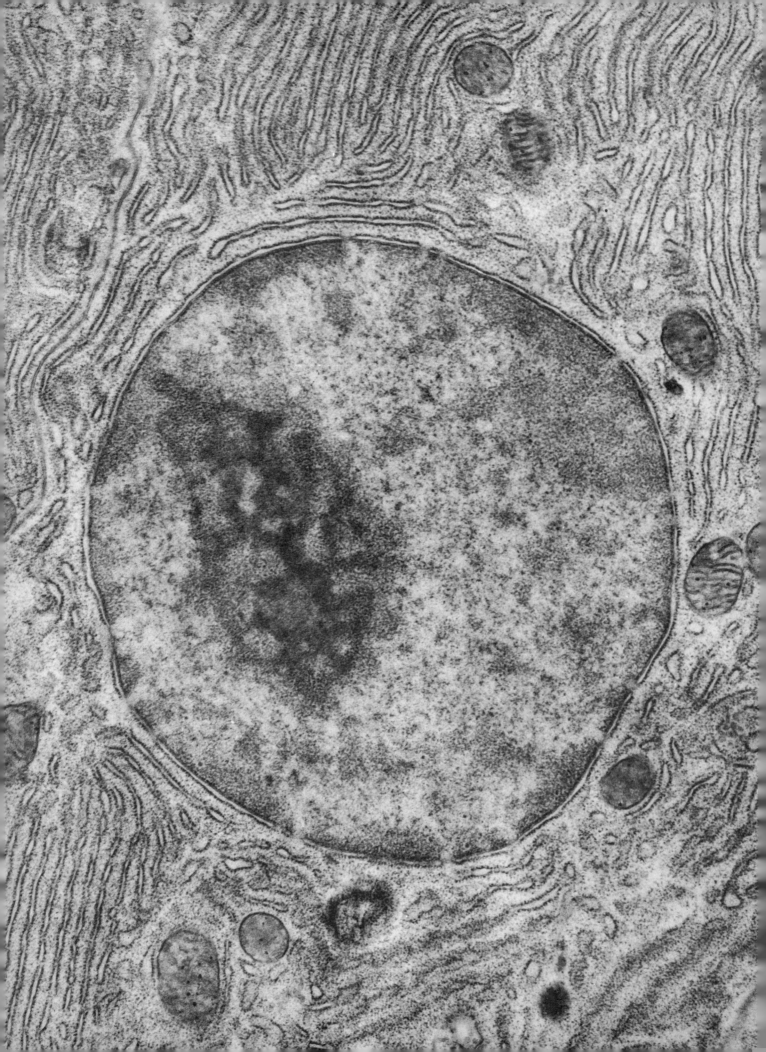

1

THE LIVING CELL

by JEAN BRACHET September 1961

The living cell is the fundamental unit of which all living organisms are made. To a reader who finds this a commonplace, it may come as a surprise that the recognition of the cell dates back only a little more than 100 years. The botanist Matthias Jakob Schleiden and the zoologist Theodor Schwann first propounded the cell theory in 1839 out of their parallel and independent studies of the tissues of plants and animals. Not long after, in 1859, Rudolf Virchow confirmed the cell's unique role as the vessel of "living matter" when he showed that all cells necessarily derive from pre-existing cells: *omnis cellula e cellula*. Since cells are concrete objects and can easily be observed, the experimental investigation of cells thereafter displaced philosophical speculations about the problem of "life" and the uncertain scientific studies that had pursued such vague concepts as "protoplasm."

In the century that followed investigators of the cell approached their subject from two fundamentally different directions. Cell biologists, equipped with increasingly powerful microscopes, proceeded to develop the microscopic and submicroscopic anatomy of the intact cell. Beginning with a picture of the cell as a structure composed of an external membrane, a jelly-like blob of material called cytoplasm and a central nucleus, they have shown that this structure is richly differentiated into organelles adapted to carry on the diverse processes of life. With the aid of the electron microscope they have begun to discern the molecular working parts of the system. Here, in recent years, their work has converged with that of the biochemists, whose studies begin with the ruthless disruption of the delicate structure of the cell. By observing the chemical activity of materials collected in this way, biochemists have traced some of the pathways by which the cell carries out the biochemical reactions that underlie the processes of life, including those responsible for manufacturing the substance of the cell itself.

It is the present intersection of the two lines of study that enables an attempt at a synthesis of what is known of the structure and function of the living cell. The cell biologist now seeks to explain in molecular terms what he sees with the aid of his instruments; he has become a molecular biologist. The biochemist has become a biochemical cytologist, interested equally in the structure of the cell and in the biochemical activity in which it is engaged. As the reader will see, the mysteries of cell structure and function cannot be resolved by the exercise of either morphological or biochemical techniques alone. If the research is to be successful, the approach must be made from both sides at once. But the understanding of life phenomena that flows from investigation of the cell has already fully ratified the judgment of the 19th-century biologists who perceived that living matter is divided into cells, just as molecules are made of atoms.

A description of the functional anatomy of the living cell must begin with the statement that there is no such thing as a typical cell. Single-celled organisms of many different kinds abound, and the cells of brain and muscle tissue are as different in morphology as they are in function. But for all their variety they are cells, and so they all have a cell membrane, a cytoplasm containing various

---

NUCLEUS OF THE LIVING CELL is the large round object in the center of the electron micrograph on the opposite page. The membrane around the nucleus is interrupted by pores through which the nucleus possibly communicates with the surrounding cytoplasm. The smaller round objects in the cytoplasm are mitochondria; the long, thin structures are the endoplasmic reticulum; the dark dots lining the reticulum are ribosomes. Actually the micrograph shows not a living cell but a dead cell: the cell has been fixed with a compound of the heavy metal osmium, immersed in a liquid plastic that is then made to solidify and finally sliced with a glass knife. The electron beam of the microscope mainly detects the atoms of osmium, distributed according to the affinity of the fixing compound for various cell constituents. The micrograph was made by Don W. Fawcett of the Harvard Medical School. The enlargement is 28,400 diameters. The cell itself is from the pancreas of a bat.

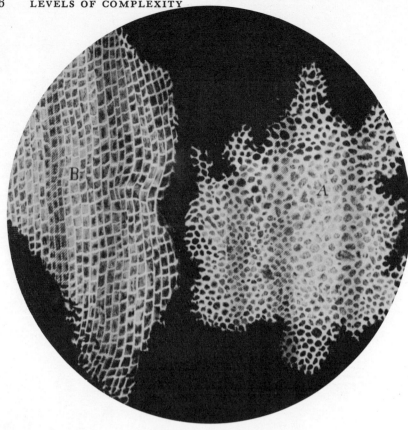

DRAWING OF CELLS in cork was published by Robert Hooke in 1665. Hooke called them cells, but the fact that all organisms are made of cells was not recognized until 19th century.

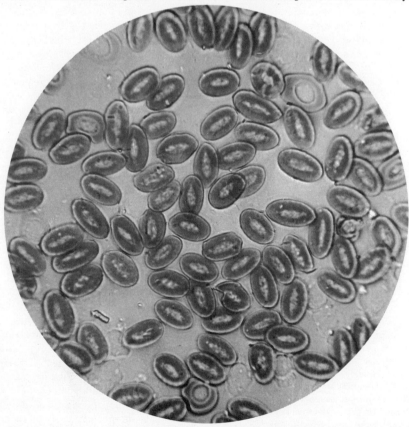

PHOTOMICROGRAPH OF CELLS in the blood of a pigeon was made by J. J. Woodward, a U.S. Army surgeon, in 1871. Woodward had made the first cell photomicrograph in 1866.

organelles and a central nucleus. In addition to having a definite structure, cells have a number of interesting functional capacities in common.

They are able, in the first place, to harness and transform energy, starting with the primary transformation by green-plant cells of the energy of sunlight into the energy of the chemical bond. Various specialized cells can convert chemical-bond energy into electrical and mechanical energy and even into visible light again. But the capacity to transform energy is essential in all cells for maintaining the constancy of their internal environment and the integrity of their structure [see "How Cells Transform Energy," page 85].

The interior of the cell is distinguished from the outer world by the presence of very large and highly complex molecules. In fact, whenever such molecules turn up in the nonliving environment, one can be sure they are the remnants of dead cells. On the primitive earth, life must have had its origin in the spontaneous synthesis of complicated macromolecules at the expense of smaller molecules. Under present-day conditions, the capacity to synthesize large molecules from simpler substances remains one of the supremely distinguishing capacities of cells.

Among these macromolecules are proteins. In addition to making up a major portion of the "solid" substance of cells, many proteins (enzymes) have catalytic properties; that is, they are capable of greatly accelerating the speed of chemical reactions inside the cell, particularly those involved in the transformation of energy. The synthesis of proteins from the simpler units of the 20-odd amino acids goes forward under the regulation of deoxyribonucleic acid (DNA) and ribonucleic acid (RNA), by far the most highly structured of all the macromolecules in the cell [see "How Cells Make Molecules," page 138]. In recent years and months investigators have shown that DNA, localized in the nucleus of the cell, presides at the synthesis of RNA, which is found in both the nucleus and the cytoplasm. The RNA in turn arranges the amino acids in proper sequence for linkage into protein chains. The DNA and the RNA may be compared to the architect and contractor who collaborate on the construction of a nice-looking house from a heap of bricks, stones and tiles.

At one or another stage of life every cell has divided: a mother cell has grown

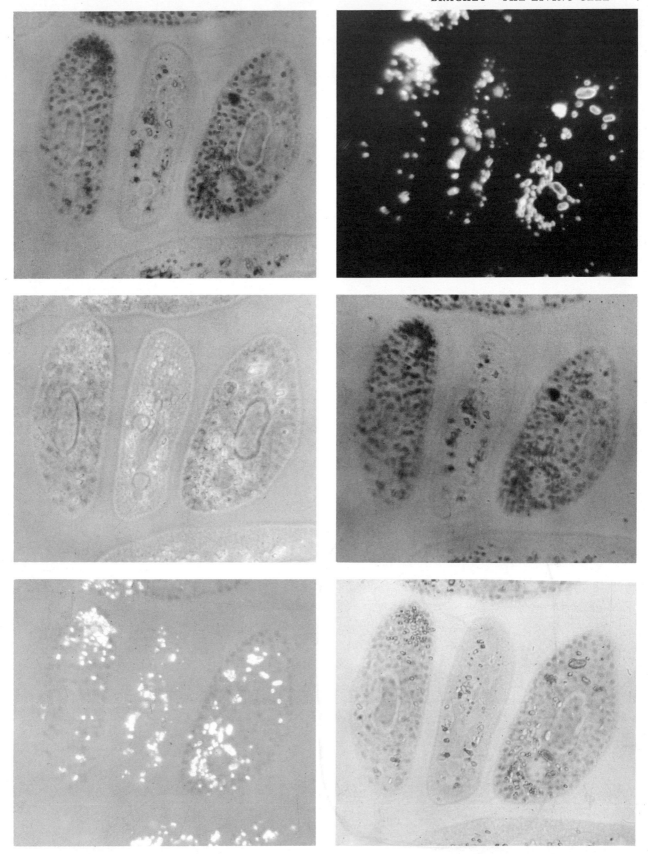

VARIOUS KINDS OF LIGHT MICROSCOPY are used to photograph the same three paramecia. The photomicrograph at top left was made with a conventional light microscope and bright-field illumination; the one at top right, with dark-field illumination. The photomicrograph at middle left was made with a phase microscope at low contrast; the one at middle right, with a phase microscope at high contrast. The photomicrograph at bottom left was made with a polarized-light microscope; the one at bottom right, with an interference microscope of the AO-Baker type. The bright spots that appear in some of the photomicrographs are small crystals that are normally present in paramecia. All the micrographs were made by Oscar W. Richards of the American Optical Company.

and given rise to two daughter cells, according to the delicate process described by Daniel Mazia [see "How Cells Divide," page 199]. Before the turn of the century biologists had observed that the crucial event in this process was the equal division of bodies in the nucleus that accepted a certain colored dye and so were called chromosomes. It was correctly surmised that the chromosomes are the agents of heredity; in their precise self-replication and division they convey to the daughter cells all the capacities of the mother cell. Contemporary biochemistry has now shown that the principal constituent of the chromosomes is DNA, and an important aim of the molecular biologist today is to discover how the genetic information is encoded in the structure of this macromolecule.

The capacity for generative reproduction is not confined exclusively to the living cell. There are in the present world macromolecules called viruses that contain nucleic acids and proteins of great complexity and specificity. When they penetrate into suitable cells, they multiply just as cells do, but at the expense of the cell. They have a heredity, since they breed true when they replicate themselves, and they synthesize their own proteins. But, lacking the full anatomical endowment of the cell, they are unable to generate the energy required for their multiplication. Viruses are thus obligatory parasites of cells and take over the enzyme system of the infected cell in order to supply the energy they need. The cell must, however, furnish exactly the right complement of enzymes. This is why tobacco mosaic virus, for example, will not multiply in human cells and so is harmless to human beings.

Such single-celled organisms as bacteria, having the capacity to make their own enzymes and so to generate the energy required for their growth and multiplication, can live and multiply in a much simpler medium than that provided by the interior of a living cell. They are, therefore, not obligatory parasites. From the viewpoint of anatomy, however, bacteria are much simpler than cells, and the various bacteria are distributed over the range of complexity from the virus upward to the cell.

In addition to the capacity for energy transformation, biosynthesis and reproduction by self-replication and division, the cells of higher organisms possess other capacities that fit them for the concerted community life that is the life of the organism. From the single-celled fertilized egg the multicelled organism arises not only by the division of the daughter cells but also by their concurrent differentiation into the specialized cells that form various tissues. In many cases when a cell has become differentiated and specialized, it does not divide any more; there is a kind of antagonism between differentiation and growth by cell division.

In the adult organism the capacity for reproduction and perpetuation of the species is left to the eggs and spermatozoa. These gametes, like all other cells in the body, have arisen by cell division from the fertilized egg, followed by differentiation. Cell division remains, however, a frequent event in the adult organism wherever cells continuously wear out and degenerate, as they do in the skin, the intestine and in the bone marrow from which the blood cells arise.

During embryonic development the differentiating cells display a capacity for recognition of others of their own kind. Cells that belong to the same family and resemble one another tend to cluster together, forming a tissue from which cells of all other kinds are excluded. In this mutual association and rejection of cells the cell membrane appears to play a decisive role. The membrane is also one of the principal cell components involved in the function of the muscle cells that endow the organism with the power of movement, of the nerve cells that provide communication lines to integrate the activity of the organism and of the sensory cells that receive stimuli from without and within.

Although there is no typical cell, one may usefully put together a composite cell for the purpose of charting the anatomical features that are shared in varying degrees by all cells. Such a cell, based largely upon what is seen in electron micrographs, is presented on the opposite page; comparison of this cell with the corresponding cell drawn from photomicrographs made by Edmund B. Wilson of Columbia University in 1922 suggests the rapid advances that have been brought about by the electron microscope.

Even the cell membrane, which is only 100 angstrom units thick (one angstrom unit is one ten-millionth of a millimeter) and appears as little more than a boundary in the light microscope, is shown by the electron microscope to have a structure. It is true that electron micrographs have not yet revealed much

DIAGRAM OF A TYPICAL CELL (although there is no such thing as a typical cell) is based on what is seen in the conventional light microscope. Diagram is based on one that appears in 1922 edition of Edmund B. Wilson's *The Cell in Development and Inheritance.*

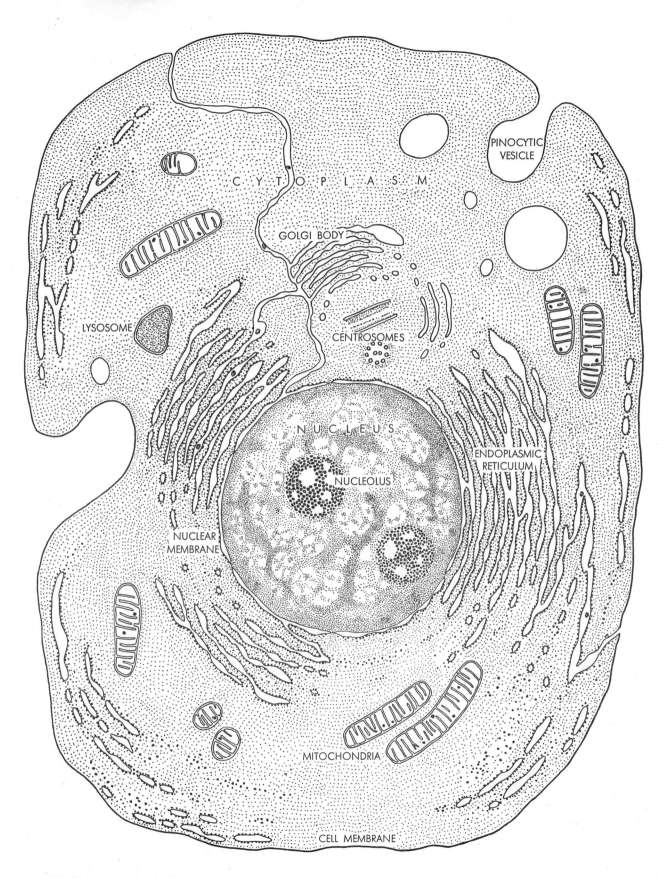

MODERN DIAGRAM OF A TYPICAL CELL is based on what is seen in electron micrographs such as the one reproduced on page 4. The mitochondria are the sites of the oxidative reactions that provide the cell with energy. The dots that line the endoplasmic reticulum are ribosomes: the sites of protein synthesis. In cell division the pair of centrosomes, one shown in longitudinal section (*rods*), other in cross section (*circles*), part to form poles of apparatus that separates two duplicate sets of chromosomes.

about this structure. On the other hand, such complexity as is shown clearly accords with what is known about the functional properties of the membrane. In red blood cells and nerve cells, for example, the membrane distinguishes between sodium and potassium ions although these ions are alike in size and electrical charge. The membrane helps potassium ions get into the cell and opposes more than a mere permeability barrier to sodium ions; that is, it is capable of "active transport." The membrane also brings large molecules and macroscopic bodies into the interior of the cell by mechanical ingestion [see "How Things Get Into Cells," Offprint #96].

Beyond the membrane, in the cytoplasm, the electron microscope has resolved the fine structure of organelles that appear as mere granules in the light microscope. Principal among them are the chloroplasts of green-plant cells and the mitochondria that appear in both animal and plant cells. These are the "power plants" of all life on earth. Each is adapted to its function by an appropriate fine structure, the former to capturing the energy of sunlight by photosynthesis, the latter to extracting energy from the chemical bonds in the nutrients of the cell by oxidation and respiration. From each of these power plants the yield of energy is made available to the energy-consuming processes of the cell, neatly packaged in the phosphate bonds of the compound adenosine triphosphate (ATP).

The electron microscope clearly distinguishes between the mitochondrion, with its highly organized fine structure, and another associated body of about the same size: the lysosome. As Christian de Duve of the Catholic University of Louvain has shown, the lysosome contains the digestive enzymes that break down large molecules, such as those of fats, proteins and nucleic acids, into smaller constituents that can be oxidized by the oxidative enzymes of the mitochondria. De Duve postulates that the lysosome represents a defense mechanism; the lysosomal membrane isolates the digestive enzymes from the rest of the cytoplasm. Rupture of the membrane and release of the accumulated enzymes lead quickly to the lysis (dissolution) of the cell.

The cytoplasm contains many other visible inclusions of less widespread occurrence among cells. Particularly interesting are the centrosomes and kinetosomes. The centrosomes, or centrioles, become plainly visible under the light microscope only when the cell approaches the hour of division, in which these bodies play a commanding role as the poles of the spindle apparatus that divides the chromosomes. The kinetosomes, on the other hand, are found only in those cells which are equipped with cilia or flagella for motility; at the base of each cilium or flagellum appears a kinetosome. Both of these organelles have the special property of self-replication. Each pair of centrosomes gives rise to another when cells divide; a kinetosome duplicates itself each time a new cilium forms on the cell surface. Long ago certain cytologists advanced the idea that these two organelles have much the same structure, even though their functions are so different. The electron

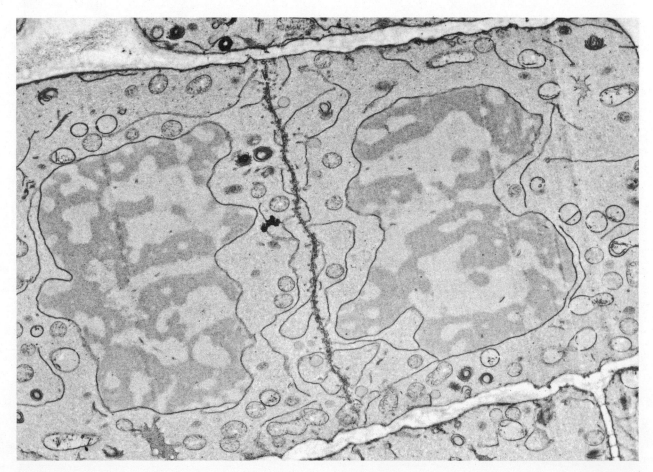

PLANT CELLS (onion root tip) are enlarged 6,700 diameters in this electron micrograph made by K. R. Porter of the Rockefeller Institute. The thin, dark line running from top to bottom of the micrograph shows the membrane between two cells shortly after the cells have divided. The large, irregularly shaped bodies to the left and right of the membrane are the nuclei of the two cells.

microscope has confirmed this suggestion. Each is a cylinder made up of 11 fibers, with two in the center and the other nine on the outside. This is the universal structure of all cilia and of flagella as well. The reason for the structure remains unknown, but it is undoubtedly related to the contractility of the cilia and flagella. It may be that the same "monomolecular muscle" principle underlies the action of the kinetosome and centrosome in their quite diverse functions.

The electron microscope has confirmed another surmise of earlier cytologists: that the cytoplasm has an invisible organization, a "cytoskeleton." Most cells show complicated systems of internal membranes not visible in the ordinary light microscope. Some of these membranes are smooth; others are rough, having tiny granules attached to one surface. The degree to which the membrane systems are developed varies from cell to cell, being rather simple in amoebae and highly articulated and roughened with granules in cells that specialize in the production of proteins, such as those of the liver and pancreas.

Electron microscopists differ in their interpretation of these images. The generally accepted view is that of K. R. Porter of the Rockefeller Institute, who has given the membrane system its name, the endoplasmic reticulum; through the network of canaliculi formed by the membrane, substances are supposed to move from the outer membrane of the cell to the membrane of the nucleus. Some investigators hold that the internal membrane is continuous with the external membrane, furnishing a vastly increased and deeply invaginated surface area for communication with the fluid in which the cell is bathed. If the membrane does indeed have such vital functions, then it is likely that the cell is equipped with a factory for the continuous production of new membrane. This might be the role, as George E. Palade of the Rockefeller Institute has recently suggested, of the enigmatic Golgi bodies, first noted by the Italian cytologist Camillo Golgi at the end of the last century. The electron microscope reveals that the Golgi bodies are made of smooth membrane, often continuous with that of the endoplasmic reticulum.

There is no doubt about the nature of the granules, which appear consistently on the "inner" surface of the membrane. They appear particularly in cells that produce large amounts of protein. As Torbjörn O. Caspersson and I showed some 20 years ago, such cells possess a high RNA content. Recent studies have revealed that the granules are exceedingly rich in RNA and correspondingly active in protein synthesis. For this reason the granules are now called ribosomes.

The membrane that surrounds the cell nucleus forms the interior boundary of the cytoplasm. There is still much speculation about what the electron microscope shows of this membrane. It appears as a double membrane with annuli, or holes, in the outer layer, open to the cytoplasm. To some investigators these annuli represent pores through which large molecules may move in either direction. Since the outer layer is often in close contact with the endoplasmic reticulum, it is also argued that the nuclear membrane participates in the formation of the reticulum membrane. Another possibility is that fluids percolating through the canaliculi of the endoplasmic reticulum are allowed to accumulate between the two layers of nuclear membrane.

Inside the nucleus are the all-important filaments of chromatin, in which the cell's complement of DNA is entirely localized. When the cell is in the "resting" state, that is, engaged in the processes of growth between divisions, the chromatin is diffusely distributed in the nucleus. The DNA thus makes maximum surface contact with other material in the nucleus from which it presumably pieces together the molecules of RNA and replicates itself. In preparation for division the chromatin coils up tightly to form the chromosomes, always a fixed number in each cell, to be distributed equally to each daughter cell.

Much less elusive than the chromatin are the nucleoli; these spherical bodies are easily resolved inside the nucleus with an ordinary light microscope. Under the electron microscope they are seen to be packed with tiny granules similar to the ribosomes of the cytoplasm. In fact, the nucleoli are rich in RNA and appear to be active centers of protein and RNA synthesis. Finally, to complete this functional anatomy of the

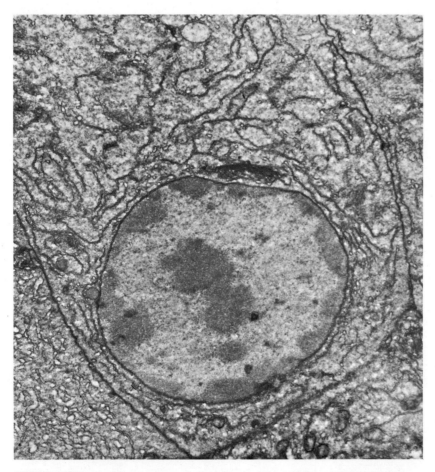

ANIMAL CELL (hepato-pancreatic gland of the crayfish) is enlarged 12,500 diameters in electron micrograph by George B. Chapman of the Cornell University Medical College. The large round object is the nucleus; the smaller dark region just above it is the Golgi body.

12 LEVELS OF COMPLEXITY

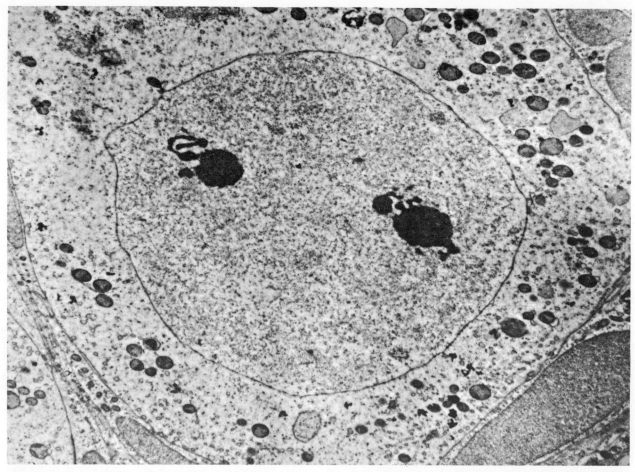

EGG CELL of rabbit is enlarged 7,500 diameters. The large round object is the nucleus; the two prominent dark bodies within it are nucleoli. Electron micrograph was made by Joan Blanchette of Columbia University College of Physicians and Surgeons.

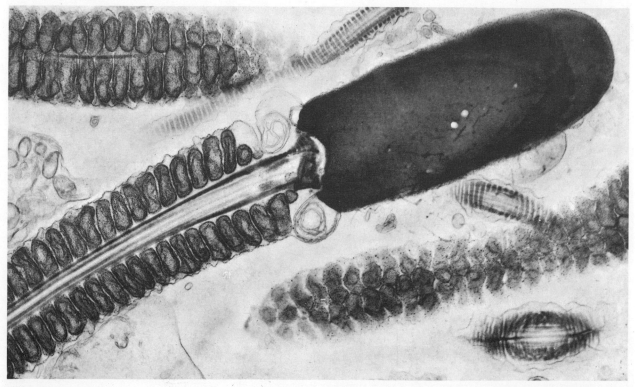

SPERM CELL of a bat is enlarged 21,500 diameters in electron micrograph by Fawcett and Susumu Ito of the Harvard Medical School. Nucleus (*top right*) constitutes almost all of sperm's head; arranged behind head are numerous mitochondria (*left*).

cell, it should be added that the chromatin and nucleoli are bathed together in the amorphous, proteinaceous matrix of the nuclear sap.

A remarkable history in the development of instruments and technique has gone into the drawing of the present portrait of the cell. The ordinary light microscope remains an essential tool. But its use in exploring the interior of the cell usually requires killing the cell and staining it with various dyes that selectively show the cell's major structures. To see these structures in action in the living cell, microscopists have developed a range of instruments—including phase, interference, polarizing and fluorescence microscopes—that manipulate light in various ways. In recent years the electron microscope, as the reader has gathered from this article, has become the major tool of the cytologist. But this instrument has a serious limitation in that it requires elaborate preparation and fixation of the specimen, which must inevitably confuse the true picture with distortions and artifacts. Progress is being made, however, toward the goal of resolving under the same high magnification the structure of the living cell.

Biochemistry has had an equally remarkable history of technical development. Centrifuges of ever higher rotation speed have made it possible to separate finer fractions of the cell's contents. These are divided and subdivided in turn by chromatography and electrophoresis. The classical techniques have been variously adapted to the analysis of quantities and volumes 1,000 times smaller than the standard of older micromethods; investigators can now measure the respiration or the enzyme content of a few amoebae or sea-urchin eggs. Finally, autoradiography, employing radioactive tracer elements, allows the worker to observe at subcellular dimensions the dynamic processes in the intact living cell.

The achievements and prospects that have been generated by the convergence of these two major movements in the life sciences furnish the subject of the articles that follow. To conclude this discussion it will be useful to consider how the two approaches have been employed to illuminate a single question: the role of the nucleus in the economy of the cell.

A simple experiment shows, first of all, that removal of the nucleus in a unicellular organism does not bring about the immediate death of the cytoplasm. The nucleate and enucleate

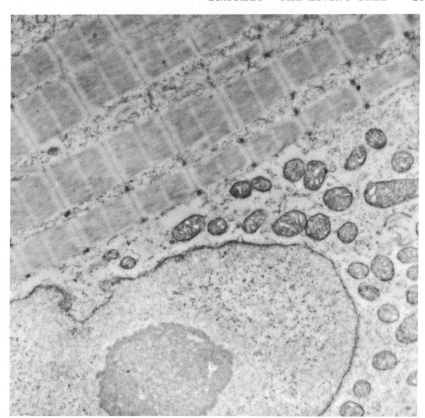

PART OF MUSCLE CELL of a salamander is enlarged 19,500 diameters in electron micrograph by George D. Pappas and Philip W. Brandt of Columbia College of Physicians and Surgeons. The nucleus is at bottom; around it are mitochondria. At top are muscle fibers.

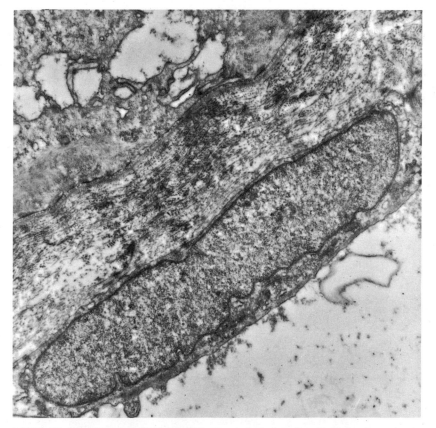

PART OF CONNECTIVE-TISSUE CELL of a tadpole is enlarged 14,500 diameters in electron micrograph by Chapman. Nucleus is oblong object; above it are fibrils of collagen.

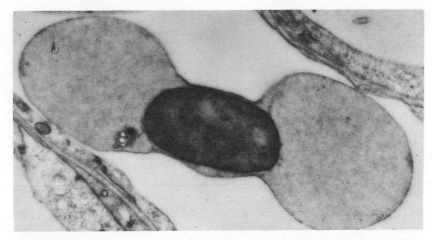

RED BLOOD CELL of a fish is enlarged 8,000 diameters in electron micrograph by Fawcett. Large dark body in center is nucleus. The mature red cells of mammals have no nuclei.

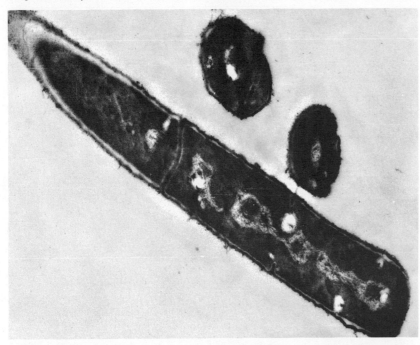

BACTERIUM *Bacillus cereus* (*long object*) is enlarged 30,000 diameters in electron micrograph by Chapman and by James Hillier of RCA Laboratories. Bacillus has several nuclei.

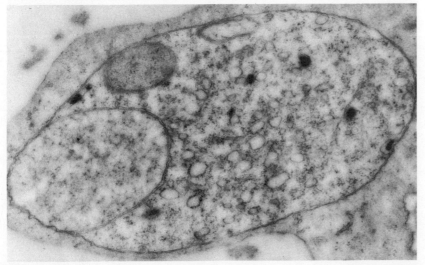

PROTOZOON *Plasmodium berghei* is enlarged 21,000 diameters in electron micrograph by Maria A. Rudzinska of the Rockefeller Institute. The nucleus is the large body at lower left.

halves of amoebae, if kept fasting, attain the same survival time of about two weeks; the cilia of an enucleate protozoon such as the paramecium continue to beat for a few days; the enucleate fragments of the unicellular giant alga *Acetabularia* may survive several months and are even capable of an appreciable amount of regeneration. Many of the basic activities of the cell, including growth and differentiation in the case of *Acetabularia*, can therefore proceed in the total absence of the genes and DNA. In fact, the enucleate pieces of *Acetabularia* are perfectly capable of making proteins, including specific enzymes, although enzyme synthesis is known to be genetically controlled. These synthetic activities, however, die out after a time. One must conclude that the nucleus produces something that is not DNA but which is formed under the influence of DNA and is transferred from the nucleus to the cytoplasm, where it is slowly used up. From such experiments—employing the combined techniques of cell biology and biochemistry—a number of fundamental conclusions emerge.

First, the nucleus is to be considered as the main center for the synthesis of nucleic acid (both DNA and RNA). Second, this nuclear RNA (or part thereof) goes over to the cytoplasm, playing the role of a messenger and transferring genetic information from DNA to the cytoplasm. Finally, the experiments show that the cytoplasm and in particular the ribosomes are the main site for the synthesis of specific proteins such as the enzymes. It should be added that the possibility of independent RNA synthesis in the cytoplasm is not ruled out and that such synthesis can, under suitable conditions, be demonstrated in enucleate fragments of *Acetabularia*. From this brief description of recently observed facts it is clear that the cell is not only a morphological but also a physiological unit.

Perhaps the reader will wonder how such knowledge of this unit helps to answer questions under the more general headings of "life" and "living." All one can venture to say is that the results of investigation invariably point in the same direction: Life, in the case of the cell and its constituents, is more a quantitative than an "all or none" concept. This dissection of cells into their constituents does not, therefore, throw much light on the questions posed by philosophy. But without this dissection, without experimentation, we would know next to nothing about the cell. And, after all, the cell is the fundamental unit of life.

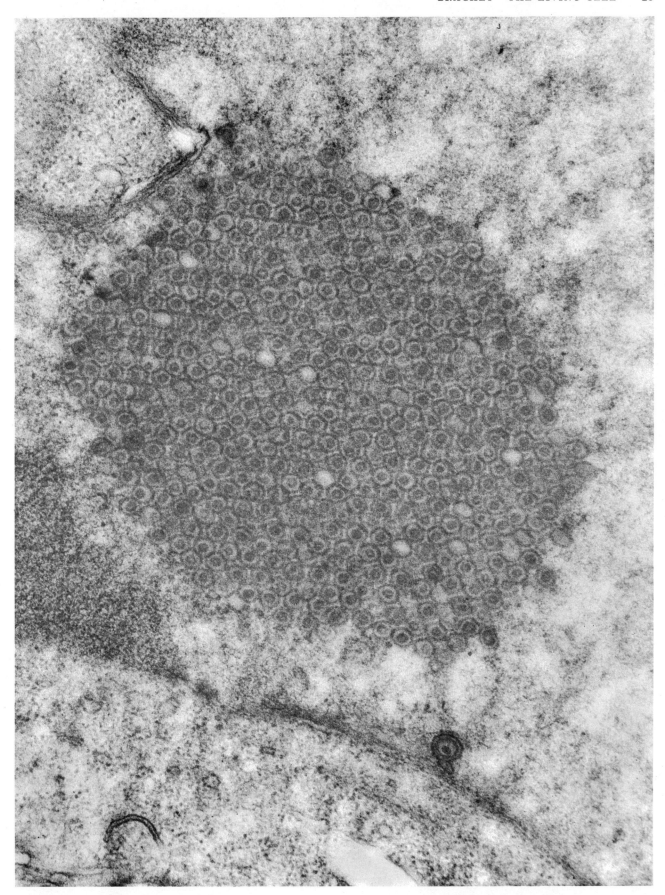

PARTICLES OF VIRUS *Herpes simplex* form a crystal within the nucleus of a cell. This electron micrograph, which enlarges the particles 73,000 diameters, was made by Councilman Morgan of the Columbia College of Physicians and Surgeons. Although viruses are exceptions to the rule that all living things are cells or are made of cells, they can reproduce only when they are inside cells.

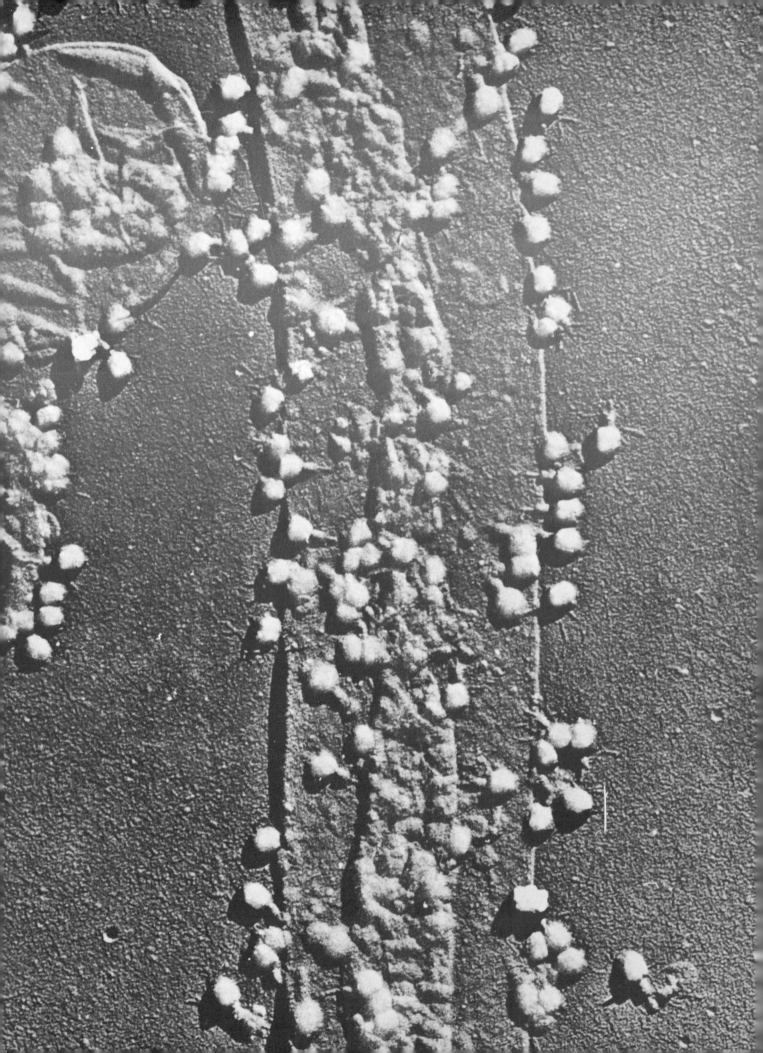

2

VIRUSES AND GENES

by FRANÇOIS JACOB and ELIE L. WOLLMAN   June 1961

Almost everyone now accepts the unity of the inanimate physical world. Physicists do not hesitate to extrapolate laboratory results obtained with a small number of atoms to explain the source of the energy produced by stars. In the world of living things a comparable unity is more difficult to demonstrate; in fact, it is not altogether conceded by biologists. Nevertheless, most students of bacteria and viruses are inclined to believe that what is true for a simple bacillus is probably true for larger organisms, be they mice, men or elephants.

Accordingly we shall be concerned here with seeking lessons in the genetic behavior of the colon bacillus (*Escherichia coli*) and of the still simpler viruses that are able to infect the bacillus and destroy it. Viruses are the simplest things that exhibit the fundamental properties of living systems. They have the capacity to produce copies of themselves (although they require the help of a living cell) and they are able to undergo changes in their hereditary properties. Heredity and variation are the subject matter of genetics. Viruses, therefore, possess for biologists the elemental qualities that atoms possess for

SCORES OF VIRUSES of the strain designated T_2 are attached to the wall of a colon bacillus in this electron micrograph. The viruses are fastened to the bacterial wall by their tails, through which they inject their infectious genetic material. (Walls of the cell collapsed when the specimen was dried by freezing. "Shadowing" with uranium oxide makes objects stand out in relief.) The electron micrograph was made by Edouard Kellenberger of the University of Geneva. The magnification is 70,000 diameters.

physicists. When a virus penetrates a cell, it introduces into the cell a new genetic structure that interferes with the genetic information already contained within the cell. The study of viruses has thus become a branch of cellular genetics, a view that has upset many old notions, including the traditional distinction between heredity and infection.

For a long time geneticists have worked with such organisms as maize and the fruit fly *Drosophila*. They have learned how hereditary traits are transmitted from parents to progeny, they have discovered the role of the chromosomes as carriers of heredity and they have charted the results of mutations—the events that modify genes. Complex organisms, however, multiply too slowly and in insufficient numbers for the high-resolution analyses needed to clarify such problems as the chemical nature of genes and the processes by which a gene makes an exact copy of itself and influences cellular activity. These detailed problems are most readily studied in bacteria and in viruses. Within the space of a day or two the student of bacteria or bacterial viruses can grow and study more specimens than the fruit-fly geneticist could study in a lifetime. An operation as simple as the mixing of two bacterial cultures on a few agar plates can provide information on a billion or more genetic interactions in which genes recombine to form those of a new generation.

It is the events of recombination, together with mutations, that model and remodel the chromosomes, the structures that contain in some kind of code the entire pattern of every organism. In recent years geneticists and biologists have clarified the nature of the hereditary message and have gained some clues as to what the letters of the code

are. The primary, and perhaps the unique, bearers of genetic information in all forms of life appear to be molecules of nucleic acid. In living organisms, with the exception of some of the viruses, these long-chain molecules are composed of deoxyribonucleic acid (DNA). In all plant viruses and in some animal viruses the genetic substance is not DNA but its close chemical relative ribonucleic acid (RNA). DNA molecules are built up of hundreds of thousands or even millions of simple molecular subunits: the nucleotides of the four bases adenine, thymine, guanine and cytosine. These subunits, in an almost infinite variety of combinations, seem capable of encoding all the characteristics that all organisms transmit from one generation to the next. RNA molecules, which are somewhat shorter in length and not so well understood, act similarly for the viruses in which RNA is the genetic material.

Ultimately the role of the genes—the words of the hereditary message—is to specify the molecular organization of proteins. Proteins are long-chain molecules built up of hundreds of molecular subunits: the 20 amino acids. The sequence of nucleotides in the nucleic acid that contains the hereditary message is thought to determine the sequence of amino acids in the protein it manufactures. This process involves a "translation" from the nucleic-acid code into the protein code through a mechanism that is not yet understood.

The Bacterial Chromosome

Before considering viruses as cellular genetic elements, we shall summarize the present knowledge of the genetics of the bacterial cell. In bacteria the hereditary message appears to be written in a single linear structure, the bacterial chromosome. For the study of this chromosome an excellent tool was discovered in 1946 by Joshua Lederberg and Edward L. Tatum, who were then working at Yale University. They used the colon bacillus, which is able to synthesize all the building blocks required for the manufacture of its nucleic acids and proteins and therefore to grow on a minimal nutrient medium containing glucose and inorganic salts. Mutant strains, with defective or altered genes, can be produced that lack the ability to synthesize one or more of the building blocks and therefore cannot grow in the absence of the building block they cannot make. If, however, two different mutant strains are mixed, bacteria like the original strain reappear and are able to grow on a minimal medium.

Lederberg and Tatum were able to demonstrate that such bacteria are the result of genetic recombination occurring when a bacterium of one mutant strain conjugates with a bacterium of another mutant strain. Further work by Lederberg, and by William Hayes in London, has shown that the colon bacillus also has sex: some individuals act as males and transmit genetic material by direct contact to other individuals that act as recipients, or females. The difference between the two mating types may be ascribed to the fertility factor (or sex factor) F, present only in males. Curiously, females can easily be converted into males; during conjugation certain types of male, called F^+, transmit their sex factor to the females, which then become males.

The Chromosome "Essay"

Our own work at the Pasteur Institute in Paris has shed light on the different steps involved in bacterial conjugation and on the mechanism ensuring the transfer of the chromosome from certain strains of male, called *Hfr*, to females. When cultures of such males and of females are mixed, pairings take place between male and female cells through random collisions. A bridge forms between the two mating bacteria; one of the chromosomes of the male (bacteria have generally two to four identical chromosomes during growth) begins to migrate across the bridge and to enter the female. In the female, portions of the male chromosome have the ability to recombine with suitable portions of one of the female chromosomes. The chromosomes may be compared to written essays that differ only by a few letters, or a few words, corresponding to the mutations. Portions of the two essays may become paired, word for word and letter for letter. Through the process known as genetic recombination, which is still very mysterious and challenging, fragments of the male chromosome, which can be anything from a word or a phrase up to several sentences, may be exactly substituted for the corresponding part of the female chromosome. This process gives rise to a complete new chromosome that contains a full bacterial essay in which some words from the male have replaced corresponding words from the female. The new chromosome is then replicated and transmitted to the daughter cell.

Perhaps the most remarkable feature of bacterial conjugation is the way in which the male chromosome migrates across the conjugation bridge. For a given type of male the migration always starts at the same end of the chromosome, which, if we represent the bacterial chromosome by the letters of the alphabet, we can call A. Then, with the chromosome proceeding at constant speed, it takes two hours before the other end, Z, has penetrated the female. After the mating has begun, conjugation can be interrupted at will by violently stirring the mating mixture for a minute or so in a blender. The mechanical agitation does not kill the cells but it disrupts the bridge and breaks the male chromosome during its migration. The fragment of the male chromosome that has entered the female before the interruption is still functional and has the ability to provide words or sentences for a chromosome [*see illustrations on pages 20 and 21*]. If conjugation is mechanically interrupted at various intervals after the onset of mating, it is found that any gene carried by the male chromosome, from A to Z, enters the female at a precise time. We have therefore been able to draw two kinds of detailed chromosome map showing the location of genes. One map, the conventional kind, is based on the observed frequency of different sorts of genetic recombination; the second is a new kind of map reflecting the time at which any gene penetrates the female cell. The latter can be compared to a road map drawn by measuring the times at which a car proceeding at a constant speed passes through various cities.

Finally, the mode of the male chromosome's migration has provided a unique opportunity for correlating genetic measurements with chemical measurements of the chromosome. In collaboration with Clarence Fuerst, who is now working at the University of Toronto, we have grown male bacteria in a medium containing the radioactive isotope phosphorus 32, which is incorporated into the DNA of the bacterial chromosome. The labeled bacteria are then frozen and kept in liquid nitrogen to allow some of the radioactive atoms to disintegrate. At various times samples are thawed and the labeled males are then mated with unlabeled females. The experiments show that the radioactive disintegrations sometimes break the chromosomes. If the break occurs between two markers, say E and F, the head part, $ABCDE$, is transferred to the female, but the tail part, $FGHIJKLMNOPQRSTUVWXYZ$, is not. Therefore the greater the number of phosphorus atoms between the A extremity of the chromosome and a given gene, the greater the chance that a break will prevent this gene from being transferred to the female. It is thus possible to draw a chromosomal map showing the location of the genes in terms of numbers of phosphorus atoms contained in the chromosome between the known genes. When we compare this map with those obtained by genetic analysis or by mechanical interruption, we find that for a given type of male all three maps are consistent.

In some types of male mutant the genetic characters have the same sequence along the chromosome but the character injected first differs from one mutant to another. The characters can also be injected either in the forward direction or in the backward direction, that is, from A to Z or from Z to A, with the alphabet capable of being broken at any point. These observations can be explained most simply by assuming that all the genetic "letters" of the colon bacillus are arranged linearly in a ring and that the ring can be opened at various points by mutation. It seems, furthermore, that the opening of the ring is a consequence of the attachment of the sex factor to the chromosome. The ring opens at precisely the point where the factor F, which is free to move, happens to affix itself. A cell with the F factor affixed to the chromosome is called an *Hfr* male, or "supermale," because it enhances the transmission of chromosomal markers. *Hfr* stands for "high frequency of recombination." When the chromosome is opened by the F factor, one of the free ends initiates the penetration of the chromosome into the female, carrying the sequence of characters after it. The other end carries the sex factor itself and is the last to enter the female. The sex factor has other remarkable properties and we shall bring it back into our story later.

The long-range objective of such stud-

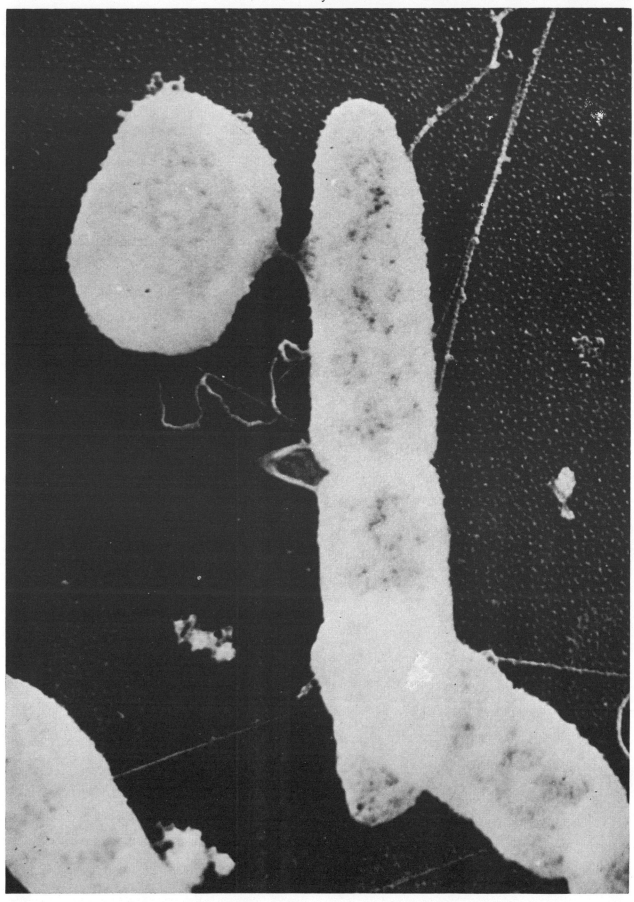

CONJUGATING BACTERIA conduct a transfer of genetic material. Long cell (*right*) is an *Hfr* "supermale" colon bacillus, which is attached by a short temporary bridge to a female colon bacillus (*see illustration on next two pages*). This electron micrograph, shown at a magnification of 100,000 diameters, was made by Thomas F. Anderson of the Institute for Cancer Research in Philadelphia.

ies is to learn how the thousands of genes strung along the chromosome control the molecular pattern of the bacterial cell: its metabolism, growth and division. These processes imply precise regulatory mechanisms that maintain a harmonious equilibrium between the cellular constituents. At any time the bacterial cell "knows" which components to make and how much of each is needed for it to grow in the most economical way. It is able to recognize which kind of food is available in a culture medium and to manufacture only those protein enzymes that are required to get energy and suitable building blocks from the available food.

At the Pasteur Institute, in collaboration with Jacques Monod, we have recently found new types of gene that determine specific systems of regulation. Mutants have been isolated that have become "unintelligent" in the sense that they cannot adjust their syntheses to their actual requirements. They make, for example, a certain protein in large amounts when they need only a little of it or even none at all. This waste of energy decreases the cells' growth rate. It seems that the production of a particular protein is controlled by two kinds of gene. One, which may be called the structural gene, contains the blueprint for determining the molecular organization of the protein—its particular sequence of amino acid subunits. Other genes, which may be called control genes, determine the rate at which the information contained in the structural gene is decoded and translated into protein. This control is exercised by a signal embodied in a repressor molecule, probably a nucleic acid, that migrates from the chromosome to the cytoplasm of the cell. One of the control genes, called the regulator gene, manufactures the repressor molecule; thus it acts as a transmitter of signals. These are picked up by the operator gene, a specific receiver able to switch on or off the activity of the adjacent structural genes. Metabolic

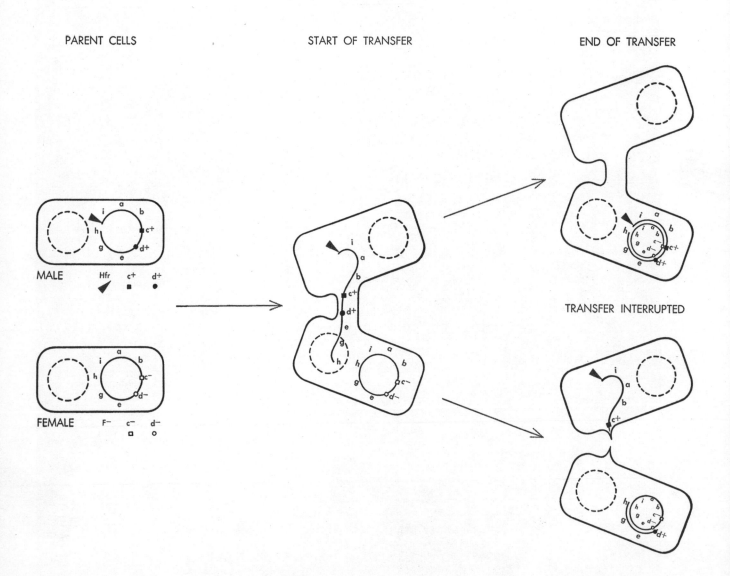

CHROMOSOMAL TRANSFER provides a primitive sexuality for colon bacillus. The bacterial chromosome, which appears to be ring-shaped, carries genetic markers (*designated by letters*), the presence or absence of which can be determined by studying cell's nutritional requirements. When the sex, or F, agent is attached to the chromosome, opening the ring, the cell is called an *Hfr* supermale. Two markers, labeled c^+ and d^+ when present and c^- and d^- when absent, can be traced from parents to daughter cells. When male and female cells conjugate, one of the male chromosomes (there are usually several, all identical) travels through the bridge.

products can interfere with the signals, either activating or inactivating the proper repressor molecules and thereby initiating or inhibiting the production of proteins.

Within the bacterial cell, then, there exists a complex system of transmitters and receivers of specific signals, by means of which the cell is kept informed of its metabolic requirements and enabled to regulate its syntheses. The bacterial chromosome contains not only a series of blueprints for the manufacture of individual molecular components but also a plan for the co-ordinated production of these components.

Let us now turn to the events that take place when a bacterial virus of the strain designated T_2 infects the colon bacillus. A T_2 virus is a structure shaped like a tadpole; by weight it is about half protein and half DNA. The DNA is enclosed in the head, the outside of which is protein; the tail is also composed of protein. The roles of the DNA and the protein in the infective process were clarified in 1952 by the beautiful experiments of Alfred D. Hershey and Martha Chase of the Carnegie Institution of Washington's Department of Genetics in Cold Spring Harbor, N. Y. By labeling the DNA fraction of the virus with one radioactive isotope and the protein fraction with another, Hershey and Chase were able to follow the fate of the two fractions. They found that the DNA is injected into the bacterium, whereas the protein head and tail parts of the virus remain outside and play no further role. Electron micrographs reveal that the tail provides the method of attachment to the bacterium and that the DNA is injected through the tail. The Hershey-Chase experiment was a landmark in virology because it demonstrated that the nucleic acid carries into the cell all the information necessary for the production of complete virus particles.

How Viruses Destroy Bacteria

A bacterium that has been infected by virus DNA will break open, or lyse, within about 20 minutes and release a new crop of perhaps 100 particles of infectious virus, complete with protein head and tail parts. In this brief period the virus DNA subverts the cell's chemical facilities for its own purposes. It brings into the cell a plan for the synthesis of new molecular patterns and the cell faithfully carries it out. The infected cell creates new protein subunits needed for the virus head and tail, and filaments of nucleic acid identical to the DNA of the invading particle. These pools of building blocks pile up more or less at random, and in excess amounts, inside the cell. Then the long filaments of virus DNA suddenly condense and the protein subunits assemble around them, creating the complete virus particle. The whole process can be compared to the occupation of one country by another; the genetic material of the virus overthrows the lawful rule of the cell's own genetic material and establishes itself in power.

A virus can therefore be considered a genetic element enclosed in a protein coat. The protein coat protects the genetic material, gives it rigidity and stability and ensures the specific attachment of the virus to the surface of the cell. As André Lwoff of the Pasteur Institute has pointed out, viruses can be uniquely defined as entities that reproduce from their own genetic material and that possess an apparatus specialized for the process of infection. The definition excludes both the cell and the specialized particles within the cell that serve its normal functions.

Another important criterion of viral growth is that of unrestricted synthesis. Infection with a virus is a sort of molecular cancer. The replication of the genetic material of the virus and the synthesis of the viral building blocks do not appear to be subject to any control system at all.

Lysogenic Bacteria

When a T_2 virus infects a bacterium, it forces the host to make copies of it and ultimately to destroy itself. Such a virus is said to be virulent, and when it is inside the cell, reproducing itself, it is said to be in the vegetative state.

There are, however, other bacterial viruses, called temperate viruses, which behave differently. After entering a cell the genetic material of a temperate virus can take two distinct paths, depending on the conditions of infection. It can enter the vegetative state, replicate itself and kill the host, just as a virulent virus does. Under other circumstances it does not replicate freely and does not kill the host. Instead it finds its way to the bacterial chromosome, anchors itself there and behaves like an integrated constituent of the host cell. Thereafter it will be transmitted for years to the progeny of the bacterium like a bacterial gene. We know that the bacterial host has not destroyed the invading particle, because from time to time one of the daughter cells in the infected line will break open and yield a crop of virus particles, as it would if it had been freshly attacked by a virulent virus. When the virus is in the subdued and integrated state, it is called a provirus. Bacteria carrying a provirus are called lysogenic, meaning that they carry a property that can lead to lysis and death.

Lysogeny was discovered in the early 1920's, soon after the discovery of the bacterial virus itself, and it remained a profound mystery for some 25 years. The mystery was explained by the fine detective work of Lwoff and his colleagues [see "The Life Cycle of a Virus," by André Lwoff; SCIENTIFIC AMERICAN, March, 1954]. Lwoff found that when he exposed certain types of lysogenic bac-

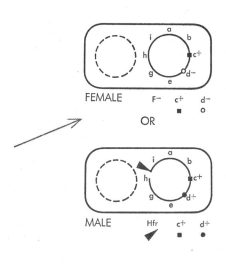

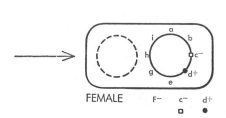

If transfer is complete, daughter cells may be male or female and carry any marker of the male. If transfer is interrupted, daughters are all female and can carry only those markers passed before bridge was broken.

22 LEVELS OF COMPLEXITY

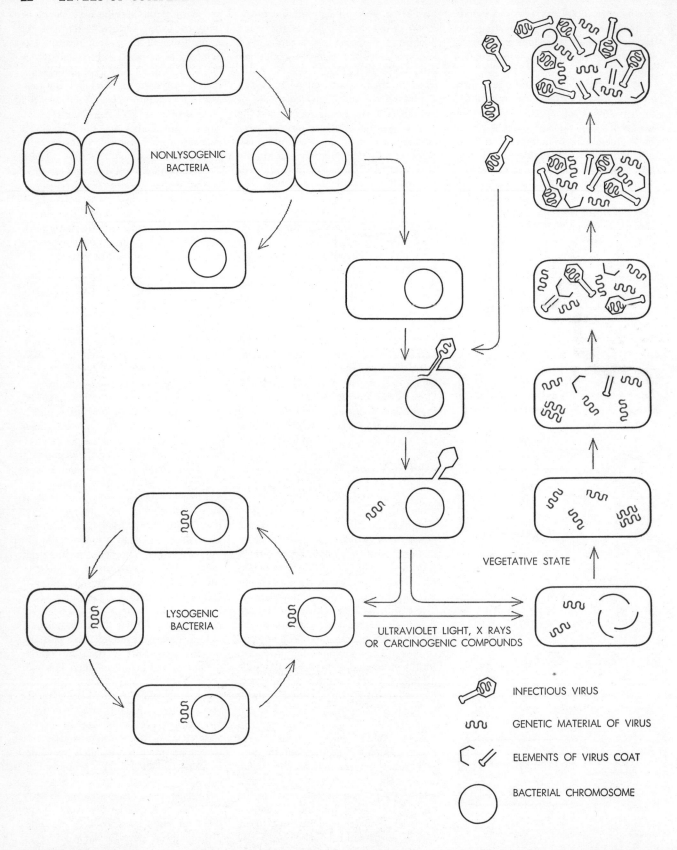

LIFE CYCLE OF BACTERIAL VIRUS shows that, for the bacterium attacked, infection and death are not inevitable. After the genes of the virus (*color*) enter a cell descended from a completely healthy line (*top left*), the cell may take either of two paths. One (*far right*) leads to destruction as the virus enters the vegetative state, makes complete copies of its infective self and bursts open the cell, a process called lysis. The other path leads to the so-called lysogenic state, in which the viral genes attach themselves to the bacterial chromosome and become a provirus; the cell lives. Exposure to ultraviolet light, however, can dislodge the provirus and induce the vegetative state. The provirus is sometimes lost during cell division, returning the cell to the nonlysogenic state.

teria to ultraviolet light, X rays or active chemicals such as nitrogen mustard or organic peroxides, the whole bacterial population would lyse within an hour, releasing a multitude of infectious virus particles. When a provirus is thus activated, or "induced," it leaves the integrated state and enters the vegetative state, eventually destroying the cell [see illustration on opposite page].

To determine the position of the provirus inside the host cell, we can apply the method of interrupting the sexual conjugation of bacteria that carry a provirus and are therefore lysogenic. In this way we can correlate the location of the provirus with that of known characters on the bacterial chromosome. Each of 15 different types of provirus takes a particular position at a specific site on the bacterial chromosome. Only one is an exception; it seems free to take a position anywhere. In the proviral state the genetic material of the virus has not become an integral part of the bacterial chromosome; instead it appears to be added to the chromosome in an unknown but specific way. However it may be hooked on, the genetic material of the virus is replicated together with the genetic material of the host. It behaves like a gene, or rather as a group of genes, of the host.

Nonviral Effects of Provirus

The presence of this apparently innocuous genetic element, the provirus, can confer on the lysogenic bacteria that harbor it some new and striking properties. It is not at all obvious why some of these properties should be related to the presence of a provirus. As one example, diphtheria bacilli are able to produce diphtheria toxin only if the bacilli carry certain specific types of provirus. The disease diphtheria is caused solely by this toxin.

In other instances the presence of a provirus is responsible for a particular type of substance coating the surface of a bacterium. The substance can be identified by various immunological tests (typically by noting if a precipitate forms when a certain serum is added). The nonlysogenic strain, carrying no provirus, will bear a different substance. In such cases the genes of the virus are responsible for hereditary properties of the host. They can scarcely be distinguished from the genes of the bacterium.

The most striking property the provirus confers on its bacterial host is immunity from infection by external viruses of the same type as the provirus. When

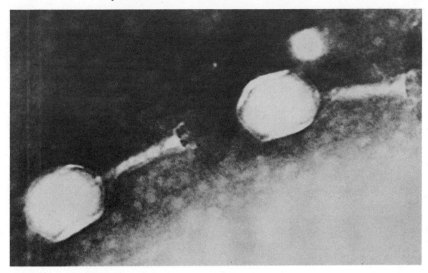

INTACT T_2 VIRUS has polyhedral head membrane and a curious pronged device at the end of its tail. The magnification is 200,000 diameters. This electron micrograph and the two below were made by S. Brenner and R. W. Horne at the University of Cambridge.

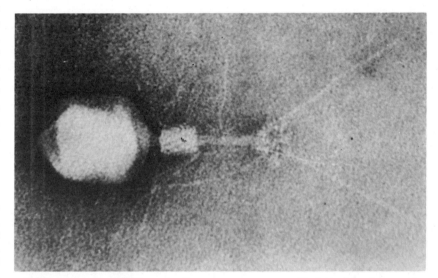

"TRIGGERED" T_2 VIRUS results from exposure to a specific bacterial substance that causes contraction of the tail sheath (stubby cylinder) and discharge of viral genes.

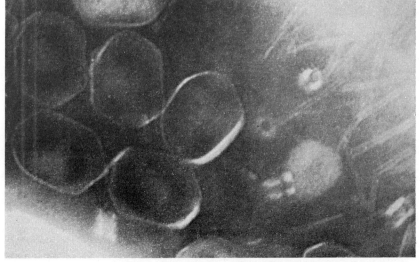

ISOLATED T_2 PARTS can be found still unassembled if host cell is forced to burst open before synthesis of virus particles is complete. Parts include head membranes and tails.

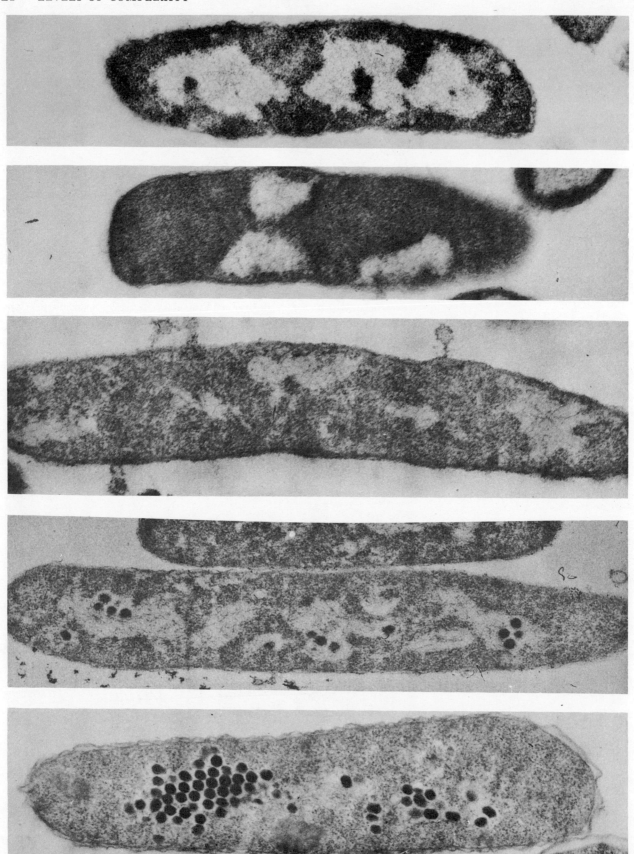

GROWTH OF T_2 VIRUS inside bacterial host is revealed in a striking series of electron micrographs by Kellenberger. Top picture shows the colon bacillus before infection. Four minutes after infection (*second from top*) characteristic vacuoles form along the cell wall. Ten minutes after infection (*third from top*) the virus has reorganized the entire cell interior and has created pools of new viral components. Twelve minutes after infection (*fourth from top*) new virus particles have started to condense. Thirty minutes after infection (*bottom*) more than 50 fully developed T_2 viruses have been produced and the cell is about ready to burst open.

lysogenic cells are mixed with such viruses, the virus particles adsorb on the cell and inject their genetic material into the cell, but the cell survives. The injected material is somehow prevented from multiplying vegetatively and is diluted out in the course of normal bacterial multiplication.

In the past two years we have attempted to learn more about the mechanism of this immunity. It seems clear that the mere attachment of the provirus to the host chromosome cannot account for the immunity of the host. The provirus must do something or produce something. We have evidence that the immunity is expressed by a substance or factor not tied to the chromosome. Remarkably enough, the system of immunity appears to be similar to the cellular systems already described that regulate the synthesis of protein in growing bacteria. It seems that the provirus produces a chemical repressor capable of inhibiting one or several reactions leading to the vegetative state. Thus immunity can be visualized as a specific system of regulation, involving the transmission of signals (repressors), which are received by an invading virus particle carrying the appropriate receptor.

Transduction

The close association that may take place between the genetic material of the virus and that of the host becomes even more striking in the phenomenon of transduction, discovered in 1952 by Norton D. Zinder and Lederberg at the University of Wisconsin [see "'Transduction' in Bacteria," by Norton D. Zinder, Offprint #106, for further information.] They found that when certain proviruses turn into infective viruses, thereby killing their hosts, they may carry away with them pieces of genetic material from their dead hosts. When the viruses infect a host that is genetically different, the genes from the old host—the transduced genes—may be recombined with the genes of the new host. The transduction process seems able to move any sort of gene from one bacterial host to another.

Lysogeny and transduction therefore represent two complementary processes. In lysogeny the genes of the virus become an integral part of the genetic apparatus of the host and replicate at the pace of the host's chromosome. In transduction genes of the host become linked to the genes of the virus and can replicate at the unrestricted viral pace when the virus enters the vegetative state.

Viruses, like all other genetic ele-

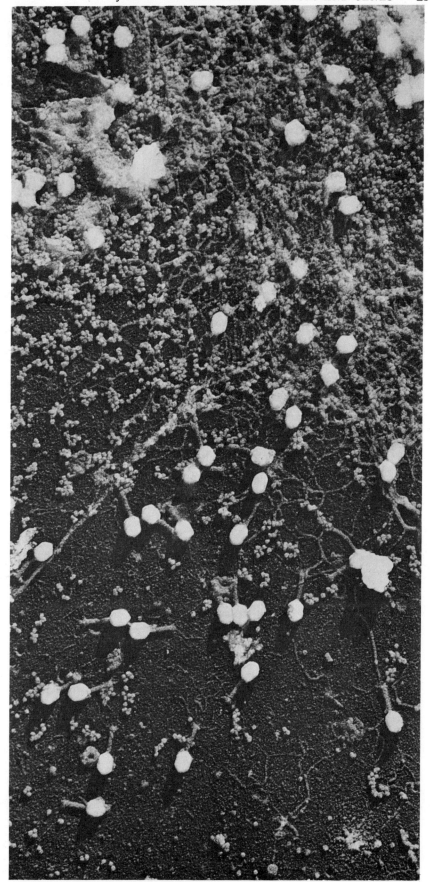

DEATH OF A BACTERIUM occurs when T_2 virus particles, having multiplied inside their host (*see sequence on opposite page*), dissolve the walls of the bacterial cell and spill out—a phenomenon called lysis. Viruses are the large white objects; the other matter is cellular debris. The electron micrograph (magnification: 50,000 diameters) is by Kellenberger.

26   LEVELS OF COMPLEXITY

ments, can undergo mutations, and these produce a variety of stable, heritable changes. The mutations of particular interest are those that prevent the formation of mature, infectious virus particles. Lysogenic bacteria in which such mutations have taken place are called defective lysogenic bacteria. These bacteria hereditarily perpetuate a mutated provirus, which is perfectly able to replicate together with the host's chromosome. If these cells are exposed to ultraviolet radiation, which activates the provirus, we observe that the defective lysogenic cells die without releasing any infectious viruses. Examination of such bacteria usually shows that virus subunits have started to appear inside the cell but have failed to reach maturity [*see illustrations on pages 28 and 29*]. Evidently some essential step in the formation of

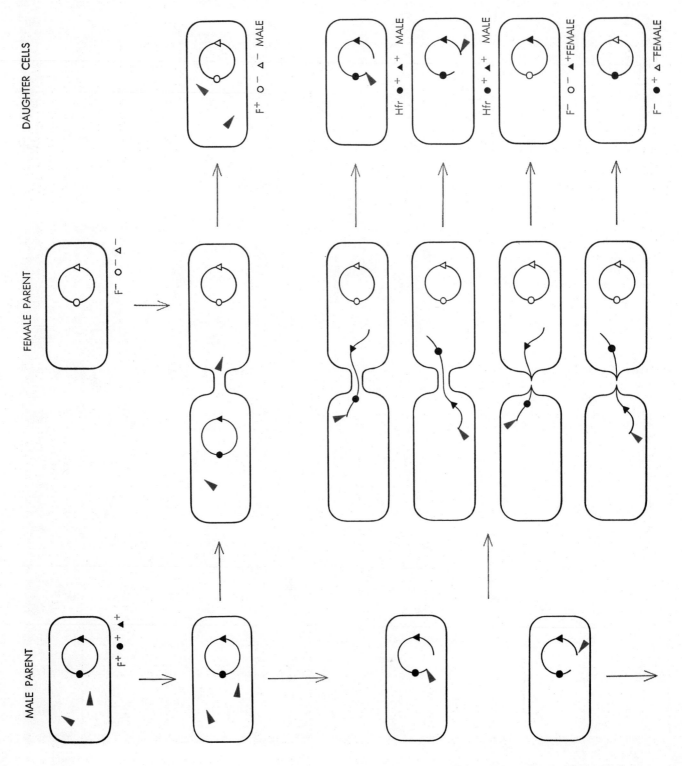

an infectious virus has been blocked by the mutation.

Just as we can study how other kinds of mutation block biochemical pathways associated with cell nutrition, we can try to identify the biochemical blockages that keep the provirus from multiplying normally. When a defective provirus turns to the vegetative state, some viral components begin to appear but the process halts. By using various biological tests, together with electron microscopy, we try to establish how far the process has gone. We have been able to identify two ways in which the process is halted and to relate the blockage to two main groups of viral genes.

One group of genes is concerned with the autonomous reproduction of the genetic material of the virus. The DNA of the provirus, which was able to repli-

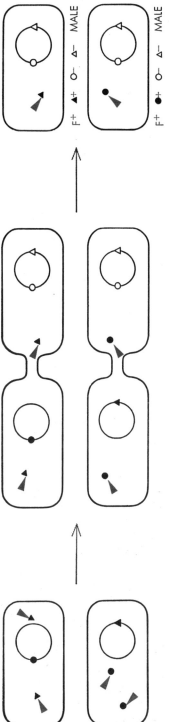

SEX-DUCTION
F AGENT NONINTEGRATED

F, OR SEX, AGENT, indicated by colored wedge, is a versatile and busy "broker" in genes. It can be attached to the bacterial chromosomes (*integrated*) or unattached (*nonintegrated*) and can alternate between the two states. When nonintegrated, it usually transmits only itself when bacteria conjugate (*top sequence*). When integrated, it opens chromosome ring and is the last marker transferred in conjugation (*middle sequence*). Daughters may inherit markers in combinations other than those shown. When F agent leaves integrated state (*bottom*), it may remove a marker and transfer it (*sex-duction*).

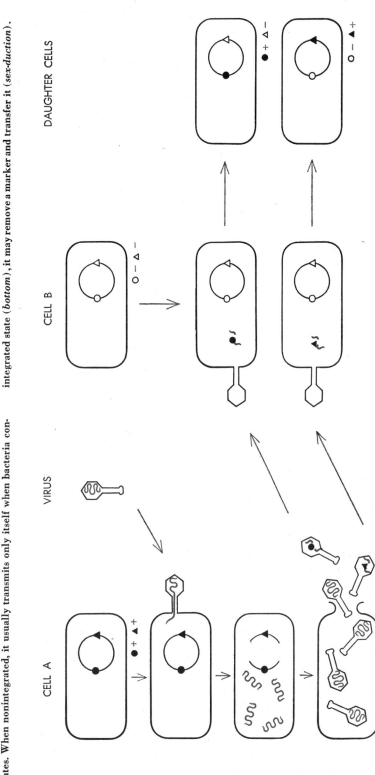

TRANSDUCTION

TRANSDUCTION is similar to sex-duction and was discovered earlier. In transduction the agent for transferring bacterial genes is a virus particle rather than an F agent. The virus injects its genes (color) into bacterial cell A and the newly formed enclose a few genes from the chromosome of the bacterial host along with a few viral genes. These imperfect viruses are able to inject their contents into another cell (cell "B") but are unable to destroy it. In this way genes (*solid black shapes*) can be transferred from cell A to the daughters of cell B. genes create new copies of the virus. Occasionally the new virus particles so

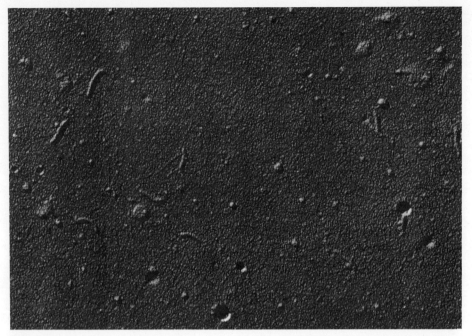

INCOMPLETE VIRUS PARTICLES are created by defective proviruses (see illustration below). The electron micrograph at left shows virus heads and tails that remain unassembled because of some defect. Occasionally (right) only heads can be found. Electron

cate when attached to the host chromosome, becomes unable to replicate on its own. A second group of genes is involved in the manufacture of the protein molecules that provide the coat and infectious apparatus of a normal virus. We have examples in which there is plenty of viral DNA, and many components of the coat material, but one or another essential protein is missing.

This study leads us to conclude that what distinguishes the genetic material of a virus from genetic elements of other types is that the virus carries two sets of information, one of which is necessary for the unrestricted multiplication of the viral genes and the other for the manufacture of an infectious envelope and traveling case.

The concept of a virus as it has emerged from the study of bacterial vi-

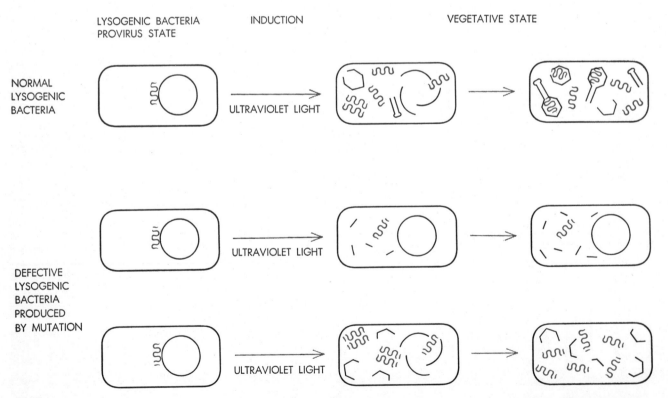

DEFECTIVE LYSOGENIC BACTERIA appear as mutations among normal lysogenic bacteria. Upon induction with ultraviolet light a normal provirus (color, top left) leaves the bacterial chromosome, replicates, produces infectious virus particles and kills its host. When defective proviruses are induced, the host cell may also be killed, but no infectious viruses appear at lysis. In

micrographs (magnification: 57,000) were made by Kellenberger and W. Arber.

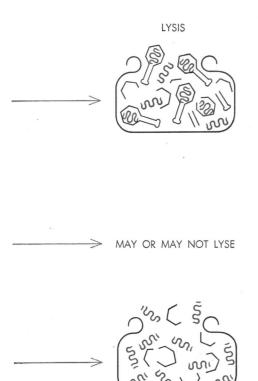

some cases (*middle*) the viral genes fail to replicate. In others (*bottom*) they replicate but the jacketing components are defective.

DNA. In the extracellular infectious state the nucleic acid is enclosed in a protective, resistant shell. The virus then remains inert like the spore of a bacterium, the seed of a plant or the pupa of an insect. In the vegetative state of autonomous replication the genetic material is free of its shell, overrides the regulatory mechanism of the host and imposes its own commands on the synthetic machinery of the cell. The viral genes are fully active. Finally, in the proviral state the genetic material of the virus has become subject to the regulatory system of the host and replicates as if it were part of the bacterial chromosome. A specific system of signals prevents the genes of the virus from expressing themselves; complete virus particles are therefore not manufactured.

The Concept of the "Episome"

Less than a decade ago there was no reason to doubt that virus genetics and cell genetics were two different subjects and could be kept cleanly apart. Now we see that the distinction between viral and nonviral genetics is extremely difficult to draw, to the point where even the meaning of such a distinction may be questionable.

As a matter of fact there appear to be all kinds of intermediates between the "normal" genetic structure of a bacterium and that of typical bacterial viruses. Recent findings in our laboratory have shown that phenomena that once seemed unrelated may share a deep identity. We note, for example, that certain genetic elements of bacteria, which we have no reason to class as viral, actually behave very much like the genetic material of temperate viruses. One of these is the fertility, or *F*, factor in colon bacilli; in the so-called *Hfr* strains of males the *F* agent is attached to one of various possible sites on the host chromosome. In the males bearing the *F* agent designated F⁺ the agent is not fixed to the chromosome and so it replicates as an autonomous unit. It bears one other striking resemblance to provirus. The integrated state of the *F* factor excludes the nonintegrated replicating state, just as a provirus immunizes against the vegetative replication of a like virus.

Another genetic agent resembling provirus is the factor that controls the production of colicines. These are extremely potent protein substances that are released by some strains of colon bacillus; the proteins are able to kill bacteria of other strains of the same or related species. The colicinogenic factors also seem to exist in two alternative states: integrated and nonintegrated. In the latter state they seem able to replicate freely and eventually at a faster rate than does the bacterial chromosome. Bacteria that lack these genetic elements—*F* agents and colicinogenic factors—cannot, so far as we know, gain them by mutation but can only receive them (by sexual conjugation, for example) from an organism that already possesses them. They may replicate either along with the chromosome or autonomously. Such genetic elements, which may be present or absent, integrated or autonomous, we have proposed to call "episomes," meaning "added bodies" [*see illustrations on page 30*].

The concept of episomes brings together a variety of genetic elements that differ in their origin and in their behavior. Some are viruses; others are not. Some are harmful to the host cell; others are not. The important lesson, learned from the study of mutant temperate bacterial viruses, is that the transition from viral to nonviral, or from pathogenic to nonpathogenic, can be brought about by single mutations. We also have impressive evidence that any chromosomal gene of the host may be incorporated in an episome through some process of genetic recombination. During the past year, in collaboration with Edward A. Adelberg of the University of California, we have shown that the sex factor, when integrated, is able to pick up the adjacent genes of the bacterial chromosome. Then this new unit formed by the sex factor and a few bacterial genes is able to return to the autonomous state and to be transmitted by conjugation as a single unit. This process, in many respects similar to transduction, has been called sex-duction [*see illustration at left on pages 26 and 27*].

Do episomes exist in organisms higher than bacteria? We do not know; but if we accept the basic unity of all cellular biology, we should be confident that the answer is yes and that mice, men and elephants must harbor episomes. So far the great precision and resolution that can be achieved in the study of bacterial viruses cannot be duplicated for more complex organisms. There is, nevertheless, evidence for episome-like factors in the fruit fly and in maize. There have been reports of two viruses in the fruit fly, transmitted through the egg to the offspring, which may exist either as nonintegrated or as integrated elements. Although it does not seem that the virus is actually located on the chromosome in the latter state, the resemblance to provirus is striking. Barbara McClintock, of

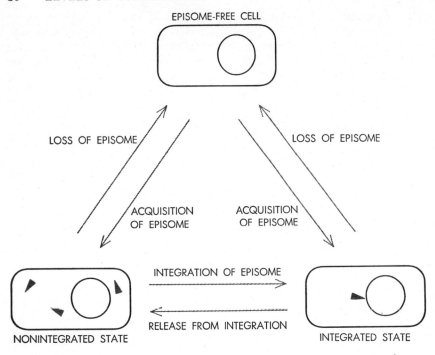

CONCEPT OF THE "EPISOME," as put forward by the authors, describes a genetic element, such as the *F* agent, that may be either attached to the chromosome or unattached. When integrated, it replicates at host's pace; nonintegrated, it replicates autonomously.

the Carnegie Institution of Washington's laboratory at Cold Spring Harbor, has discovered in maize "controlling elements" that are able to switch a gene off or on. (A gene responsible for a reddish color in corn may be switched on and off so fast that a single kernel may turn out speckled.) The controlling elements in maize are not always present, but when they are, they are added to specific chromosomal sites and can move from one site to another or even from one chromosome to another. These elements, therefore, act like episomes.

The discovery of proviruses and episomes has brought to light a phenomenon that biologists would scarcely have considered possible a few years ago: the addition to the cell's chromosome of pieces of genetic material arising outside the cell. The bacterial episomes provide new models to explain how two cells that otherwise possess an identical heredity can differ from each other. The episome brings into the cell a supplementary set of instructions governing additional biochemical reactions that can be superimposed on the basic metabolism of the cell.

The episome concept has implications for many problems in biology. For example, two main hypotheses have been advanced for the origin of cancer. One assumes that a mutation occurs in some cell of the body, enabling the cell to escape the normal growth-regulating mechanism of the organism. The other suggests that cancers are due to the presence in the environment of viruses that can invade healthy cells and make them malignant [see "The Polyoma Virus," by Sarah E. Stewart, Offprint #77, for further information]. In the light of the episome concept the two hypotheses no longer appear mutually exclusive. We have seen that proviruses, living peacefully with their hosts, can be induced to turn to the vegetative, replicating state by radiation or by certain strong chemicals—the very agents that can be used to produce cancer experimentally in mice. If defective, the provirus will not even make viral particles. Malignant transformation involves a heritable change that allows a cell to escape the growth control of the organism of which it is a part. We can easily conceive that such a heritable change may result from a mutation of the cell, from an infection with some external virus or from the action of an episome, viral or not. Thus in the no man's land between heredity and infection, between physiology and pathology at the cellular level, episomes provide a new link and a new way of thinking about cellular genetics in bacteria and perhaps in mice, men and elephants.

3

THE SMALLEST LIVING CELLS

by HAROLD J. MOROWITZ AND MARK E. TOURTELLOTTE

March 1962

What is the smallest free-living organism? The most likely candidate for this niche in the order of nature was discovered by Louis Pasteur when he recognized that bovine pleuropneumonia, a highly contagious disease of cattle, must be caused by a microbial agent. But Pasteur was unable to isolate the microbe: he could not grow it in nutrient broth nor could he see it under the microscope. Apparently it was too small to be seen.

Then, in 1892, the Russian investigator D. Iwanowsky succeeded in demonstrating that certain infectious agents were so small that they could pass easily through the porcelain filters used to trap bacteria. The size of the microbes postulated by Pasteur was comparable to that of Iwanowsky's organisms, which were subsequently named viruses. All viruses, however, are parasites of the living cell. The pleuropneumonia agent, on the other hand, is not. In 1898 Pasteur's successors E. I. E. Nocard and P. P. E. Roux were able to grow the pleuropneumonia agent in a complex, but cell-free, medium. In this respect the agent seemed more like a bacterium than a virus. In 1931 W. J. Elford of the National Institute for Medical Research in London, who developed the first filters in which pore size could be precisely determined, showed that cultures of the pleuropneumonia agent contained viable particles only .125 to .150 micron (.0000125 to .000015 centimeter) in diameter. Thus the particles were smaller than many viruses. Yet, as subsequent investigations have shown, the particles fully satisfy the definition "free-living": they have the ability to take molecules out of a nonliving medium and to give rise to two or more replicas of themselves.

More than 30 strains of this tiny organism have now been isolated from soil and sewage, as contaminants from tissue cultures and from a number of animals,

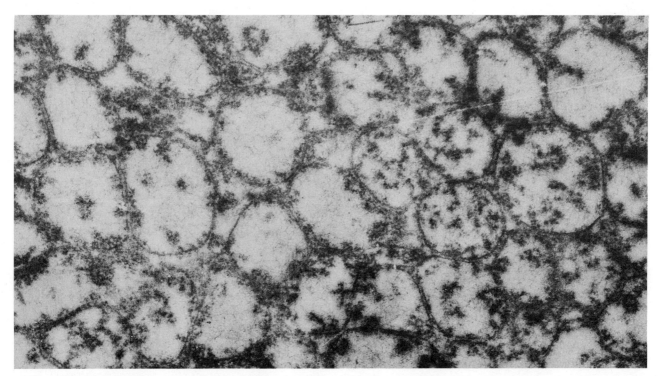

CELLS OF PLEUROPNEUMONIA-LIKE ORGANISM, abbreviated PPLO, are seen in cross section in this electron micrograph made by Woutera van Iterson of the University of Amsterdam. The cells, which are enlarged 72,000 diameters, are not the smallest PPLO's that have been observed. Nevertheless, they are only about 50 per cent larger than the vaccinia virus. Unlike the virus, however, these cells and smaller PPLO's meet a biologist's criterion for life: they are able to grow and reproduce in a medium free of other cells.

32 LEVELS OF COMPLEXITY

including man. In veterinary medicine one or another of them has been identified as the cause of a respiratory disease in poultry, of a type of arthritis in swine and of an udder infection in sheep. Although a pleuropneumonia organism was implicated in cases of human urethritis (inflammation of the urethra), it was not until January of this year that one of them was positively identified as an agent of disease in man. Robert M. Chanock and Michael F. Barile of the National Institutes of Health and Leonard Hayflick of the Wistar Institute of Anatomy and Biology then published their finding that an organism called the Eaton agent, first isolated in 1944, is actually a member of the pleuropneumonia group and is the cause of a common type of pneumonia. Because these organisms pass through filters (like viruses) and grow in nonliving media (like bacteria) they are considered by some workers to be a bridge between these two large classes of organism, and because they show obvious differences from both bacteria and viruses they have been accorded the status of a separate and distinct order: *Mycoplasmatales*. Because of their similarity to the original pleuropneumonia

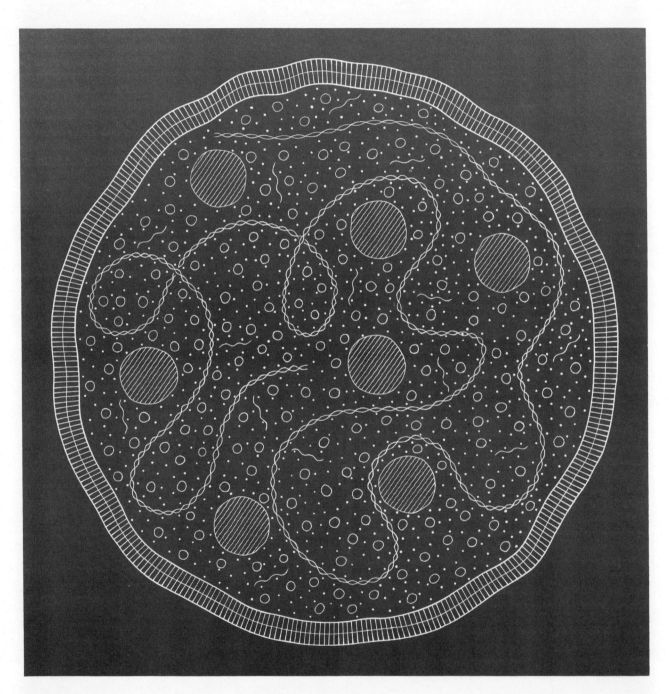

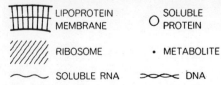

SCHEMATIC REPRESENTATION of a single cell of a PPLO is based on the authors' chemical analysis of *Mycoplasma gallisepticum*, which causes a respiratory disease in poultry. Deoxyribonucleic acid (DNA) and ribonucleic acid (RNA), found both in the ribosomes and in soluble particles, constitute 12 per cent of the total weight of the cell. The soluble proteins are similar to those in larger cells. The delicate cell membrane is composed of successive layers of protein, lipid, lipid and protein.

organism they are usually referred to as pleuropneumonia-like organisms, abbreviated to PPLO.

Although some very small bacteria are smaller than the larger PPLO, none is as small as the smaller PPLO: .1 micron (.00001 centimeter) in diameter. This is a tenth the size of the average bacterium; it is only a hundredth the size of a mammalian tissue cell and a thousandth the size of a protozoon such as an amoeba. But as the British mathematical biologist D'Arcy Wentworth Thompson observed some years ago, a major factor in any comparison of living things is mass, and mass varies as the cube of linear dimension. By such reckoning a protozoon is a billion times heavier than a PPLO. This vast gap in size gains vividness in the mind's eye from the reckoning that a laboratory rat is about a billion times heavier than a protozoon. A protozoon weighs .0000005 gram; a PPLO weighs a billionth as much: 5×10^{-16} gram.

In terms of linear dimensions the smallest PPLO is as close in size to an atom as it is to a 100-micron protozoon. A hydrogen atom measures one angstrom unit (.0001 micron) in diameter; a PPLO cell .1 micron in diameter is only 1,000 times larger. The existence of such a small cell raises intimate questions about the relationship of molecular physics to biology. Does a living system only a few orders of magnitude larger than atomic dimensions possess sufficient molecular equipment to carry on the full range of biochemical activity found in the life processes of larger cells? Or does the minuscule amount of molecular information it can carry compel it to operate in a simpler way? What biological or physical factors place a lower limit on the size of living cells?

In our laboratory at Yale University we have cultured 10 distinct strains of PPLO, clearly distinguished from one another by their metabolic behavior and by the antibody responses they produce in rabbits. Our work so far has been concentrated primarily on two of these strains: *Mycoplasma laidlawii*, a strain that is normally free-living in nature, and *Mycoplasma gallisepticum*, which causes chronic respiratory disease in poultry. In the first, which contains the smallest cells we have thus far studied, we have been able to follow the life cycle. In the second we have been able to determine details of chemical composition and structure.

At many stages in its life cycle the individual PPLO cell is too small to be

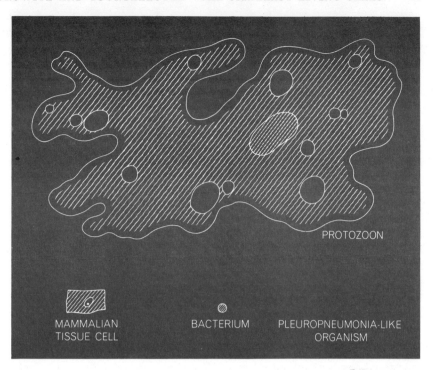

SIZES OF VARIOUS CELLS are compared. A protozoon, with a diameter of .01 centimeter, is 10 times bigger than a tissue cell, 100 times bigger than a bacterium and 1,000 times bigger than the smallest PPLO, with a diameter of .1 micron, or .00001 centimeter.

seen in the light microscope. In the electron microscope, however, we have been able to examine at least four different types of cell in *M. laidlawii*. One, called an elementary body, is a small sphere between .1 and .2 micron in diameter. A second is somewhat larger than this. A third is still larger: up to a full micron in diameter, about the size of a bacterium. A fourth type, which is of similar size, contains inclusions that are about the size of elementary bodies. In addition to observing the cell sizes directly, we measured them by forcing the cultures through filters with pores of various sizes and then examining in the electron microscope the material that had gone through the filters [see illustration on page 34]. To determine the size of the smallest PPLO cells we calibrated our filters by performing filtrations on two viruses of known size: the influenza virus, which is .08 to .1 micron in diameter, and the vaccinia virus, which is .22 by .26 micron in size. The smallest PPLO cells lie between these two; they are larger than the influenza virus but smaller than the vaccinia virus.

To separate the smallest cells of the strain from the others we had to employ the method of density-gradient centrifugation [see bottom illustration on next two pages]. This technique derives its effectiveness from the fact that cells as small as the PPLO vary in density as well as in size as they go through their life cycle. The density at each phase depends on the changing chemical composition of the cell and closely approximates the mean of the densities of its constituents. Salt solutions of different concentration are layered in a centrifuge tube, and the cell culture is added at the top. When the tube is inserted in the centrifuge and spun at high speed, cells of various sizes settle in the layer of salt solution that has a density equal to their own. Centrifuging a 72-hour culture of *M. laidlawii* in solutions that varied in density from 1.2 to 1.4 (the density of water is 1.0) revealed three bands. Examination of these bands in the electron microscope showed the bottom band contained large cells; the top band, elementary bodies; and the middle band, cells of intermediate size and large cells with inclusions.

Starting with elementary bodies thus isolated from a culture, we have been able to follow a culture of *M. laidlawii* through its life cycle. Our method is to sample the culture at periodic intervals and inspect the samples in the electron microscope. Young cultures—about 24 hours old—are primarily composed of large cells. Cultures about six days old, on the other hand, are predominantly elementary bodies. Samples

34 · LEVELS OF COMPLEXITY

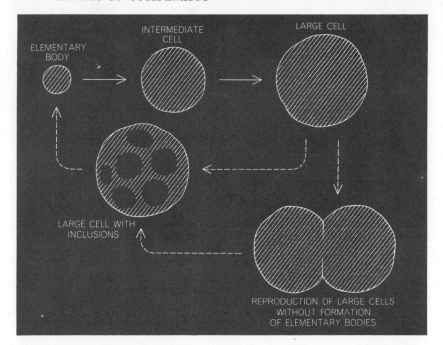

LIFE CYCLE of the PPLO *Mycoplasma laidlawii* is outlined. Elementary bodies grow to intermediate cells and then large ones. The large cells may divide, some developing inclusions released as elementary bodies, or may develop and release inclusions directly.

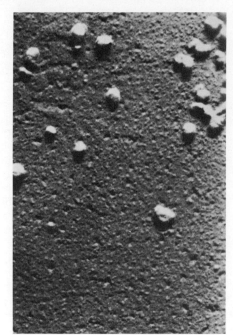

FOUR TYPES OF CELL in the PPLO *M. laidlawii* are seen in these electron micrographs. First micrograph (*far left*)

taken over the course of the five-day interval suggest that this strain has two methods of reproduction. In both cases the organism goes through a cycle in which elementary bodies are transformed first into intermediate cells, then into large cells and then back into elementary bodies again. Differences in composition between young and old cultures show, however, that the organism can probably adopt one of two courses once it has reached the large cell stage. In one cycle the large cells develop inclusions, which are apparently released as elementary bodies. In the second the large cells seem to reproduce by binary fission. Thereafter it appears that some of them form inclusions from which new elementary bodies are liberated. In either case the new elementary bodies begin the life cycle all over again.

We have not so far been able to establish the mode of reproduction in *M. gallisepticum*. None of our cultures has

SEPARATION OF PPLO CELLS BY SIZE AND TYPE is achieved by density-gradient centrifugation. This method can be used because small cells have different densities at different times in their life cycle.

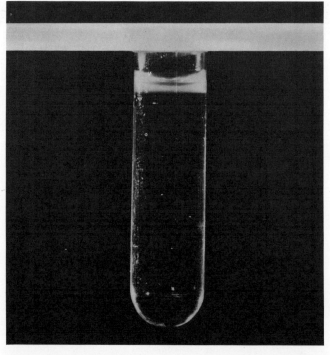

In photograph at far left two densities of salt solution are layered in a centrifuge tube. In the second photograph a PPLO culture has been added at the top of the tube. In the third photograph the

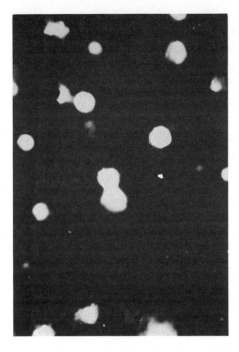 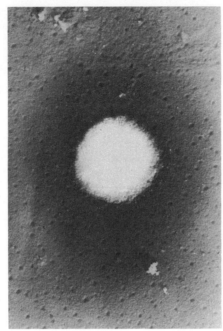 

shows elementary bodies about .1 micron in diameter. The second shows intermediate cells. The third shows large cells about 1 micron in diameter, and fourth shows large cells that have developed inclusions. The inclusions may be released as elementary bodies to begin the life cycle again. All the micrographs, which enlarge the cells 17,750 diameters, were made by the authors.

revealed either elementary bodies or large cells. All the cells we have seen appear uniformly spherical and all appear to be about .25 micron in diameter [*see illustration on page 32*]. Our work with *M. gallisepticum* has helped, however, to settle the question of whether or not these tiny organisms conduct the same biochemical processes as larger cells.

Chemical analysis shows that the *M. gallisepticum* cell has the full complement of molecular machinery. In the first place, the nonaqueous substance of the cell contains 4 per cent deoxyribonucleic acid (DNA) and 8 per cent ribonucleic acid (RNA). These large molecules have been identified in larger cells as the bearers of the genetic information that governs the synthesis of the other components of a cell. Moreover, we find that in the tiny cells the composition of these molecules, the so-called base ratios, falls within the normal range. The DNA ap-

 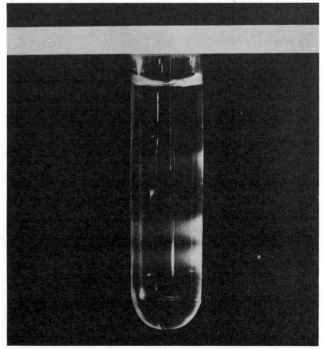

tube is ready to be placed in a container (*left*) and fastened to a rotor, part of which is seen at right. The rotor is then placed in a centrifuge. The last photograph shows the tube after centrifugation. The PPLO's have settled in three bands. The bottom band contains large cells; the middle band, intermediate cells and large cells with inclusions; the top band, elementary bodies.

36 LEVELS OF COMPLEXITY

COLONIES OF THE EATON AGENT, now known to be a PPLO, are seen at a magnification of 600 diameters in this light micrograph. Discovery that the Eaton agent causes a type of pneumonia in man is the first proof that a PPLO can produce human disease.

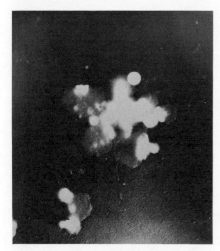

PPLO CELLS from a rat strain are contrasted in size with a .26-micron sphere. The large cell seen at the center contains inclusions.

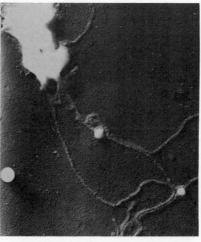

FILAMENTS are observed in many strains of PPLO. The cells shown here are the same strain as those in micrograph at left.

HUMAN PPLO is seen in this electron micrograph. All three micrographs, magnified 16,000 diameters, were made by the authors.

pears as the familiar double-stranded helix found in the chromosomes of larger cells, and most of the RNA appears to be in the form of particles resembling ribosomes, the organelles that are believed to conduct protein synthesis in larger cells. The soluble proteins in the cell seem to have the usual range of size and variety, and the amino acid units of which they are composed occur in the expected ratios.

In several respects this PPLO cell appears to resemble animal cells more than it does plant cells or bacteria. The composition of its fatty substances, including cholesterol and cholesterol esters as an essential element, is characteristic of animal cells. More important, it has no rigid cell wall but has instead a flexible membrane that, in other strains of PPLO, permits the cells to assume a great variety of shapes. In spite of its delicacy the PPLO membrane is able to fulfill the functions of a cell membrane. It effectively distinguishes the cell from its environment and it is firm enough to contain the cell's internal structures in a coherent way. Indirect measurement of its electrical properties shows that they fall within the normal range. At a sufficiently high magnification the membrane can be seen in the electron microscope. It measures about 100 angstrom units (.01 micron) in thickness, which is typical of many animal cells.

Thus far we have been able to demonstrate more than 40 different enzymatic functions in *M. gallisepticum*. These include the entire system of enzymes necessary for the metabolism of glucose to pyruvic acid, one of the processes by which cells extract energy from their nutrients. Therefore the evidence points to a considerable biochemical complexity in these organisms. In spite of their size they seem to compare in structure and function with other known cells.

The demonstration that these tiny cells are indeed free-living compels a further question: Can there be other cells, even smaller than the PPLO and as yet undiscovered, that possess the capabilities for growth and reproduction in a cell-free medium? The mere detection of such cells presents a challenge to the ingenuity and technical resources of the biologist. If the cells happen to be pathogens, they might be discovered by the diseases they produced. If they are harmless to other forms of life, they might put in a visible appearance by causing turbidity in a culture medium through mass growth, or they might

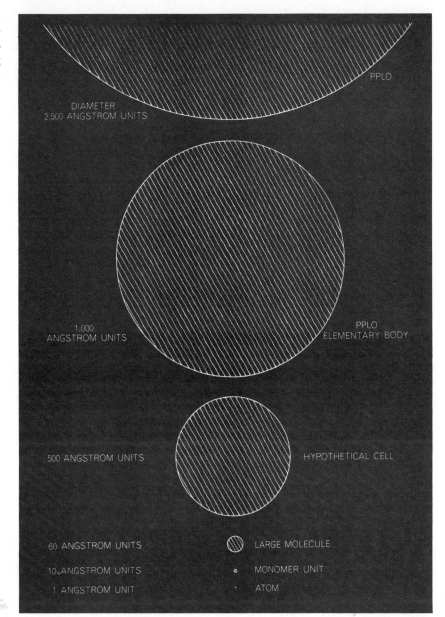

	NUMBER OF ATOMS IN DRY PORTION OF CELL	MOLECULAR WEIGHT OF DNA (DNA = 4 PER CENT OF CELL CONTENT)	NUMBER OF MONOMER UNITS (AMINO ACIDS AND NUCLEOTIDES)	NUMBER OF LARGE MOLECULES
PLEUROPNEUMONIA-LIKE ORGANISM (DIAMETER: 2,500 ANGSTROM UNITS)	187,500,000	45,000,000	9,375,000	18,750
PLEUROPNEUMONIA-LIKE ORGANISM ELEMENTARY BODY (DIAMETER: 1,000 ANGSTROM UNITS)	12,000,000	2,880,000	600,000	1,200
HYPOTHETICAL CELL (DIAMETER: 500 ANGSTROM UNITS)	1,500,000	360,000	75,000	150

PPLO CELLS AND ATOMS can be shown on the same scale; a PPLO elementary body is only 1,000 times larger than a hydrogen atom. Table shows number of atoms, molecular weight of DNA, number of monomer units (the repeating units of a large molecule) and number of large molecules anticipated in PPLO cells and in smallest theoretical cell.

show up in electron microscope preparations. A very small cell, however, might escape detection by any of these methods. If the population it formed grew to a concentration of only 100,000 cells per cubic centimeter of the growth medium, the chances of finding it would be slight. There is no assurance that the proper growth medium could be found to culture such cells. It may well be, therefore, that the PPLO is not the smallest living cell.

Yet there are lower limits, in theory at least, to the size of a living organism. Biological considerations suggest one such limit. A cell must have a membrane, if only to provide coherence for its structure. Since all cell membranes so far studied appear to be on the order of 100 angstrom units (.01 micron) in thickness, it would seem that no cell could exist that had a diameter less than 200 to 300 angstrom units (.02 to .03 micron), or about a tenth the diameter of the *M. gallisepticum* cell. Biochemistry suggests that the smallest cell would have to be somewhat larger in size. The complexity of function necessary to growth and reproduction indicates that the minimal organism must be equipped to conduct at least 100 enzymatically catalyzed reactions. If each reaction were mediated by a single enzyme molecule, the molecules would require a sphere 400 angstrom units in diameter to encompass them and the raw materials on which they operate.

Biophysics suggests another limit to the smallest size. In his little book *What Is Life?* the physicist Erwin Schrödinger pointed out that a cell has to survive against the ceaseless internal deterioration caused by the random thermal motion of its constituent molecules. In a very small cell even small motions are large in proportion to the entire system, and small motions are statistically more likely to occur than large ones. With only one or at most a few molecules of each essential kind present in the smallest conceivable cell, the most minute dislocation might be enough to disable the cell.

The foregoing reasoning seems to set 500 angstrom units as the minimum diameter of a living cell. A cell of this size would have, in its nonaqueous substance, about 1.5 million atoms. Combined in groups of about 20 each, these atoms would form 75,000 amino acids and nucleotides, the building blocks from which the large molecules of the cell's metabolic and reproductive apparatus would be composed. Since these large molecules each incorporate about

FILTRATION of a PPLO culture requires filters calibrated for viruses. At left is a tube holding an unfiltered culture of PPLO's. In middle, PPLO culture is forced through a filter with .22-micron pores. Tube at right contains filtered PPLO elementary bodies.

500 building blocks, the cell would have a complement of 150 large molecules. This purely theoretical cell would be delicate in the extreme, its ability to reproduce successfully always threatened by the random thermal motion of its constituents.

The smallest living organism actually observed—the .1-micron, or 1,000-angstrom, elementary body of *M. laidlawii*—is only twice this diameter. Its mass, of course, is eight times larger, and calculation from its observed density shows that it may contain 1,200 large molecules. This is a quite finite number, and since the organism grows to considerably larger size in the course of its reproductive cycle it cannot be said that 1,200 large molecules constitute its complete biochemical equipment. In the case of *M. gallisepticum*, however, we know that a diameter of .25 micron—only five times the theoretical lower limit—does encompass an autonomous metabolic and reproductive system. Our chemical analysis shows that the entire system is embodied in something less than 20,000 large molecules [*see chart on page* 37].

This is still an exceedingly small amount of material to sustain the complexity of biochemical function necessary to life. In fact, the portion of it allotted to the genetic function seems inadequate to the task. The 4 per cent of its dry substance that is DNA has a total molecular weight of 45 million. Since, according to current views of genetic coding, it takes an amount of DNA with a molecular weight of one million to encode the information for the synthesis of one enzyme molecule, *M. gallisepticum* would seem to contain enough genetic material to encode only a few enzymes beyond the 40 we have identified so far. That is far short of the 100 enzymes thought to be the minimum for cellular functions. It may be that the enzymes of very small cells are less specific and hence more versatile in their action than the enzymes of larger ones. On the other hand, it may prove necessary to re-examine prevailing ideas about the way information is encoded in the genetic material.

Questions of this kind suggest the principal challenge of very small cells. If they are indeed simpler than other cells, they can tell much about the basic mechanisms of cell function. If, on the other hand, they are functionally as complicated as other cells, they pose the fundamental question of how such functional complexity can be carried in such tiny pieces of genetic material.

Part II

ORGANELLES

Organelles

II

INTRODUCTION

Soon after the cellular nature of organisms was understood, the interior of the living cell became the object of a scrutiny limited only by the resolution of the available optical tools. The history of this increasingly intimate examination has been marked by a progressive shift away from the concept of the cell as a sac of relatively homogeneous aqueous solution, and toward the view that nucleus and cytoplasm are highly structured systems. First, a number of large and obvious inclusions—nuclei, chloroplasts, mitochondria—were recognized. Then it was discovered that even the remaining cytoplasm, far from being weak and watery, could undergo changes of state suggestive of substantial internal structure. Most recently, the electron microscope has revealed a pervasive system of internal membranes—a "cytoskeleton"—as well as a complex ultrastructure in many of the classic organelles.

This section begins with a treatment of the lipoprotein boundary of the cell. In "The Membrane of the Living Cell," J. David Robertson demonstrates an approach to general biological problems that is often productive: concentration upon a specific system in which the component of interest shows unusual elaboration. For studies on membrane structure, the myelin sheath of vertebrate nerve fibers, which consists of the rolled-up membrane of a satellite Schwann cell, has been such a system. The importance of the external membrane in regulating the lively commerce that takes place between the interior of a cell and its environment has long been realized; indeed, before the advent of electron microscopy, studies on the permeation of various materials were the primary means of analyzing the composition of the membrane. The major surprise provided by the higher-resolution techniques has been the similarity of—and the continuity between—the cell surface and a system of previously unknown internal membranes comprising the endoplasmic reticulum, Golgi apparatus, and nuclear envelope. Membranes enveloping mitochondria and other organelles have also been shown to have the same "unit" structure revealed in the myelin sheath.

The membrane is responsible for the regulation of exchange; but in certain complex cells, membrane-associated organelles may have important integrative functions. In "Cilia," Peter Satir describes the most important class of these organelles—a class that is remarkably ubiquitous in unicellular organisms as well as in metazoan cells. Among the important questions raised by the distribution and structure of cilia are these: How does transfer of energy occur along the protein chains of an essentially fibrillar system? How is it that rows of cilia in a single cell are coordinated so that they beat in a successive wave? How is the same coordination achieved in a row of ciliated cells in a multicellular organism? Finally, why have cilia apparently been evolved into sensory organelles in a number of metazoa—notably, the sensory hairs of the inner ear and the photosensitive rods and cones of the retina? Can one suppose that the fibrillar organization of the cilium has properties uniquely appropriate for the conduction of excitation?

Of the strictly cytoplasmic organelles, the mitochondrion is the largest and, partly for this reason, has had a particularly long experimental history. Mitochondria were visible (and stainable with certain vital dyes) well before the days of electron microscopy. Later, they were the objects of the first correlations between cytology and biochemistry, when it was shown that a fraction of particles having the size and properties of mitochondria could be centrifuged out of a homogenate of cells, and that most of the metabolic activity (the oxygen uptake) of the homogenate was concentrated in that fraction. This important discovery established the relationship between the mitochondrion and cellular energetics, and began a series of experimental probings at increasingly high resolution into the relationship between mitochondrial structure and function. In "The Mitochondrion," David E. Green describes the state of that search. Whatever uncertainties remain in the assignment of specific events to specific places, it is nevertheless evident that most of the chemistry of the mitochondrion takes place not in solution but in ordered structural components of the organelle itself. Because these structures are within the resolving range of the electron microscope, the concerns of anatomy and of biochemistry are blended at this new level of inquiry. Green concludes with a theme strikingly similar to Robertson's: a suggestion that the mitochondrial membrane may be the functional as well as structural equivalent of other biological membrane systems.

The final article in this section, "The Lysosome" by Christian de Duve, demonstrates that the search for new subcellular particles may not yet be ended. It describes an organelle, which was fully defined for the first time only within the past several years, whose biochemical function is altogether different from that of the similarly sized mitochondrion. These particles contain a battery of hydrolytic enzymes that function not only in the intracellular digestion of nutrient molecules but also in the postmortem chemical breakdown of cell contents.

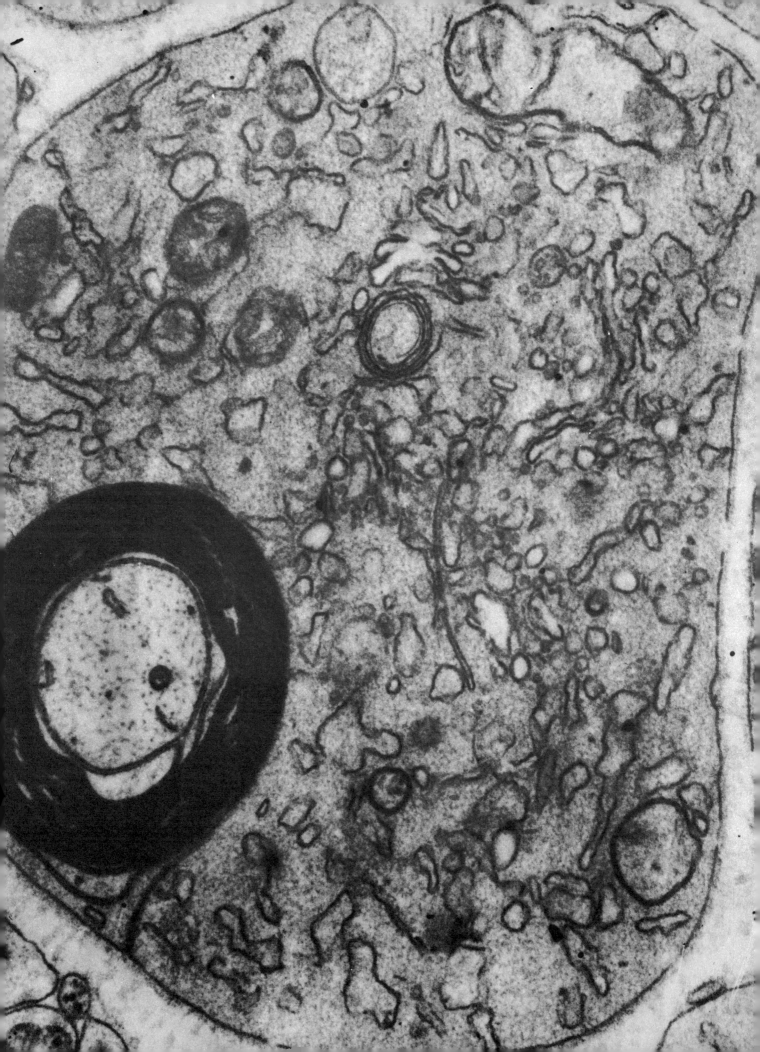

4

THE MEMBRANE OF THE LIVING CELL

by J. DAVID ROBERTSON   April 1962

Almost everyone who has looked at a cell through a microscope has assumed that it is surrounded by a membrane. Some sort of coating seems necessary to maintain the integrity of the soft, yet far from shapeless, bit of protoplasm in its liquid environment and to control the constant exchange of material between inside and outside. But the nature of the membrane—its thickness and composition—was until quite recently a matter of conjecture. The membrane is far too thin to be visible under the most powerful light microscope.

Now, with the help of the electron microscope, the membrane has been seen. Its thickness has been measured. By combining the new direct evidence with the results of older studies, the general features of its molecular structure have been deduced and a fundamental constant pattern of organization common to all cells has been defined. Moreover, the cell membrane has proved to be much more than an outer coat: in most cells it forms an essential part of the internal structure. The present understanding is the culmination of a fascinating search, extending over many years and following several pathways that have at last converged.

UNIT MEMBRANE surrounds a Schwann cell protecting the axon of mouse sciatic nerve. Axon (*lower left*) is encased in its membrane and in a myelin sheath, composed of layers of Schwann cell membrane. Round object at top center is a growth spiral, where new membrane may be produced. Mitochondria, seen as oblong object at upper right and larger round objects throughout cell, are composed in part of paired unit membranes. Other membranous forms are endoplasmic reticulum. Micrograph enlarges structures 75,000 diameters.

The indirect evidence for a definite cell membrane structure has long been highly persuasive, if not conclusive. Studies of the traffic between the interior of the cell and its surroundings show molecules of various kinds passing back and forth in an orderly sequence. Often they move against strong concentration barriers that can be overcome only by expenditure of energy [see "How Things Get into Cells," by Heinz Holter, Offprint #96, for further information]. For example, sodium ions are continuously pumped out of the cell while a high concentration of potassium ions (with respect to the surrounding liquid) is maintained within the cell. In nerve fibers this activity is related to an electric potential difference between inside and outside, which forms the basis for the transmission of nerve impulses. It is possible to imagine how such electrochemical processes could be carried out without a membrane, but they are much more readily understood by supposing that there is a separate, specialized structure at the boundary, capable of converting and expending energy.

However strongly biologists may have believed in the cell membrane, they could not hope to see it directly in the light microscope. At best this instrument has a resolving power of about a thousandth of a millimeter. Assuming, as seemed likely, that the membrane is only a few molecules thick, it would be expected to measure less than a hundred-thousandth of a millimeter across. Nevertheless, light microscopists had fallen into the habit of talking about a cell membrane, by which they meant the thinnest dense line they could see next to the cytoplasm (the part of the cell that lies between the membrane and the nucleus). But there was little assurance that this line represented any particular constant structure. After all, structures radically different from one another could look the same.

With the advent of the electron microscope the nature of the problem changed completely. Now the microscopist has a device with about 1,000 times more resolving power. The thinnest dense line seen in the light microscope stands out as a complex structure containing several denser and lighter layers. How many of them constitute the membrane proper? The former stratagem of defining it as the thinnest dense line will not do. Present-day electron microscopes are quite capable of resolving a layer one protein molecule thick, and so this approach would lead to the definition of the cell membrane as a monomolecular layer. From the physiological point of view, however, it seems unlikely that a layer one molecule thick could carry out the functions of cell membranes.

The solution to the problem of defining the limits of the membrane has come chiefly from studies of long nerve fibers, or axons. These are threads of protoplasm, extending from nerve cells located mainly in the spinal cord or brain, that carry messages to and from the central nervous system. In human beings axons are commonly a few thousandths of a millimeter in diameter and up to several feet long. They are accompanied along their entire length by a succession of specialized satellite cells, named Schwann cells after Theodor Schwann, one of the great 19th-century cell biologists.

Under the light microscope two different types of nerve fiber could be made out. One is enclosed in a thick sheath of fatty material called myelin; the other is not. Whether the myelin is part of the axon or of the Schwann cells was not clear. It could be seen, however, that the sheath is broken up into segments, one for every Schwann cell. Almost always each axon in myelinated fibers has its own set of Schwann cells. In unmyelinated fibers several axons usually share one set. It appeared to light microscopists that the unmyelinated fibers were completely surrounded by the substance of their Schwann cells.

The first views of nerve fibers in the electron microscope, obtained by H. Fernandez-Moran in Venezuela and Fritiof Sjöstrand in Sweden, told a different story. In the first place the myelin sheath proved to be part of the Schwann cell, not the axon. Seen in cross section, the sheath turned out to consist of many repeating layers, each 110 to 140 angstrom units thick (an angstrom unit is a ten-millionth of a millimeter). In each repeating stratum there is a "major dense line" about 30 angstroms thick followed by a lighter zone 80 to 110 angstroms across. This zone is bisected by another dense layer, called an intraperiod line [see illustrations at bottom of these two pages].

The stratified myelin structure held the key not only to delimiting the cell membrane but also to determining its molecular organization. These problems were solved by combining the results of two lines of investigation: (1) older biochemical and biophysical studies of myelin, which I shall discuss shortly, and (2) electron microscope studies that showed how the myelin sheath is formed. The latter investigation began early in the 1950's, when Herbert S. Gasser of the Rockefeller Institute first examined unmyelinated nerve fibers by electron microscopy. He soon discovered that, contrary to what had been thought, the fibers are not completely enclosed within the cytoplasm of the Schwann cells. Instead they lie close to each satellite cell or are sometimes embedded in its surface. No matter how deep the intrusion is, the overhanging lips of Schwann cytoplasm do not quite touch. They are always separated by a gap 100 to 150 angstroms wide. Topologically the fiber is outside the Schwann cell. Gasser named the almost closed pair of lips a mesaxon.

Shortly afterward Betty Ben Geren-Uzman of the Harvard Medical School began to study the development of the myelin sheath in embryonic chick nerves. She found some unmyelinated fibers arranged like those Gasser had described but with only one axon per Schwann cell, like myelinated fibers. She noted that some of these fibers had long mesaxons drawn loosely into a simple spiral running around the axon in several loops. It occurred to her that the myelin sheath might be nothing more than a mesaxon, greatly drawn out and wound around the axon in a tightly packed spiral.

In 1955 I published an electron micrograph of an adult myelin sheath that confirmed Mrs. Uzman's conjecture [see illustrations on page 48]. Although technically inferior to present-day micrographs, the picture displayed for the first time the outer and inner end of the postulated mesaxon loop. Further-

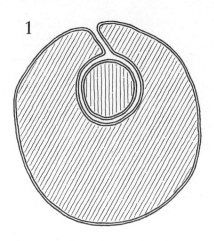

FORMATION OF MYELIN SHEATH from Schwann cell membrane is depicted. The axon, shown as the circle with vertical

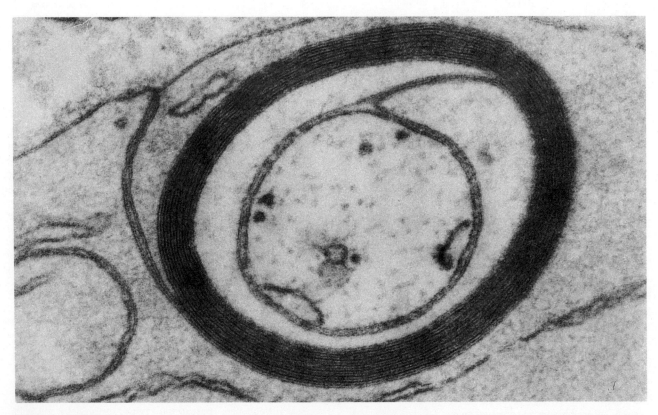

MYELIN SHEATH is enlarged 160,000 diameters in this micrograph. The dark lines represent the inner sides of the membranes that come together to form the sheath. The fainter lines are the outer sides of these membranes. The two membranes composing the sheath can be distinguished clearly. The diagram on the opposite page shows the left-hand section of the sheath schematically.

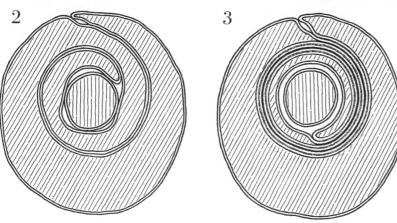
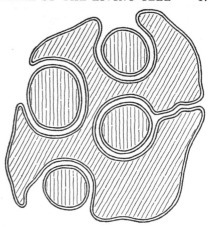

hatching, is surrounded by its own membrane and embedded in a Schwann cell (*1*). The myelin sheath is produced when membranes separating overhanging lips of Schwann cell cytoplasm come together and elongate (*2*) to form a tightly packed spiral around the axon (*3*).

UNMYELINATED AXONS are also embedded in Schwann cells. But the membranes remain separated and do not elongate.

more, it was clear enough to suggest, although not to prove, that the turns of the spiral are tightly packed, with neither additional material from the Schwann cytoplasm nor any extracellular material between them. In other words, it appeared that the myelin sheath might be nothing but a winding of a double layer of the Schwann cell membrane, just as Mrs. Uzman had postulated. The idea was exciting because a great deal was already known about the molecular structure of myelin, and it now seemed possible to apply this knowledge directly to the cell membrane.

Research on myelin goes back to the latter part of the 19th century. At that time it had already been learned that a large part of the myelin sheath is dissolved away by fat solvents. This fact, plus the staining characteristics of intact myelin, made it clear that it contains a large proportion of fatty substances, or lipids. In addition, it was soon realized that the material remaining after the lipids are dissolved consists largely of protein.

The next major advance came in the 1930's through studies of myelin with the polarizing microscope, particularly by the noted German biophysicist W. J. Schmidt. With this instrument it is possible to discover something about the internal molecular architecture of a translucent material. If it is composed of long, thin molecules with their long axes parallel to one another and a beam of plane-polarized light is sent through the material, the direction of polarization of the emerging beam is rotated. From the amount and direction of the rotation the orientation of the molecules can be deduced.

A myelinated nerve fiber, examined in the polarizing microscope, alters the polarized beam in such a way as to indicate that the sheath contains long molecules lined up radially with respect to the axis of the fiber. Moreover, pure lipid extracted from myelin can be shown to produce a similar effect under the same optical conditions. Therefore it seems reasonable to suppose that it is the lipid molecules in myelin that have the radial orientation. As has been mentioned, the residue remaining after the extraction of lipid from myelin consists largely of protein. This residual material also alters polarized light, but in a way opposite to that caused by intact myelin. The protein molecules, then, must be aligned at right angles to the lipid molecules; they lie parallel to the fiber axis. Schmidt therefore concluded that myelin consists of alternating layers of lipid and protein molecules stacked at right angles to each other.

As for the lipid fraction, it had also

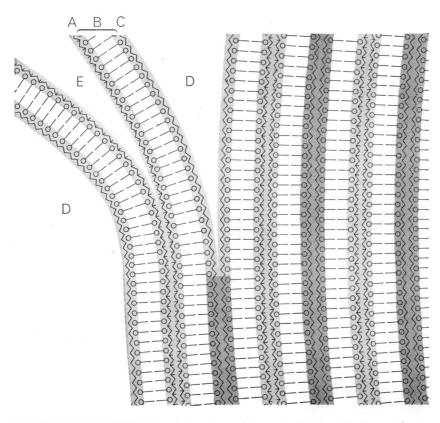

MOLECULAR STRUCTURE of unit membranes forming myelin sheath is diagramed. *A* is a layer of protein and carbohydrate on outer side of membrane. *B* is its lipid core. *C* is the protein layer of the inner side. *D* is Schwann cell cytoplasm, *E* extracellular matter.

been studied intensively by itself and a good deal was known about it. The phospholipid molecules found in myelin consist of two parts: a long, straight chain of carbon and hydrogen atoms attached to a more complicated structure containing atoms of carbon, hydrogen, oxygen, nitrogen and phosphorus. The two portions can be split apart and examined separately. The hydrocarbon chains are soluble in organic solvents such as ether and chloroform but are not soluble in water; the more complex end of the molecule, containing phosphorus and nitrogen, is not soluble in these substances but dissolves readily in water. The difference in solubility is explained by the electrical characteristics of the two fractions. In the water-soluble material the negatively charged electron clouds in the atoms are slightly displaced with respect to the positive nuclei, so the group as a whole is electrically polarized: one end is positive and the other negative. Water molecules are also strongly polarized, so the substances attract each other. In the hydrocarbon chain, on the other hand, the centers of negative and positive charge are symmetrically disposed, and there is no such electrical polarization.

Intact phospholipid molecules, then, have a "hydrophilic" end, which is attracted to water, and a "hydrophobic" end, which is not. Under certain conditions pure phospholipids take on an orderly arrangement in which the molecules lie side by side in sheets, the hydrophilic ends at one surface and the hydrophobic ends at the other. The sheets themselves pair up, with the hydrophobic, or nonpolar surfaces, facing each other. Such a pair is known as a bimolecular leaflet. The over-all ordered structure is a stack of many leaflets.

Like any other regular molecular array, a stack of bimolecular leaflets will diffract a beam of X rays in a meaningful pattern. Long before the electron microscope became available, X-ray diffraction techniques had revealed a good deal about the architecture of ordered lipids. From the angle through which the rays were bent, each bimolecular leaflet, containing mixed lipids of the kind readily extracted from nerve fibers, was computed to be about 65 angstroms thick.

X-ray diffraction was also applied to whole myelin, particularly by Francis O. Schmitt, Richard S. Bear and G. L. Clark, then at Washington University in St. Louis. They found a repeating unit 170 to 185 angstroms thick in the radial direction. Since the material consists of protein as well as lipid, the repeating interval could obviously contain no more than two bimolecular leaflets. It would also have room for a few layers of protein molecules with their long axes perpendicular to the radial direction. In this position they would be only about 15 angstroms thick. Schmitt's group concluded that myelin is made up of bimolecular leaflets alternating with protein layers.

Here matters rested until the development of the electron microscope. Now the structures of pure phospholipid and of myelin, which had been inferred from X-ray analysis, could be examined directly. The first electron micrograph of ordered phospholipids, made by Schmitt and Mrs. Uzman, displayed a series of alternating dense and light strata. Each of the strata was narrower than the known length of the phospholipid molecule, and so it was concluded that each dense layer must represent one part of the molecule and each light layer the other part. Then the question arose: Is the dense layer the hydrophilic or the hydrophobic part of the molecule? To find out I studied samples of phospholipid treated with water. When water penetrates the material, it runs between the polar surfaces of the individual molecular sheets but not between the nonpolar, or hydrophobic, surfaces. In the electron microscope it could be seen that the water had split each dense stratum into two layers and had left the light strata unchanged. Therefore the dense material represents the polar surfaces of the bimolecular leaflets and the lighter regions the hydrophobic carbon chains. This conception has recently received support from the work of Walther Stoeckenius of the Rockefeller Institute.

The appearance of whole myelin under the electron microscope was described earlier in this article. To recapitulate, it is made up of alternating major dense lines and light zones, the pattern repeating in an interval of 110 to 140 angstroms. Bisecting the light

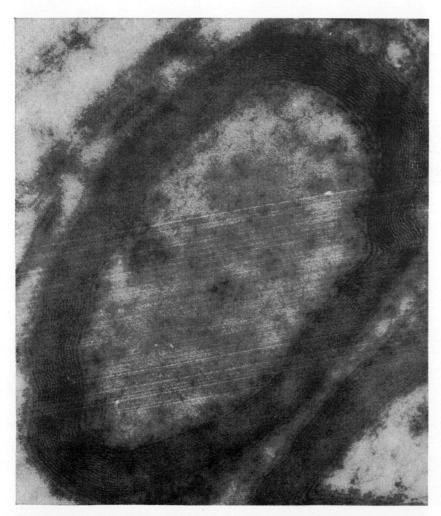

ELECTRON MICROGRAPH published by the author in 1955 showed that myelin is made of layers of Schwann cell membrane. Both outer and inner ends of myelin sheath can be seen in upper right-hand section of micrograph, which enlarges the structure 81,250 diameters.

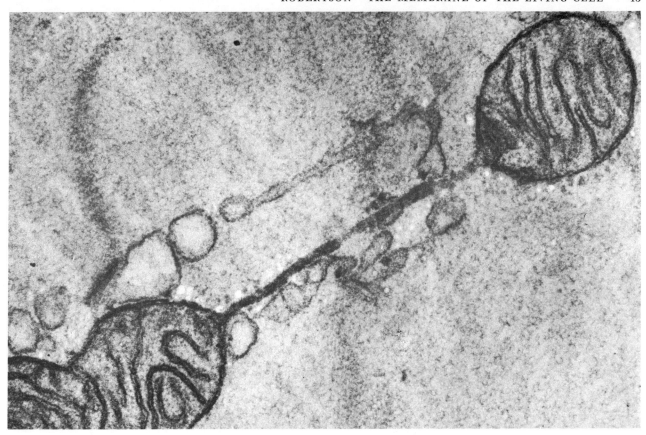

MITOCHONDRIA in frog muscle are seen here connected by a narrow tubular neck. Mitochondria, which usually appear separate from one another, are composed of a pair of unit membranes. In some electron micrographs the outer membrane is connected with the endoplasmic reticulum. As such micrographs and this one suggest, mitochondria may be formed when cytoplasm fills a cavity of the endoplasmic reticulum, which then pinches off to produce discrete organelles. The structures are enlarged 77,500 diameters.

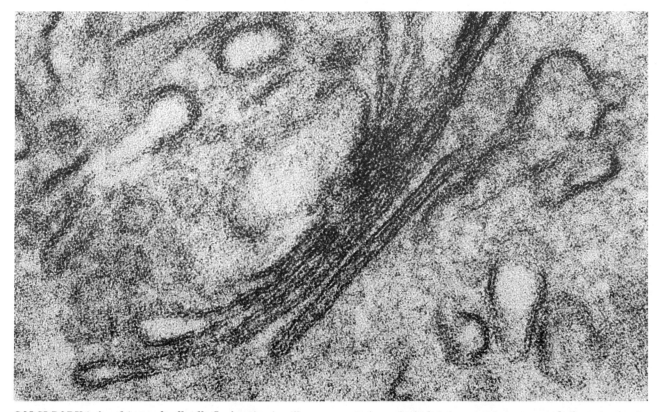

GOLGI BODY is found in nearly all cells. Its function is still unknown but its structure can be seen clearly. Like the endoplasmic reticulum, of which it is a part, it is composed of a pair of unit membranes. In this micrograph it is enlarged 240,000 diameters.

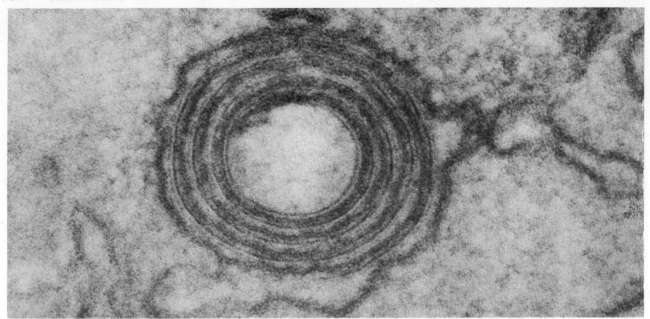

GROWTH SPIRAL is an elongated, cigar-shaped structure formed of a flattened sac bounded by a pair of unit membranes wrapped up in a spiral. It is often found in developing Schwann cells. The shape of the growth spiral may indicate that its outer membrane is the seat of new membrane synthesis for the growing cell. This micrograph shows the growth spiral in the sciatic nerve of a young mouse at an enlargement of 265,000 diameters. Like all the other micrographs accompanying the article, it was made by the author.

zones is a darker intraperiod line, which in the early micrographs was broken into short segments or dots.

In trying to correlate this picture with the earlier results, one must take into account the method used to prepare materials for electron microscopy. First they are "fixed" by treatment with certain chemicals; then they are embedded in a block of plastic and cut into very thin slices. J. B. Finean in England and Fernandez-Moran considered the effects of fixing and embedding on the myelin structure. They showed that the material shrinks just enough to reduce the size of the repeating interval from the value deduced from X-ray diffraction (170 to 185 angstroms) to the width seen in electron micrographs (110 to 140 angstroms).

As was pointed out above, the polar groups in lipid bimolecular leaflets were found to produce dense areas in an electron micrograph. (They do so because they combine readily with the fixing chemical, which contains heavy metal atoms that strongly scatter electrons.) Moreover, it was known that pure protein also forms dense lines or spots on fixation. Therefore the dense lines in the electron micrographs of myelin were interpreted as representing combined protein monolayers and lipid polar surfaces.

Much of this had been learned by about 1955. Considering what was also known about the origin of myelin from the Schwann cell, it was tempting to trace the laminated structure of myelin directly to the Schwann cell membrane. The evidence was not yet quite good enough, however. The main trouble was that the membrane itself could not be seen clearly.

Then in 1956 John H. Luft at the Rockefeller Institute introduced a new fixing technique using potassium permanganate instead of osmium tetroxide, which had been the only effective agent up to that time. When the method was tried on preparations of axons and Schwann cells, the hazy line that had bounded their cell bodies in earlier micrographs stood out clearly as a layered structure containing three distinct strata. Next to the cytoplasm is a thin dense line measuring about 20 angstroms in thickness. This borders a light central zone about 35 angstroms thick that is in turn bounded externally by another thin dense line about 20 angstroms thick. The whole structure measures about 75 angstroms across and is called a unit membrane.

Micrographs of developing myelin sheaths fixed with potassium permanganate show clearly that they consist simply of a double layer of Schwann cell unit membranes wound into a spiral. The major dense lines represent the intimate apposition of the inside, or cytoplasmic, surfaces of the membrane. The intraperiod lines are formed by the outside layers coming together as each layer wraps around the previous one. Since the new micrographs also demonstrate the absence of any material other than the unit membrane, the structure of myelin layers can now be related directly to that of the membrane. The membrane must consist of a core containing a single bimolecular leaflet of lipid, its polar surfaces pointing outward and covered by monomolecular films of nonlipid.

Until the introduction of the potassium permanganate technique the nonlipid material was assumed to be almost all protein. This turns out not to be so. Osmium tetroxide produces a dense line at the inside surface of the membrane but usually does not produce such a line

EVOLUTION OF CELL may have followed the pattern diagramed here. First drawing shows matrix material bounded by a unit

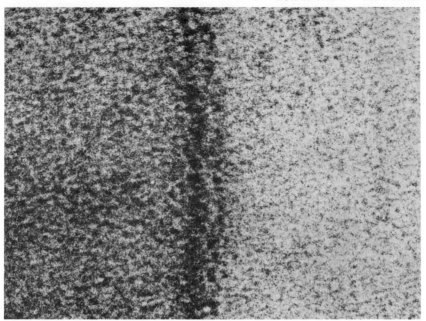

MEMBRANE of human red blood cell is seen as a double line separating the cell body (*at left*) from extracellular matter. The red blood cell is composed only of cytoplasm bounded by a unit membrane. Although it originates from a more complex cell, its simplicity suggests the form the earliest cells may have taken. Micrograph enlarges the cell 925,000 diameters.

at the outside. Potassium permanganate, on the other hand, gives a density at both surfaces. Obviously the inner layer must differ chemically from the outer layer. The difference also carries over to myelin. Treated with osmium tetroxide, the substance displays a heavy major dense line (formed by the inside surfaces of the membrane), but the intraperiod line (representing the outside surfaces) is broken into irregular dots or granules. In contrast potassium permanganate yields a solid intraperiod line as thick but not quite as dense as the major dense line.

Evidently osmium tetroxide does not fix the outer half of the Schwann cell membrane as well as potassium permanganate does. Recently it has been found that osmium tetroxide reacts poorly with many sugars and more complex carbohydrates. It therefore seems likely that the outer surface of the membrane is mainly carbohydrate in nature and the inside mainly protein.

The same kind of 75-angstrom unit membrane found at the surface of the Schwann cell can apparently be demonstrated at the surfaces of all cells. A survey of a great many different kinds of animal cell, plant cell and bacterium has failed to turn up a single case where the structure was lacking. It would appear to be a biological constant common to the surfaces of all kinds of cell.

In the past 10 years electron microscopists have found that practically all cells are filled with numerous membranes folded in complicated ways. Keith R. Porter, George E. Palade and Sanford L. Palay at the Rockefeller Institute and Sjöstrand have studied the arrangement of the internal membranes and have produced strong evidence that many of them are linked together. One reason for thinking so is that the membranes always appear in pairs, forming the lining of extensive cavities, sinuses and canals that run throughout the cell. This extensive system is known as the endoplasmic reticulum. Michael L. Watson of the University of Rochester has shown that the nuclear membrane is in fact simply a system of sacs of the endoplasmic reticulum arranged around a spherical bit of cytoplasm containing the principal genetic material. There is direct continuity between nuclear and cytoplasmic matrix maintained by openings in the nuclear membrane. Many workers have discovered continuities between the sacs of the endoplasmic reticulum and the surface membrane of the cell. Therefore it seems reasonable to think of the cell as essentially a three-phase system. First, there is a cytoplasmic phase. Second, there is a phase made up of contents of the cavities in the endoplasmic reticulum. Third, there is a membrane phase separating the first and second phases.

This concept implies that the material

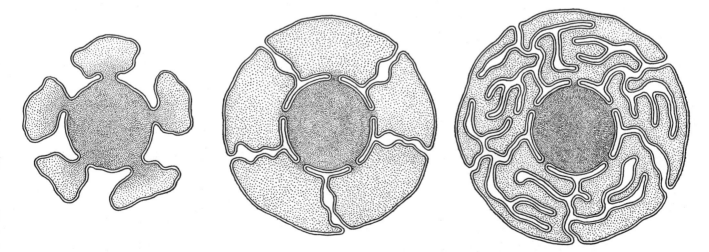

membrane. In second, matrix material has pushed out to form pseudopodia. In third, evaginated membranes have folded back to make paired nuclear membranes and a primitive endoplasmic reticulum. Fourth drawing shows further development of endoplasmic reticulum. Now material that was at first completely outside cell is included in its volume, although remaining outside topologically.

of the second phase is in a sense outside the cell. It also implies that the membranes of the endoplasmic reticulum are an extension of the surface membrane. Recent electron micrographs clearly demonstrate that the unit membrane structure does appear in various parts of the endoplasmic reticulum. Although for technical reasons the structure cannot always be seen, there seems good reason to suppose that it occurs throughout the system of interior membranes.

If the three-phase idea is literally true, it must also accommodate the various membranous bodies, or organelles, inside the cell. They are surrounded by membranes but seem to be distinct from the endoplasmic reticulum. Mitochondria are the most abundant of these organelles. Electron micrographs of mitochondria show that their membranes too exhibit the familiar unit structure. In fact, it now seems that mitochondria are formed when cytoplasm pushes into a cavity bounded by an internal membrane, which then pinches off and separates from the continuous system.

Recently I have found a new membranous organelle, about the size of mitochondria, in developing Schwann cells. This long, cigar-shaped body also consists of unit membranes, which are rolled up in a tight helix around a core of cytoplasmic matrix filled by the small particles known as ribosomes. The coiled arrangement could result if one membrane were growing faster than another. In my opinion this is just what is happening. I have named the body a growth spiral and visualize it as a structure actively spinning off new membranes of the endoplasmic reticulum.

The three-phase view implies that the simplest possible cell might consist of cytoplasm bounded by a purely surface membrane. In that case the second phase would be completely outside the cell. An example of such an organism is the human red blood cell, which has a surface membrane but no organelles. If the simple primordial cells were of this type, more complex cells could have evolved by the invagination, or drawing inward, of the surface membrane. Conversely, they could have evolved by the evagination, or pushing outward, of sections of the cytoplasmic matrix together with its membrane. The hypothesis means that any piece of membrane in the cell must have come from pre-existent membranes; it was not formed normally *de novo* in the cytoplasm or elsewhere. Membranes make membranes.

If this rule is true, one is led to ask: "Where did the first membranes come from?" In speculation about the origin of life the emphasis has been on protein and nucleic acid. There has so far been little interest in the origin of lipids. Actually the production of lipid molecules may have been crucial to the origin of life. This is so because lipids in a watery medium tend to aggregate into thin continuous sheets. The primitive membrane would have served both as a container for the various large molecules and as a surface on which they could be anchored and arranged in the integrated patterns that are as essential to life as the existence of the large molecules themselves. It seems highly probable that only after membranes appeared could the first living things be organized from the soup brewing in the ancient seas.

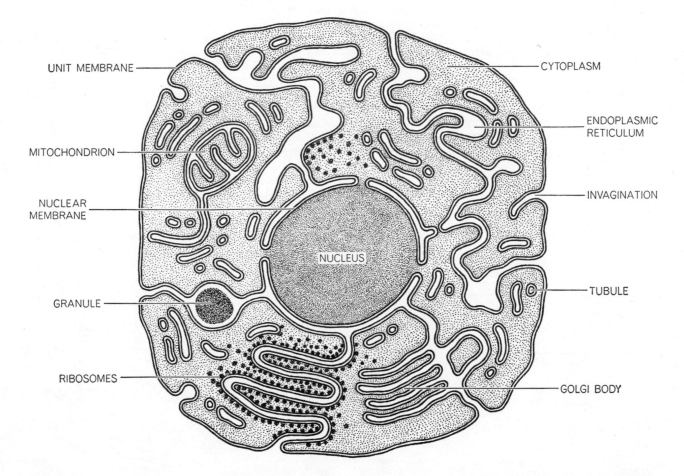

SCHEMATIC DIAGRAM, based on the author's concept of cell structure, shows the extensive distribution of membrane within the cell and some of the many membranous organelles so far identified. Since the membranes within cells always have the unit membrane pattern, it may be that most of these organelles are formed from the unit membrane structure bounding all cells.

5

CILIA

by PETER SATIR   February 1961

In 1676 Antony van Leeuwenhoek, the Dutch naturalist and microscope maker, described an animalcule he had been watching through one of his instruments: its "belly is flat," he wrote, and "provided with diverse incredibly thin feet, or little legs, which were moved very nimbly." The nimble legs that Leeuwenhoek saw were cilia. After he described them they were discovered on a wide variety of living cells, from protozoa to the cells that line the human trachea.

These appendages, which really look more like fine hairs than legs, are one of the most easily seen "organelles"—specialized parts of single cells that correspond to the organs of more complex living forms. When the appendages are less numerous, they are called flagella, but there is no essential difference between the two types. In most of the cells where cilia or flagella occur their primary function is obvious: they move back and forth like oars. When attached to a movable, boatlike object such as a protozoon, they propel it through the liquid around it. On a stationary object, for example a tracheal cell, they move the surrounding liquid over the surface of the cell. They either bring the cell into new environments or bring new environments to the cell.

Until about 10 years ago there was little more to be said on the subject. Since then the study of cilia, as of other parts of the cell, has undergone a major revolution, brought about by the electron microscope. It is now possible to examine the detailed substructure of cilia, mitochondria, microsomes and other organelles and to compare them in many different types of cell. The method of "comparative cytology" is providing new insights into the design and evolution of the cell, playing a role similar to that of comparative anatomy for larger living structures.

Perhaps the most remarkable part of these investigations is the correlation they reveal between form and function. Cellular structure, down to its minute details, remains constant as long as function is constant. When the structure of an organelle changes from one type of cell to another, the difference usually corresponds directly to a change in its function. Taking advantage of this fact, one can interpret the features seen in electron micrographs. By comparing the

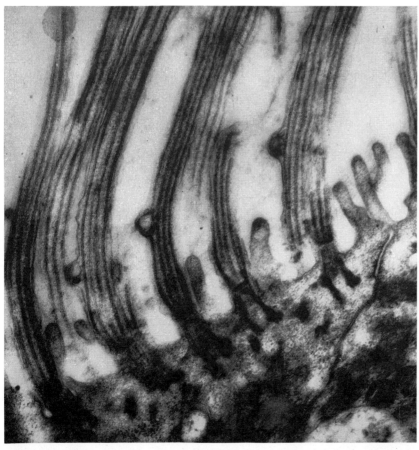

GROUP OF CILIA projecting from the gill cells of a fresh-water mussel is magnified 34,000 diameters in this electron micrograph. Long dark lines are filaments; dark areas at bottom of cilia are basal plates and bodies. (These and other ciliary features are detailed in illustrations on opposite page and on page 74). The smooth protuberances on the cell surface (*slope*) are microvilli. Irregular line (*right center to bottom*) marks cell boundaries. The micrographs on the first six pages of this article were made by the author.

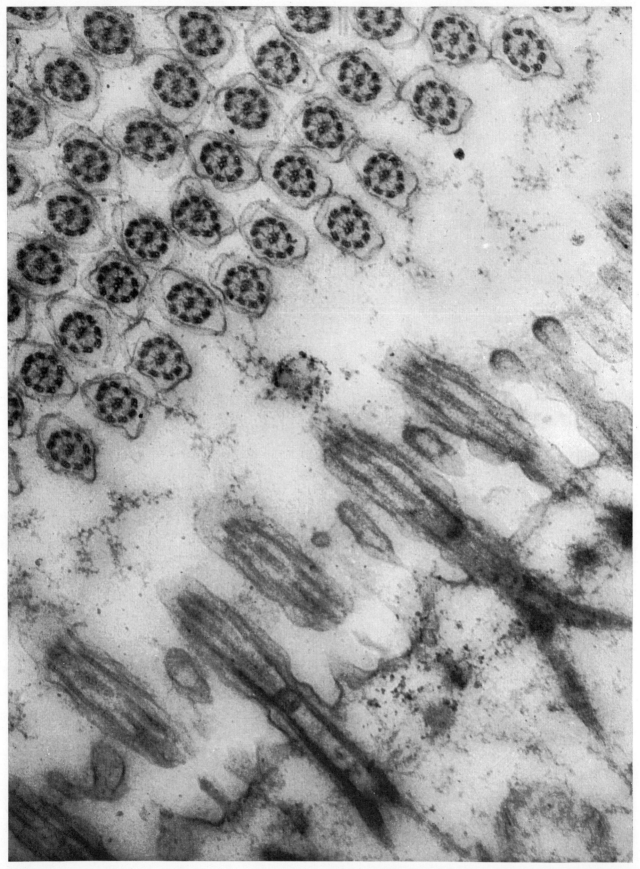

PRINCIPAL FEATURES OF CILIA of mussel-gill epithelium are magnified 51,000 diameters. Cross sections of cilia (*upper left*) reveal common pattern of two central filaments surrounded by nine peripheral ones. Longitudinal sections of several ciliary stalks (*large gray projections*) show filaments (*thick gray lines*) and stalk membrane, an extension of cell membrane (*rough sloping line*). Bar of elongated H (*bottom*) is basal plate; lower legs, basal body. Rootlets (*bifurcation at right*) extend from basal body.

DETAILS OF CILIARY STRUCTURE are depicted somewhat schematically in longitudinal section of cilium (*left*). As shown, cilium has been sectioned parallel to the plane of ciliary motion (*i.e., left and right in the plane of this page*); rootlet at bottom right generally does not appear in actual section. In electron micrographs at right cross sections corresponding to different levels (*broken lines*) of cilium are magnified 49,000 diameters.

structures of homologous bodies with different functions, the roles of the various features in the over-all workings of the organelle can be determined.

What functions, then, do most cilia perform? Two are obvious: contraction and conduction. Any structure that can move must contain some component that contracts in response to an impulse conducted through it. In this respect a cilium is not much different from a muscle: an impulse conducted to a muscle by a nerve, and through it by a system still imperfectly understood, causes the muscle fibers to contract. There must be elements in the ciliary structure that perform the same tasks. The third function is rather more special: it is duplication. Since cells reproduce by direct division, producing two structures where there was only one, their parts must duplicate themselves, or be duplicated, during each generation. Cilia are self-duplicating and therefore they must have a component that is concerned with this function as well.

Cilia are very small. The stalk extending from the body of the cell is about .0002 millimeter in diameter and .005 to .015 millimeter long. This is not the complete structure, however. There is a part, the "basal body" and its associated "rootlet fiber," extending deep into the cell. Also connecting cilium and cell body is the common membrane that encloses both.

Not surprisingly, ciliary stalks fail to function when their connection with the rest of the organelle is disrupted. This was demonstrated some 30 years ago by the English zoologist James (now Sir James) Gray. He studied the cilia in the gill of the mussel. Normally their beating is timed so that a wave motion, called the metachronal wave, appears to pass down the row of stalks [*see illustration on page 58*]. When Gray stripped some of the stalks away from their cell bodies, the wave traveled as far as the stripped region and then stopped. The separated stalks did not move at all. Recently the German investigator H. Hoffman-Berling has refined Gray's observation. Working with sperm flagella, he found that isolated stalks are motionless, but he made them move by adding adenosine triphosphate (ATP). This substance, which supplies energy for almost all biological reactions, was known to cause the proteins in muscle to contract. Evidently the proteins of a flagellar stalk also have inherent contractile ability that is usually controlled through the connection with the rest of the cell.

For a more detailed picture of the cil-

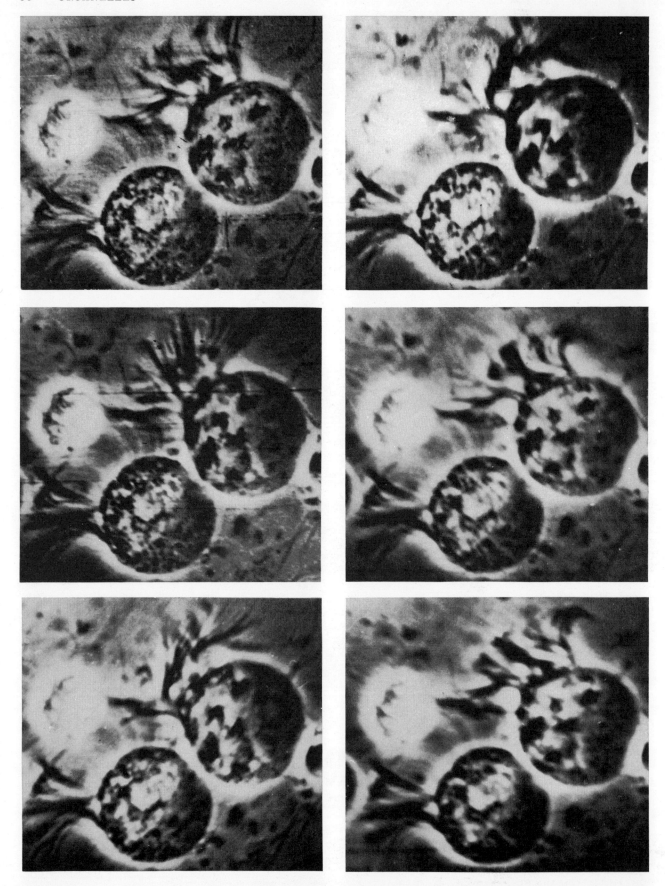

CILIARY BEAT involves effective stroke (*sequence at left*) and recovery stroke (*sequence at right*). Cilia of single mussel-gill cell (*top right in each frame*) are in motion. Effective stroke can be followed by noting successive positions of cilium that stands nearly vertical (*top left*). Cilium bends stiffly to position 45 degrees from vertical (*middle left*), then to horizontal position (*bottom left*) at end of stroke. In recovery stroke cilium straightens from horizontal (*top right*) to vertical (*bottom right*). Cilium is limp during stroke. The frames were selected from a motion-picture sequence. Magnification is 1,800 diameters.

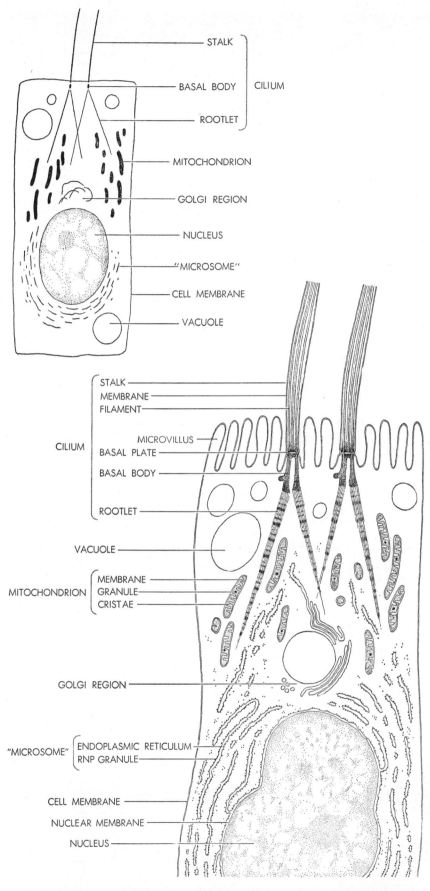

CILIATED CELL is depicted schematically as it would look under light microscope (*top*) and in electron micrograph (*bottom*). Organelles depicted (*e.g.*, cilium and mitochondrion) are found throughout plant and animal kingdoms. Electron microscope reveals subunits of organelles. In mitochondrion the cristae are the light internal finger-like projections.

iary apparatus let us turn to the electron microscope studies. To begin with, we find that all movable cilia or flagella, no matter from what type of cell they come, are built on an identical plan. The cytoplasm of the stalk is enclosed in an extension of the cell membrane, and in it are embedded 11 filaments running up from the basal body. They are distributed in a distinctive manner: two central filaments are surrounded by nine others, an arrangement that has come to be known as the 9 + 2 pattern. The nine peripheral filaments are not simply round rods; in cross section they appear as figure eights with hooks at one end. The central filaments, on the other hand, have a circular cross section. Some fibrous connections can be seen between the central pair and the nine peripheral filaments and occasionally between the outer filaments and the ciliary membrane.

At their tips cilia are narrow and probably closed. Some of the filaments in the stalk come to an end before the tip, so that the 9 + 2 pattern is lost in the cross section of the outer region. There the remaining filaments no longer have a complex structure but appear as simple black dots.

The 11 ciliary filaments run down to the surface of the cell, where the ciliary membrane becomes the cell membrane, ending at this level in a plate. Opposite it is a second plate, which forms the end of the basal body. This body is a cylindrical object about .0004 millimeter long, containing nine filaments that are continuations of the nine peripheral filaments of the stalk but with somewhat more variable and complex form. From the bottom of the basal body the rootlet fiber, striated along its length, extends deeper into the cytoplasm of the cell, where it may become part of a complex network of fibers.

Since the electron microscope does not work with living material, the foregoing description undoubtedly does not apply exactly to the intact cell. Some of the details are surely "artifacts" resulting from reactions between the material of cilia and osmium tetroxide, the chemical used to kill and fix the tissue. Nevertheless, the universal prevalence of the 9 + 2 pattern, even down to the form and arrangement of the filaments, in all electron micrographs of all motile cilia or flagella is convincing evidence of a common molecular arrangement in the live organism.

Which structures correspond to which functions? The question cannot yet be answered in full, but certain facts have been established. The 9 + 2 pattern, for

58 ORGANELLES

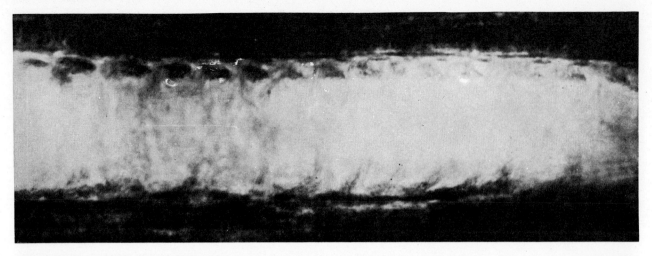

METACHRONAL WAVE consists of repeating patterns of crests and troughs that move along a row of ciliated cells. In this light micrograph of mussel gill, the cilia are magnified 860 diameters. Electron micrograph of metachronal wave appears below.

ELECTRON MICROGRAPH OF METACHRONAL WAVE shows that pattern of crests and troughs results from cilia in different phases of their "beat" at the same time. At crest cilia are beginning the effective stroke; trough occurs at or near beginning of the recovery stroke. In mussel gill the cilia beat at right angles to the direction of wave motion. Magnification is 6,500 diameters.

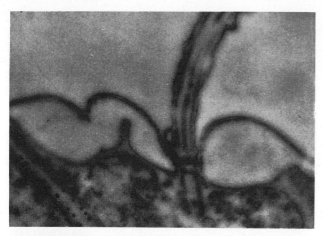

PARAMECIUM CILIA propel organism. In cross section (*left*) cilium shown above has central filaments characteristic of motile cilia. A. W. Sedar and K. R. Porter of the Rockefeller Institute made electron micrographs.

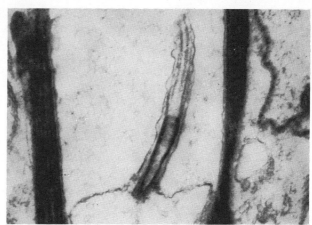

CILIA OF INSECT EAR are nonmotile. Cross section (*left*) of cilium shows two central filaments missing, a characteristic of nonmotile cilia. Micrographs were made by E. G. Gray of University College London.

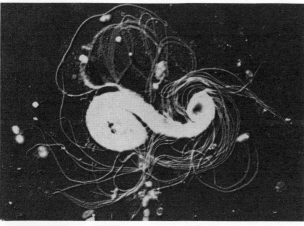

FERN SPERMATOZOID (*white area*) moves by means of cilia (*long threadlike strands*). Four of these cilia are shown in cross section at left. Micrographs were made by Irene Manton of the University of Leeds.

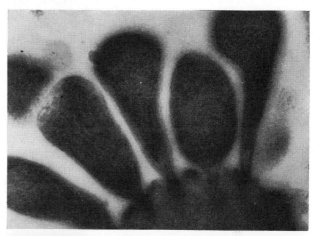

CROWN CELL OF FISH BRAIN has modified cilia. Central filaments are missing in cross section (*left*); such cilia usually perform a sensory function. K. R. Porter of Rockefeller Institute made the micrographs.

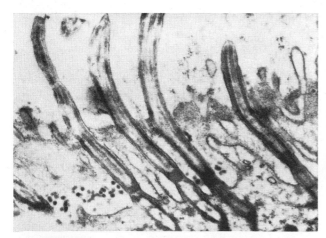

CILIA IN RESPIRATORY ORGAN of tubed polychaete (marine worm) drive current of water over the surface of cells (*thin horizontal gray area*). Samuel Dales of the Rockefeller Institute made both of these micrographs.

WHORL ON RETINAL CELL of scallop eye is composed of modified cilia. Cross section (*left*) of ciliary stalk shows typical nine-filament pattern. W. H. Miller of the Rockefeller Institute made these electron micrographs.

example, must play a fundamental part in contraction. It appears in the motile cilia of all protozoa, higher animals and plants. Some organisms contain cilia that have been modified so that they conduct impulses but do not move. And in these the central pair of filaments is usually missing.

A striking example is found in the eye of vertebrates. Its sensory elements—rods and cones—have evolved from cilia, as is demonstrated by their possession of basal bodies and striated rootlet fibers. The stalks, however, have been radically altered. Instead of terminating in a narrow tip they now lead to bulbous stacks of light-sensitive rhodopsin molecules. The only connection between these molecules and the inner part of the cell (and hence the nerves and the brain) is a piece of cilium-like stalk about .001 millimeter long. The stalk does not move but merely transmits impulses to the sensory nerves. And although it contains the usual nine peripheral filaments, the central pair is missing. Similarly modified cilia are found in the eyes of scallops, the ears of insects and the brains of fishes. These cilia can transmit impulses but so far as we know cannot contract.

Another kind of modification is seen in the ray, a fish related to the shark. In its labyrinth (an organ concerned with hearing and balance) are cilia fixed with respect to the surface of the cells to which they are attached. They can be moved only by an external force. When a force moves them, an impulse is conducted to the cells and sense centers. The movement is a necessary preliminary to the conduction; the two functions are linked together. Hence their operation is the reverse of that in normal cilia: contraction produces an impulse instead of vice versa. But the important point is that these cilia do move, and they possess the pattern of filaments always associated with motion: 9 + 2.

If the filaments, and especially the two in the center, are involved in motion, what part of the structure has to do with the conduction of impulses? Considerable evidence implicates the ciliary membrane. Cilia that conduct but do not move always have a membrane that links the stalk to the cell. In rods and cones, at least, the membrane is known to contribute to the development of the sensory portion of the organelle. Moreover, the cell membrane plays an important part in the transmission of impulses, especially in nerve and muscle cells. Since the ciliary membrane is an extension of the cell membrane, it can logically be assigned the same function.

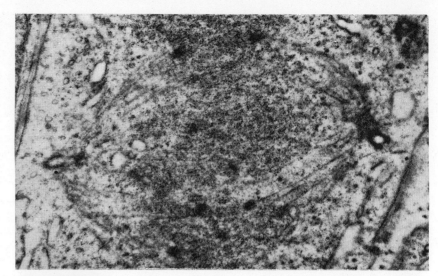

CENTRIOLE closely resembles the basal body of a cilium in structure and function. Two centrioles (*small dark C-shaped areas at left and right*) appear in a chicken-spleen cell during division. In cross section of centriole (*left*) the central filaments are missing. The micrographs were made by W. Bernhard of the Institute for Scientific Research on Cancer of the University of Paris.

There is also experimental evidence for this conclusion. Cutting into a cell near the bottom of ciliary stalks destroys the co-ordination of their motion. Although there may be other explanations for the result, the cell membrane is clearly disrupted by the cuts, and this is a likely cause of loss of co-ordination. Secondly, a number of experiments indicate that impulse conduction in cilia is electrical in nature, as is conduction by the membranes of nerve fibers. In particular, the beating cilia of an isolated mussel-gill have recently been shown to produce an electrical effect much like that associated with the passage of impulses along a cell membrane.

Finally, we should like to account for the self-duplication of cilia. Here we turn to the basal body. Electron microscope studies of various cells show that the basal body has the same structure as another cellular body, the centriole. Both are short cylinders with nine sets of

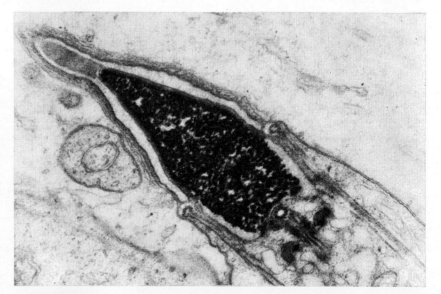

TWO CENTRIOLES are present in developing sperm of cat. Ring with bright center below dark area is one centriole. Other, from which flagellum (*short gray lines at lower right*) grows, does not appear. Section of guinea-pig sperm tail (*left*) shows extra filaments. M. H. Burgos and D. W. Fawcett of the Harvard Medical School made micrograph above, Fawcett that at left.

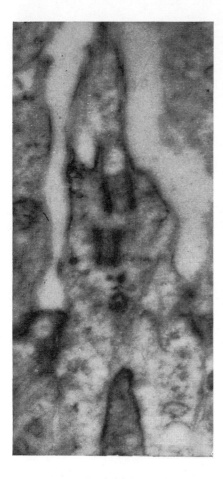
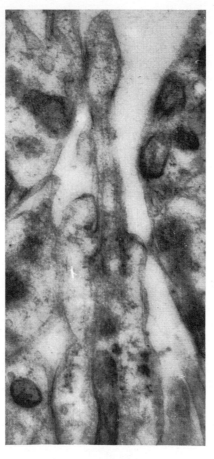
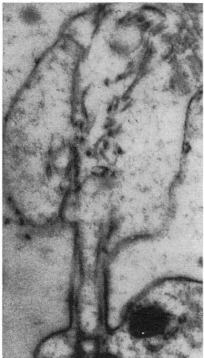
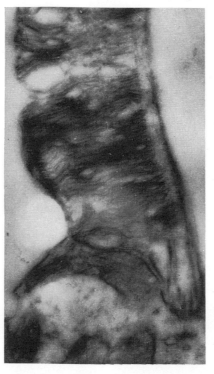

filaments. In certain cells their functions apparently overlap as well. As the sperm cell matures, for example, a centriole gives rise to a flagellum and becomes a true basal body. The normal function of the centriole is to manufacture, or at least to help organize, the paraphernalia of cell division—the filamentary assembly called the mitotic apparatus. In the process the centriole precisely duplicates itself. Similarly, the basal body is a self-duplicating organelle concerned with synthesizing or organizing the cilium. This is not to say that the basal body plays no part in the contraction or conduction processes, but what part it plays, if any, is still unknown.

Still less is known about the function of the rootlet fibers. From the fact that some fully functioning cilia do not even have a rootlet fiber, we judge it to be of secondary importance. There is no lack of guesses as to its purpose, but so far there is little agreement.

In cytology, as in other branches of biology, comparative studies are intimately connected with the theory of evolution. At this point we can only speculate on the origin of cilia. Judging from the importance of the centriole and its near universality, it seems reasonable to suppose that this organelle emerged first. Then, sometime still near the dawn of evolution, a centriole may have become linked to a cell membrane, an alteration that somehow conferred an advantage in responding to the environment. A little later one of its descendants may have synthesized two extra filaments, a sort of mistaken mitotic apparatus, in the center of the already existing ring of nine. If these filaments had the power of contraction the cell had acquired a new motile element, which was then conserved and extended by the process of selection.

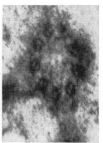

VERTEBRATE RETINAL ROD is a modified cilium. In developing visual cell of newborn kitten, one centriole (*top pair of short dark lines at top left*) attaches itself to cell membrane, becoming a basal body. Stalk grows from basal body (*top right*). Outer portion of cilium widens (*bottom left*) and assumes primitive shape of outer segment of rod. Unchanged lower section of stalk is "connecting cilium." Micrograph of outer segment of adult rabbit rod (*second from bottom left*) shows connecting cilium (*light gray margin*) and filaments (*thin double lines near bottom*). In cross section rod stalk of rat (*third from left*) shows nine-filament ring characteristic of modified cilia, as does basal body of rabbit rod (*right*). The micrograph of the rat stalk was made by K. R. Porter of the Rockefeller Institute. The other five micrographs shown here were made by Eichi Yamada of Kyushu University in Japan.

6

THE MITOCHONDRION

by DAVID E. GREEN January 1964

Although we do not know what the first forms of life on the earth were like, we do know the nature of the equipment that started them on the road to survival. Life was no more than an experiment of nature, in danger of being snuffed out at any time, until the proto-organisms developed dependable machinery to perform two basic functions: (1) reproduce themselves and (2) generate energy in a form usable for an organism's various requirements. It is clear what that machinery must have been even in the first successful living creatures some two billion years ago, because all forms of life on our planet have basically the same systems for these two purposes. They are summed up in the familiar initials DNA and ATP.

DNA stands for the genetic apparatus that is responsible for the replication of the key substances of life: proteins and the nucleic acids themselves—deoxyribonucleic acid (DNA) and ribonucleic acid (RNA). Fully as important as DNA is adenosine triphosphate, or ATP. This molecule supplies the energy for all the processes of life, including replication; in fact, as it is shaped by evolution, the replicating process is designed to be energized by ATP.

The energy-generating system of living things depends on two related actions: oxidation and the synthesis of ATP. The function of oxidation is to release electrons, which act as the agents for the storage and transformation of energy. In the absence of oxygen other methods of electron release can serve, and there are many cells and organisms that carry out "oxidation" in these substitute ways. In all cases, however, the goal of the process is the formation of ATP, which acts as a kind of storage

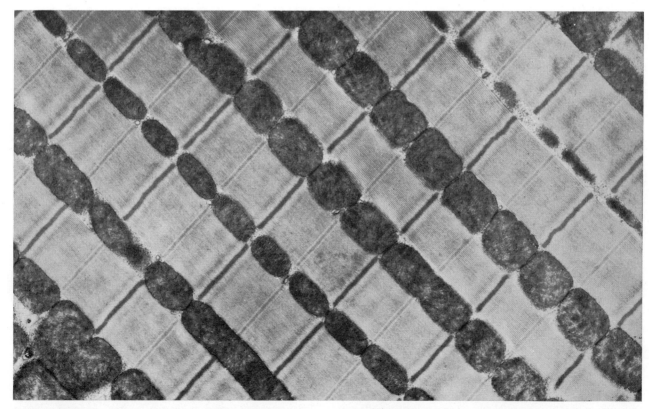

MITOCHONDRIA are the dark, oblong objects lined up end to end from top left to bottom right in this electron micrograph of a wasp's flight-muscle fiber, made by David S. Smith of the University of Virginia. Magnification is approximately 9,500 diameters.

battery feeding energy to the cell as it is needed.

Living organisms have developed three different systems for generating energy in the form of ATP: the various enzymes that catalyze the process of glycolysis and the intracellular particles known as the chloroplast and the mitochondrion. The most primitive of these is glycolysis; it is the breakdown of sugar by enzymes in the absence of free oxygen. The system is comparatively inefficient, yielding only one molecule of ATP for each pair of electrons released. The mitochondrion produces three molecules of ATP for each pair of electrons released by oxidation; the chloroplast possibly produces the same number of ATP molecules for each pair of electrons released by light. This threefold gain in efficiency makes the chloroplast and the mitochondrion the favored power plants of living things—the chloroplast in the world of green plants and the mitochondrion in the world of animal cells. In principle both are variations on the same theme.

The mitochondrion is often called the powerhouse of the cell. It is a good deal more than that; it carries out functions other than generating energy. Indeed, its basic design seems to be copied by all other systems in the cell that have to do with the transformation or use of energy, such as those responsible for the transport of materials through membranes, for the contraction of muscle and so on. Thus the mitochondrion can serve as a Rosetta stone for deciphering the *modus operandi* of all the energy-transforming processes in cells. This article will discuss the mitochondrion's chemical structure and operations, now known in considerable detail, in that light.

The Structure of the Mitochondrion

Let us first look at the structure of this tiny body. In the average cell there are several hundred mitochondria. An average mitochondrion would be a sausage-shaped object about 15,000 angstrom units long and 5,000 angstroms in diameter. (An angstrom unit is a ten-millionth of a millimeter.) Rather like a Thermos bottle, it has a two-layered wrapping: an outer and an inner membrane with a watery fluid filling the space between them. Extending from the inner membrane into the interior of the sausage are a number of sacs called cristae. The surfaces of both membranes are sprinkled with thousands of smaller particles; they are anchored to the outside surface of the outer membrane and the inside surface of the inner membrane. These particles are the elementary units that carry out the chemical activities of the mitochondrion. The fluid between the membranes also participates, providing communication between the layers and supplying the enzymes in the membranes with the auxiliary catalysts known as coenzymes.

The two membranes serve as structural backbones for the mitochondrion. From the structural standpoint they have three notable properties: good tensile strength, stability and flexibility. (The mitochondrion is by no means a rigid body.) One of the fascinating developments in the analysis of the mitochondrion was the discovery of the chemical arrangement that accounts for these properties.

Each membrane is made up of two materials apart from its attached particles. The principal material, accounting for about four-fifths of the weight of the membrane, is a structural protein that has recently been identified in our laboratory at the University of Wisconsin. In its usual form the protein is completely insoluble in water. Analysis of its constituent amino acids showed that half of them have paraffin-like side chains, that is, side chains insoluble in water. When such chains join together, they do so by virtue of what is known as a hydrophobic bond. The bond is fairly weak, but the large number of bonds involved in the joining of chains gives it a high degree of stability. As we shall see, the hydrophobic bond is a key to the structure and functions of the mito-

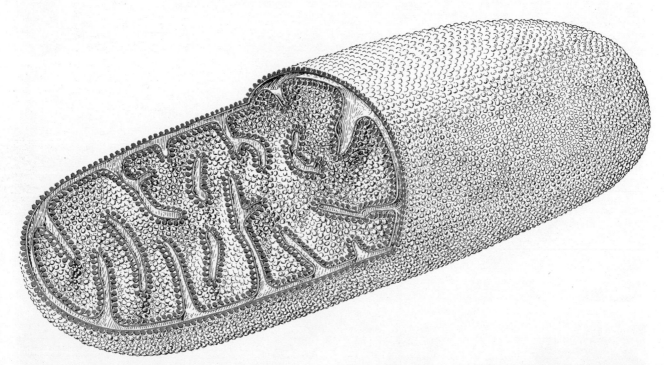

CUTAWAY DRAWING of a typical mitochondrion shows the two membrane layers separated by a fluid-filled space called the intrastructure space. The space within the inner membrane is called the interstructure space. The invaginations of the inner membrane are the cristae. The stalkless particles distributed over the outer surface of the outer membrane are involved in various oxidation reactions that supply electrons to the interior of the mitochondrion. The particles extending inward on short stalks from the inner surface of the inner membrane transfer the electrons along a chain of complexes that synthesize molecules of adenosine triphosphate (ATP).

chondrion; we can call it one of the motifs of the mitochondrial theme.

The structural protein can be broken down into monomers, or subunits, by various reagents such as lauryl sulfate, alkali or acetic acid. They weaken the hydrophobic bond, either by increasing the number of electrically charged groups on the molecules or by a detergent action. The monomer, whose molecular weight is 22,000, is soluble in water. If the reagents are removed from the solution, the smaller units spontaneously rejoin to form the insoluble polymer.

The second material in the mitochondrial membranes, constituting about a fifth of their weight, is lipid, or fatty material, almost entirely in the form of the molecules known as phospholipids. We can picture a phospholipid molecule as having the shape of a clothespin, the head consisting of an electrically charged group of atoms and the two legs made up of long-chain fatty acid [see upper illustration on next page.] By itself this molecule is insoluble in water. In a structural combination known as a micelle, however, a group of phospholipid molecules becomes soluble. Imagine two rows of the clothespin-shaped molecules—hundreds or thousands of them—lined up back to back so that the heads of the clothespins all face outward. On the opposite, or inward, side of the rows the fatty-acid chains forming the legs of the molecules nest together and are joined by hydrophobic bonds. The resulting micelle is a stable structure, and the behavior of the phospholipid is determined by the properties of the micelle as a whole rather than by those of the individual molecule. By virtue of the fact that in water the electrically charged heads of all the molecules in the micelle face the water, the micelle is soluble.

It is a combination of structural protein and phospholipid micelles, then, that forms the membranes of the mitochondrion. We can picture the membrane as a network consisting of alternating protein and micelle units linked by hydrophobic bonds [see lower illustration on next page]. This model would account for the relative proportions of protein and phospholipid in the membrane and for the membrane's physical properties; its tensile strength presumably is derived from the structural protein, and its flexibility from the phospholipid micelle. The outer and inner membranes of the mitochondrion are not quite alike; the outer one shows some properties the inner one does not have. Probably components other than

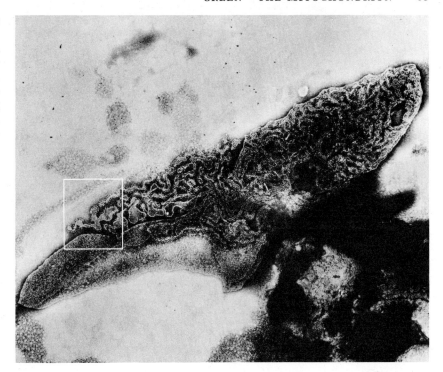

LONGITUDINAL CROSS SECTION of a beef-heart mitochondrion is magnified about 22,-000 diameters in this electron micrograph made by Humberto Fernández-Morán, now at the University of Chicago. This particular mitochondrion is about 29,000 angstrom units long and 7,000 angstrom units in diameter, after having been flattened out on a grid for viewing.

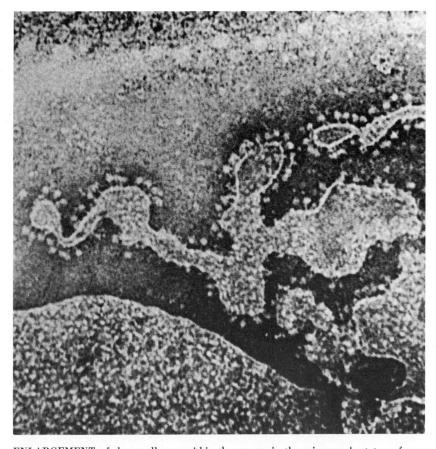

ENLARGEMENT of the small area within the square in the micrograph at top of page shows the electron-transfer particles in greater detail. The base pieces of the particles form a continuous layer around the crista. Magnification is approximately 250,000 diameters.

66 ORGANELLES

structural protein and phospholipid are responsible for these properties.

The Mitochondrial Particles

Let us now examine the particles that perform the various functions of the mitochondrion. Electron micrographs show that each mitochondrion has many thousands of these particles; they are distributed, as we have noted, over the outside surface of the outer membrane and the inside surface of the inner membrane. The particles have three different functions: (1) carrying out the oxidation reactions that supply electrons, (2) transferring the electrons along a chain of complexes that synthesize ATP, (3) catalyzing synthetic reactions that are powered by ATP.

The available evidence indicates that the particles concerned with the first and third of these functions are all located on the outer membrane, and that those concerned with electron transfers leading to the synthesis of ATP are on the inner membrane. (For example, bombardment of mitochondria with ultrasonic radiation does not disturb the electron-transfer particles but does dislodge most of the others, presumably because they are on the outer and more fragile surface.) It appears that the electron-transfer job is carried out by a particle of a particular kind that contains the entire set of catalysts that make up the chain.

The general picture of electron transmission seems to be somewhat as follows. On the outer membrane certain particles responsible for providing "energetic" electrons (some four to six different particles are involved) do so by implementing oxidation reactions such as those of the so-called citric acid cycle and the oxidation of fatty acids. They pass on the released electrons to the coenzyme diphosphopyridine nucleotide, or DPN, which shuttles them across the liquid-filled space between the membranes to the particles on the inner membrane. In accepting the electrons the DPN molecule is reduced (and is then called DPNH); in turning them over to the inner-membrane particles it is oxidized.

The existence of a unit in which a single, complete electron-transfer chain is contained was implicit in studies conducted in our laboratory over the past

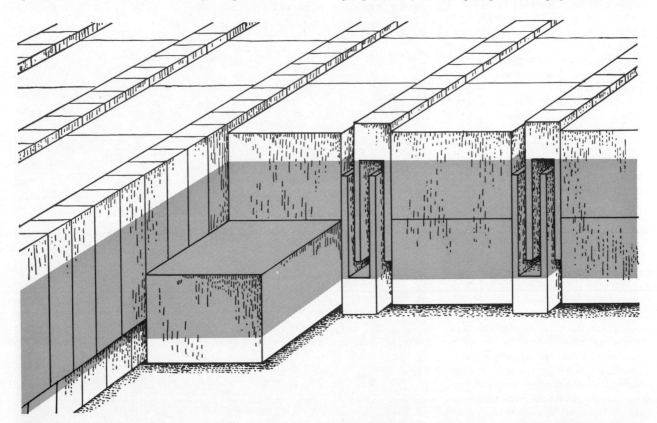

PHOSPHOLIPID MOLECULE, one of the two materials that make up the membranes of the mitochondrion, is shaped somewhat like a clothespin. The head (*left*) consists of an electrically charged group of atoms; the two legs (*right*) are long hydrocarbon chains.

MEMBRANES OF MITOCHONDRION contain networks of alternating protein and phospholipid units. The phospholipid molecules are lined up back to back with their heads facing outward in long rows known as micelles. The network is two protein molecules thick. The hydrophobic, or water-insoluble, bonding faces of both the phospholipid and the structural protein molecules are in color. This hypothetical model would account for the relative proportions of protein and phospholipid in the mitochondrial membranes.

10 years. The physical unit, however, was first discovered by Humberto Fernández-Morán, now at the University of Chicago, during examination of mitochondria in the electron microscope. The electron-transfer particle has proved to be a momentous discovery, opening the door to investigation of the activities of the mitochondrion at the molecular level.

The chain traveled by the electrons in this particle consists of a series of catalysts arranged in a certain sequence, each of which can undergo oxidation and reduction. We can think of the chain as a scale of notes that must be struck in regular order, except that the notes are struck not individually but in groups. There are 11 different components in

the sequences (10 of them are proteins), and they are grouped in four complexes, or clusters. Complexes I and II are alternates: the chain starts with Complex I if the electrons come from the coenzyme DPNH; it starts with Complex II if the electrons are donated, as they may be, by succinate, a product of the citric acid cycle. In either case the electrons travel

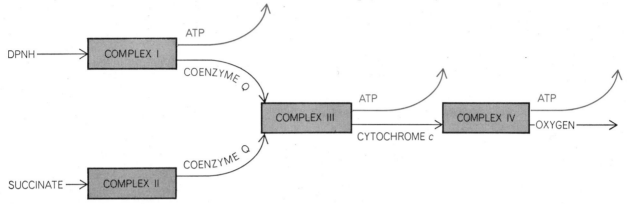

ELECTRON TRANSMISSION through a series of complexes inside the mitochondrion results in the formation of ATP, which stores the energy produced in the mitochondrion and supplies it to the cell when needed. Three molecules of ATP are formed in this process. DPNH and succinate are produced during the operation of the so-called citric acid cycle and also by the oxidation of fatty acids. Black arrows accompanying colored arrows denote oxidation reactions that donate electrons toward the formation of ATP.

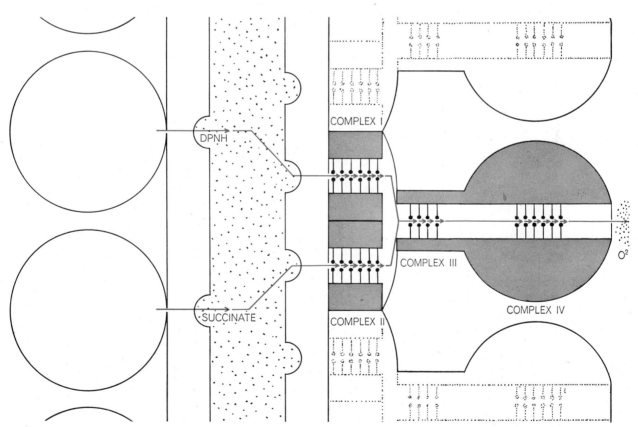

HYPOTHETICAL ARRANGEMENT of the four energy-generating complexes within the electron-transfer particle would reconcile the disparity between the weight requirements of the proposed electron-transmission system and the observed weight of the particle. According to this hypothesis Complexes I and II are in the base piece, Complex III is in the stalk and Complex IV is in the headpiece. Electrons are supplied by the stalkless outer particles (left) to molecules of DPNH and succinate at the inner surface of the outer membrane. These coenzyme molecules then shuttle the electrons across the intrastructure space to the outer surface of the inner membrane, where they are donated to Complexes I or II. The gray parts of the electron-transfer particle at right are protein and the white parts are lipid. The fine structure of the electron-transfer mechanism within a single complex is depicted on pages 68 and 69.

on from Complex I or II to Complexes III and IV, and eventually at the end of the chain they are carried off by molecules of oxygen.

The point of this arrangement is that each of the successive steps from Complex I to the end produces ATP. The transfer of a pair of electrons from DPNH to Complex I, then to Complex III and then to Complex IV results in the formation of a molecule of ATP at each site, or three molecules altogether. When ATP is synthesized, energy is stored in the molecule in the form of a high-energy bond, and this energy is provided by the transferred electrons because of differences in potential between the steps in the chain.

The four complexes are stationary and are separated by lipid. From Complex I (or II) electrons are shuttled across a lipid layer to Complex III by a catalyst known as coenzyme Q, and from Complex III to IV the electrons are carried by cytochrome c [*see upper illustration on page 67*]. The complexes are arranged in a certain order; it has been found that if the four complexes are isolated and then dissolved in water, they will reassemble themselves in precisely the same arrangement they have in the mitochondrion.

This general scheme shows how electrons are transported from one complex to the next, but it is not easy to see how the traveling electrons can move within a complex. The proteins making up a complex are locked in fixed positions, which would seem to allow them no means of transferring electrons from one protein to another. Robert M. Bock of the University of Wisconsin and Richard S. Criddle of the University of California at Berkeley have proposed a brilliant explanation that seems to be borne out by the evidence so far, namely that the proteins make contact with one another by means of swinging groups of atoms, mounted on the respective proteins by flexible arms, that transfer and accept the electrons [*see illustration on these two pages*].

Models of the Complexes

We now know all the components of the electron-transfer chain (all except coenzyme Q are proteins), the number of molecules of each component, the amount of lipid associated with the chain (about 30 per cent of the total mass), the approximate molecular weight of each complex (between 250,000 and 600,000) and the approximate molecular dimensions of the various parts of the system and of the electron-transfer particle as a whole. With this information one can work out molecular models of the complexes and the entire particle. It appears that each complex, containing four to six protein molecules, must have roughly the shape of a doughnut, with the protein around the periphery and phospholipid in the center [*see illustration on these two pages*]. The proteins are linked to one another and to the phospholipid by hydrophobic bonds, as in the over-all membranes of the mitochondrion. The functional—that is, electron-transferring—groups in the protein molecules are probably directed toward the hydrophobic interior of the lipid core, which presumably is in the form of phospholipid micelles. Lipid also permeates the entire particle, providing a medium for the shuttling of electrons from one complex to another.

Sidney Fleischer and Gerald P. Brierley of our laboratory have elegantly demonstrated the essential role played by the lipid in the transfer of electrons. They delicately removed the lipid from electron-transfer particles, using a sol-

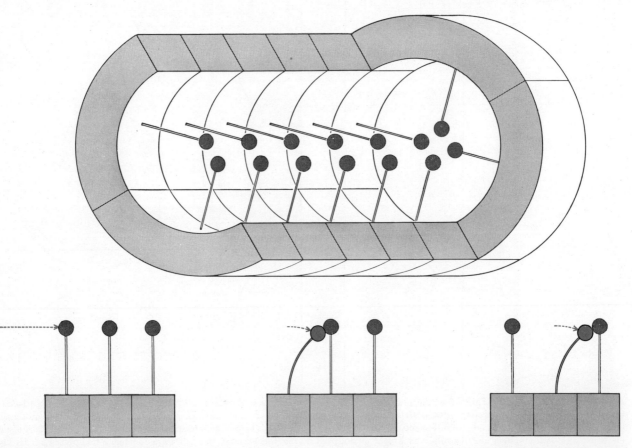

ELECTRONS ARE PROPELLED through a single complex (in this case Complex I) from one protein to another by means of swinging groups of atoms (*black balls*) that extend into the lipid core of the complex from the surrounding protein sheath (*top*). The five-stage drawing across the bottom shows how an electron is transferred and accepted by three of these protein groups; the

vent composed of 90 per cent acetone and 10 per cent water at low temperature to avoid damage to the functional proteins. With the lipid extracted, the particles lost their ability to transfer electrons. Fleischer and Brierley then restored the lipid by supplying phospholipid micelles in a form that readily entered the particles. Thus supplied, the particles recovered full activity in electron transfer. In short, lipid is an active partner, and there is accumulating evidence that phospholipid micelles take part in many catalytic activities in the mitochondrion, in addition to serving as a component of its membranes.

It is interesting to see that, structurally speaking, the mitochondrion has a striking resemblance to a virus. The virus particle consists of a protein coat and a nucleic acid core. When these two parts are separated, they are inactive; when they are combined again, the virus completely recovers its capacity for infection. In an analogous way the mitochondrion completely loses its ability to transfer electrons when its protein and phospholipid parts are separated and regains this ability when they recombine. In each case the partners are incompetent when alone and become active when coupled.

The Formation of ATP

How do the electron-transfer particles produce ATP? Essentially their function is to form ATP by joining a phosphorus atom in a high-energy bond to adenosine diphosphate (ADP). This is done in several steps by a rather intricate process. Let us examine how the process is carried out in the interaction of Complex IV and cytochrome c, where the details are best known.

The crux of the process is the formation of the high-energy bond. Energy for the bond is provided, as we have noted, by virtue of a drop in potential as electrons pass from one component in the chain to the next. In Complex IV these components are six protein molecules. The difference in potential between any one protein and the adjacent one is small—too small to supply enough energy for the high-energy bond. The total potential drop through the six molecules, however, is large enough. The set of six molecules does in fact function as if they were acting in unison, thereby assembling an energy package sufficient for the bond, as six men pulling on a rope together can move a weight that none of them could move singly.

How is the flow of electrons translated into a high-energy bond in Complex IV? George C. Webster of our laboratory has shown that the electron-bearing messenger, reduced cytochrome c, forms a compound with Complex IV; the two are joined by a high-energy bond, probably attached to the copper-containing group in the Complex IV protein molecule. Apparently one molecule of reduced cytochrome c donates its spare electron to the complex, and the reduced complex then interacts with another cytochrome c molecule and receives a second electron, which so to speak seals the bond. After the compound is formed it interacts with a comparatively small protein molecule called the coupling factor, which splits off Complex IV and forms a compound with cytochrome c, still using the same high-energy bond. Cytochrome c is in turn replaced by an inorganic phosphate group that attaches itself to the coupling factor by means of the high-energy bond. Then follows a final reaction in which a molecule of ADP bumps away the coupling factor and joins up with the phosphate group, still retaining the high-energy bond, to form ATP. In short, ATP is the end product of a series of reactions in which cytochrome c combines with Complex IV, the coupling factor replaces Complex IV, inorganic phosphate replaces cytochrome c, ADP replaces the coupling factor and ADP and the phosphate addition at last come together with a high-energy bond that has survived all the substitutions.

Imagine that the high-energy bond is a coiled spring, stretched by a molecule pulling at each end. Once it has been stretched, the tension can be maintained by other molecules replacing those pulling at the two ends, and the combination may wind up with a pair of partners entirely different from the one that originally did the stretching. This analogy essentially describes the process by which ATP is formed not only at Complex IV but also at the other complexes. Gifford B. Pinchot at Johns Hopkins University and Archie L. Smith and Marc F. Hansen in our laboratory have found that DPNH and Complex I form a compound with a high-energy bond; it can be assumed that coenzyme Q and Complex III likewise form a high-energy intermediate compound as a preface to the production of ATP.

The virtue of this roundabout method is that the series of substitutions serves to separate the enzyme systems taking part in the set of interactions so that they do not get in one another's way. As we shall see, the mitochondrion uses the same tactic (forming a high-energy intermediate compound) in generating energy for processes other than the synthesis of ATP, notably for moving substances through its membranes.

Before we leave the electron-transfer particle, let us note briefly the progress that has been made toward identifying it and locating its position in the mitochondrion. The particle, as first discovered by Fernández-Morán, seemed to be a minute spherical body about 100 angstroms in diameter, present in large numbers in the mitochondrion's inner membrane. A curious contradiction soon emerged. On any reasonable assumptions about its contents, a particle of that size would have a molecular weight of no more than 400,000. Two kinds of evidence later showed, however, that the electron-transfer system must be a great deal larger than that. In the first place, according to calculations based on the known components of the system their total molecular weight must be in the neighborhood of 1.3 million. This figure received strong experimental support when in our laboratory we isolated a particle that carried the electron-transfer activity and proved to have a molecular weight of about 1.4 million. It seemed that this unit could not be the same as the 100-angstrom particle seen with the electron microscope on the mitochondrion's inner membrane. How could a particle with a molecular-weight capacity of no more than 400,000 accommodate a system with a weight

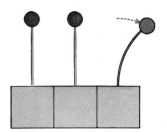 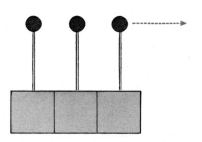

colored group in each stage is carrying the electron. This system of electron transfer within the complexes was first proposed by Robert M. Bock of the Enzyme Research Institute of the University of Wisconsin and Robert S. Criddle of the University of California at Berkeley.

of about 1.4 million?

The situation was rescued from total mystery when electron micrographs of mitochondria were made at very high resolution, first by Fernández-Morán, and later by others, particularly David S. Smith, now at the University of Virginia. These showed that the particles in the inner membrane were actually made up of three parts: a spherical headpiece (the tiny sphere that had been seen earlier), a cylindrical base piece and a cylindrical stalk connecting the base and head. Together the three parts could total about 1.3 million in molecular weight.

The electron-transferring unit that our group had isolated did not have this shape; it was a sphere about 150 angstroms in diameter. It would not be surprising, however, to find that these spheres had lost their natural shape on being isolated; left alone, biological material, particularly lipid, has a tendency to round itself into a ball. Then Takuzo Oda of the University of Okayama, examining electron micrographs of mitochondria in the cells of human heart-muscle tissue, made the exciting discovery that there the particles are spherical and about 150 angstroms in diameter. Our micrographs had been of mitochondria in beef-heart cells. Apparently, then, the electron-transfer particles can take different shapes in different cells: extended in some, compressed into a ball in others. The three-part shape is not inconsistent with the four-complex picture of the electron-transfer chain: Complexes I and II may be in the base piece, Complex III in the stalk and Complex IV in the headpiece.

The Larger Meaning of the Particle

Let us now see what the electron-transfer particle has to offer as a Rosetta stone—that is, what it can tell us about other processes in the dynamics of the cell. Consider, for instance, the movement of substances through cell membranes, an act that calls for the application of propulsive energy. This activity has been extremely difficult to study at the chemical level; now the discoveries concerning the electron-transfer particle make it accessible to both theoretical and experimental investigation.

Among the materials that penetrate the mitochondrion and that play a role in its activities are magnesium, calcium and manganese. They make their entry into the mitochondrion, through its outer membrane, as divalent ions (Mg^{++}, Ca^{++} and Mn^{++}). In each case the entry is effected with a phosphate ion as an escort; in other words, the movement is by sets—a phosphate ion with a pair of magnesium ions or calcium ions or manganese ions. It turns out that this movement through the membrane is energized by a high-energy intermediate compound, of the same kind as the intermediates that lead to the synthesis of ATP. One molecule of the compound energizes the passage of one set of ions into the mitochondrion. Once the ions are there the high concentration they attain leads to their precipitation as metal phosphate, and as a result two hydrogen ions are released for each molecule deposited. Gerald P. Brierley of our laboratory, Albert L. Lehninger of Johns Hopkins University and J. Brian Chappell and Guy D. Greville of the University of Cambridge have independently documented this transport of ions through the membrane.

The synthesis of ATP and the movement of ions both require the same high-energy intermediate. What determines which of the two activities will have priority? Under ordinary conditions synthesis wins by overwhelming odds. For all practical purposes the movement of ions stops cold when synthesis of ATP is going on. Hector F. De Luca and Howard Rasmussen of the University of Wisconsin have made the exciting discovery that one of the hormones of the parathyroid gland can switch off synthesis and turn on the movement of ions. We have here an instance of regulation by a hormone at the level of the mitochondrial membrane.

How, exactly, is the set of ions propelled through the membrane? The most attractive hypothesis put forward so far likens the propelling agent to a gun placed in the outer membrane. We may give this agent, presumably an enzyme, the name "translocase." To begin with, the translocase must be loaded with the right kind of projectile; only certain pairs of ions will fit this gun. When it is properly loaded, the translocase is fired by means of a detonating charge supplied by the high-energy compound. This firing causes the translocase to twist around, face the interior of the mitochondrion and propel its ion-pair shell through the membrane.

The features of the system have been investigated in a number of laboratories. These facts emerge. It has been shown that the high-energy compound that provides the charge can be generated either by the chain of electron transfers leading toward the synthesis of ATP or from ATP itself by reversal of the process. It has been shown further that the ion-moving agent (the assumed translocase) is in the outer membrane, because when the mitochondrion is broken into fragments by ultrasonic radiation, which destroys the outer membrane, the fragments are no longer able to move the ions, although they can still synthesize ATP.

What is translocase? Evidently it is a contractile substance, probably designed specifically for the propulsion of ions rather than for developing tension or the muscular type of contraction. Several laboratories have succeeded in extracting from mitochondria a contractile protein that on exposure to ATP undergoes changes in the shape and size of the molecule, which can catalyze the

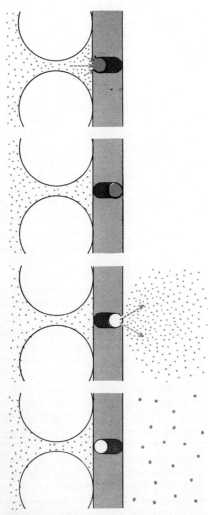

TRANSLOCASE "GUN" transports metallic ions and phosphates across the outer membrane of the mitochondrion from the external medium to the intrastructure space. The translocase molecule (*oblong shape inside membrane*) is loaded with pairs of ions (*top*) and is detonated, causing it to twist around and face the interior of the mitochondrion. It then fires the ion pair and returns to its original position (*bottom*). The ions combine with the phosphate groups in the intrastructure space to form crystals.

breakdown of ATP to ADP and which is greatly aided in this breakdown by calcium—the ion that is most easily propelled through the mitochondrial membrane. It remains to be determined if this protein is actually translocase.

On all sides evidence is piling up that the mitochondrion represents a general blueprint that is characteristic of all membrane systems—in fact, of all the energy-transforming systems of the cell. Membrane systems generally show the same features as those exemplified by the mitochondrion: two membranes separated by a space, a membrane structure composed of structural protein and phospholipid in a network arrangement, the presence of elementary functional particles in the membrane layers, the presence of a contractile protein system in the outer membrane and the possession of a system that can move ions. The basic elements—a double membrane with active, specialized giant molecules attached to the membranes—seem to be universal in all living systems.

Now that the energy-generating system of living things can be described in molecular terms, we can expect an accelerating tempo and revolutionary developments in all the fields of investigation bearing on this central problem of biology.

THE LYSOSOME

by CHRISTIAN de DUVE   May 1963

The study of the living cell has in recent years established an increasingly complete catalogue of its working parts and identified these with their functions. The new understanding has come from a collaborative effort of, on the one hand, the cell anatomist, whose electron micrographs portray the internal structures of the cell in almost molecular detail, and, on the other, the biochemist, who disrupts and fractionates the cell so that he can observe the activity of the cellular organelles and their molecular components in isolation from one another. This concurrent study of structure and function has shown, for example, that the organelles called mitochondria conduct the primary energy transformations of the cell and that the smaller organelles called ribosomes are the centers of enzyme manufacture. The latest addition to the list of organelles is the lysosomes. They serve a function more comprehensible in terms of the grosser life processes of multicelled organisms. The lysosomes are tiny bags filled with a droplet of a powerful digestive juice capable of breaking down most of the constituents of living matter, much as these constituents are fragmented in the gastrointestinal tract of higher animals. In point of fact, the lysosomes function in many ways as the digestive system of the cell.

First identified in rat liver cells in 1955, lysosomes are now known to occur in many—possibly in all—animal cells. (It remains to be shown if they are present in plant cells.) It is significant that they are particularly large and abundant in cells, such as the macrophages and the white blood cells, that are called on to perform especially important digestive tasks. Lysosome function and malfunction appear to be involved in such vital processes as the fertilization of the egg and the aging of cells and tissues and in certain diseases. Challenging questions are presented by the properties of the membrane of the lysosome, which enable the organelle to contain enzymes that, on liberation, are capable of digesting the entire cell. Indeed, the death and dissolution of the cell following rupture of the membrane may play a part in the developmental processes of some animals and in a number of degenerative phenomena. This suggests the possibility that cell "autolysis" might be deliberately promoted or retarded for therapeutic purposes by the use of substances affecting the stability of the lysosome membrane.

Although lysosomes are frequently above the lower limit of visibility in the light microscope and are well within the range of the electron microscope, they were not discovered by optical methods. They were undoubtedly seen many times, but their nature and function were not recognized until they had been characterized chemically. The first clue was provided by a chance observation in our laboratory at the Catholic University of Louvain in 1949.

We had just begun to use the then newly developed technique of centrifugal fractionation, in which cells are disrupted in a homogenizer and then spun in a centrifuge at successively higher speeds to yield a number of fractions containing organelles of different types. When isolated in this manner, the organelles still maintain many of their functional properties, which can then be explored by means of biochemical methods. Our object was to localize in such fractions certain enzymes involved in the metabolism of carbohydrates in the liver of the rat and thereby to determine with which cellular structures these enzymes are associated. The standard procedure in this work is first to assay the homogenate of the disrupted cells for the presence of a given enzyme and then to look for the activity of the enzyme in the fractions. Among the enzymes included in our routines was the enzyme called acid phosphatase. This enzyme, which splits off inorganic phosphate from a variety of phosphate esters, is not directly connected with carbohydrate metabolism. We included it largely for control purposes.

To our surprise the acid-phosphatase activity in the homogenate was only about a tenth of what we had come to expect from previous assays of preparations that had been subjected to the more drastic homogenizing action of a Waring Blendor. The total of the activities found in the fractions, about twice that observed in the homogenate, was still only a fifth of the expected value. When the assays were repeated five days later on the same fractions (they had been kept in the icebox), the enzyme activity was much greater in all the particulate fractions, especially in the fraction containing mitochondria. The total activity was now within the expected range.

Fortunately we resisted the temptation to discard the first series of results as being due to some technical error, and a few additional experiments quickly gave us the clue to the mystery. In living cells the enzyme is largely or entirely confined within little baglike particles; the surface membrane of these particles is able not only to retain the enzyme inside the particle but also to resist the penetration of the small molecules of phosphate esters used in the assay. What we measured in our assays was only the amount of enzyme that

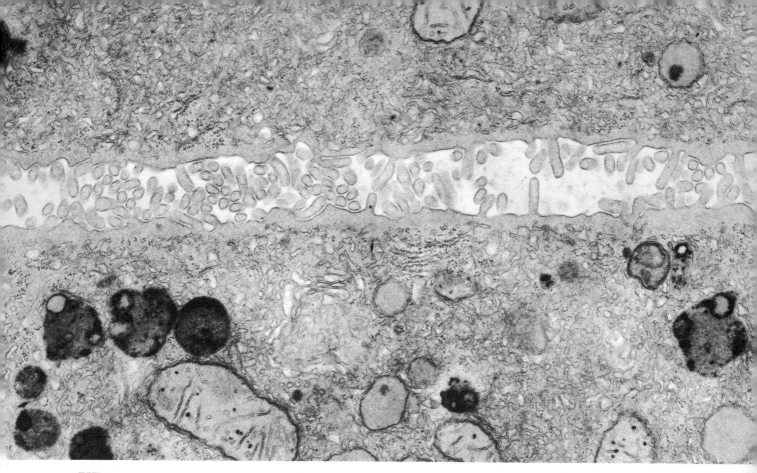

LYSOSOMES appear as relatively large dark objects in the electron micrograph above, which shows parts of two rat liver cells separated by a bile canaliculus. The canaliculus is the light strip running horizontally through the micrograph; the protuberances in the canaliculus are microvilli. The oblong body near the six lysosomes at bottom left is a mitochondrion. The micrograph was made at the Rockefeller Institute by Henri Beaufay of the Catholic University of Louvain. The magnification is 26,000 diameters.

TWO TYPES OF LYSOSOMES in a nephrotic rat kidney cell are magnified 60,000 diameters in the electron micrograph below: the kidney-shaped "digestive vacuole" at upper right and the two round "residual bodies" near the center and at lower left. Layered structures in the latter are "myelin figures," probably consisting of undigested fats. Minute black areas in the lysosomes are lead phosphate precipitated in staining. The micrograph was made by Alex B. Novikoff of the Albert Einstein School of Medicine.

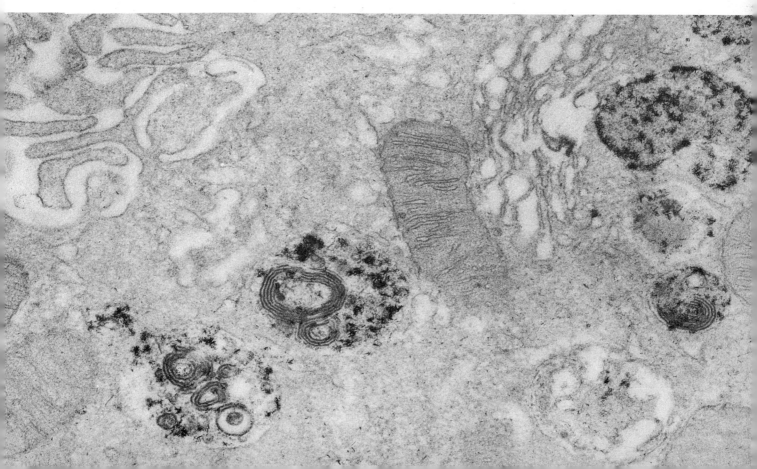

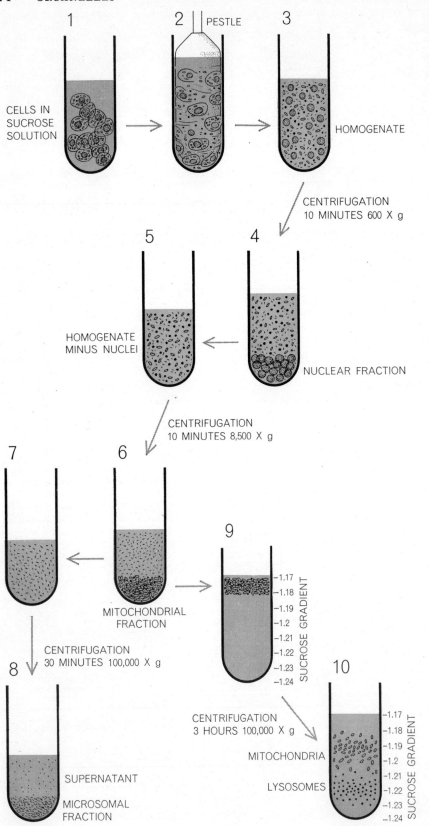

CENTRIFUGAL FRACTIONATION separates cells into fractions containing various cell components. Rapid mechanical rotation of the pestle ruptures the cells, setting the intracellular particles free in the medium. Successive centrifugations of the resulting homogenate produce fractions in which certain cell particles predominate. Steps *1* through *8* represent a method developed by W. C. Schneider of the National Institutes of Health. Steps *9* and *10* show a modification developed by the author and his co-workers; the mitochondrial fraction (Step *6*) is sedimented by centrifugation for 10 minutes at 25,000 times gravity. Numbers associated with the sucrose gradient give density in grams per cubic centimeter.

either was free in the cell or had escaped from particles injured by our manipulations. Whereas the Waring Blendor disrupts essentially all the particles, the gentler homogenizing procedure we had been using in our fractionation work ruptured only about 10 per cent of the particles, thus accounting for the low result obtained in the original homogenate. Further fractionation released an additional 10 per cent of the total activity from the fractions; the remainder came out as a result of the aging of the particles for five days in the refrigerator.

When these observations were transposed to the living cell, they suggested an interesting means of control of the enzyme activity. Living cells contain numerous phosphate esters of great importance to cellular function. Most of these phosphate compounds can be broken down by acid phosphatase, and investigators had often wondered why this breakdown does not occur in cells where the enzyme is present in large amounts. It now appeared from our results that the protective agent preventing the enzyme from acting indiscriminately on all the compounds might be simply the particle membrane that segregated the enzyme from the rest of the cell. This possibility, opened up by chance, was so interesting that we decided to make it the primary objective of our work.

At first it was believed that the particles containing acid phosphatase were the mitochondria, but later experiments indicated that they formed a distinct group, different from both the mitochondria and the microsomes on which most biochemists had been working. It took several years to establish the identity of the new particles as a separate group. In the meantime the list of enzymes contained within them began to grow. The number now stands at more than a dozen. In common with acid phosphatase, each new enzyme has demonstrated its ability to split important biological compounds in a slightly acid medium. Ultimately all the major classes of biologically active compounds, including proteins, nucleic acids and polysaccharides, were shown to be susceptible to action by the enzymes contained in these particles. As the spectrum of activity broadened, we became the more impressed with the significance of the new particles and of their surrounding membrane. Considered as a group, the enzymes present in the particles could have but one function: a lytic, or digestive, one. Hence the name "lysosome" (meaning lytic body) that

```
        RIBONUCLEIC ACIDS  ←
     DEOXYRIBONUCLEIC ACIDS  ←                    RIBONUCLEASE
  RIBONUCLEASE                                    DEOXYRIBONUCLEASE
  DEOXYRIBONUCLEASE    PHOSPHATE ESTERS  ←        PHOSPHATASES
  PHOSPHATASES              PROTEINS  ←           CATHEPSINS
  CATHEPSINS                                      GLYCOSIDASES
  GLYCOSIDASES                                    SULFATASES
  SULFATASES         POLYSACCHARIDES  ←
                     AND GLYCOSIDES

                       SULFATE ESTERS  ←

         LYSOSOME                                 INJURED LYSOSOME
```

LYSOSOME CONCEPT developed by the author is that of a minute "bag" filled with powerful digestive enzymes. So long as the lysosome membrane remains intact, digestion of the substrates on which these enzymes act is confined within the lysosome. But when the membrane is ruptured, the enzymes leak out and digestion takes place externally, often resulting in digestion of the cell.

we gave to the particles. As for the membrane, it must act as a shield between this powerful digestive juice and the rest of the cell. The digestive processes, we deduced, must be confined within the limits of the membrane, and the substances to be digested must somehow be taken up in the particles. Conversely, we were alerted to look for those pathological or normal conditions that might lead to the release of the enzymes inside the cell and the dissolution of the cell.

It was not until 1955 that the electron microscope made its contribution to the identification of the lysosomes. Working in collaboration with Alex B. Novikoff of the Albert Einstein College of Medicine in New York, we obtained our first electron micrographs of cell fractions containing partially purified lysosomes. In addition to known particles, mostly mitochondria, the pictures showed large numbers of characteristic bodies that had occasionally been observed in intact liver cells and that had been named "pericanalicular dense bodies." Their function was quite unknown; their name signified only their preferential location in cells along the bile canaliculi—the smaller bile ducts—and their high electron density, or opacity to the beam of the electron microscope. The identification of the lysosome activity with the dense bodies, made provisionally at that time, has since been confirmed by a variety of techniques.

We hoped that the identification of the liver lysosomes would lead quickly to the recognition of the lysosomes in other cells—much as the characteristic structure of the mitochondrion makes it

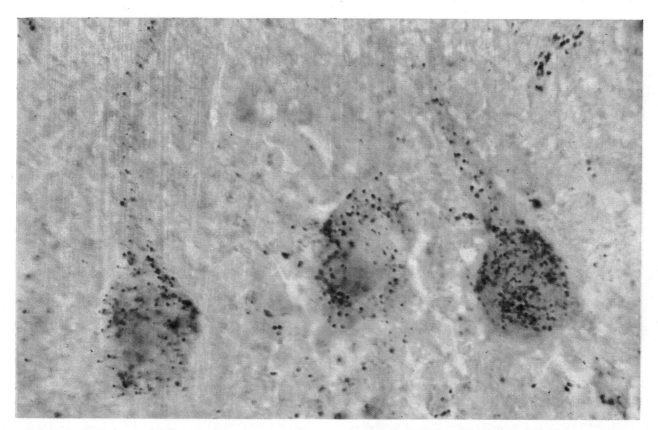

PURKINJE CELLS in the cerebellum of the pigeon contain lysosomes, which appear as tiny dark brown dots. Most of the lysosomes are located in the body of the three cells seen here. Single dendritic processes extending upward from the neurons at left and right also contain a few lysosomes. The magnification of this micrograph, which was made by Novikoff, is 1,600 diameters.

76 ORGANELLES

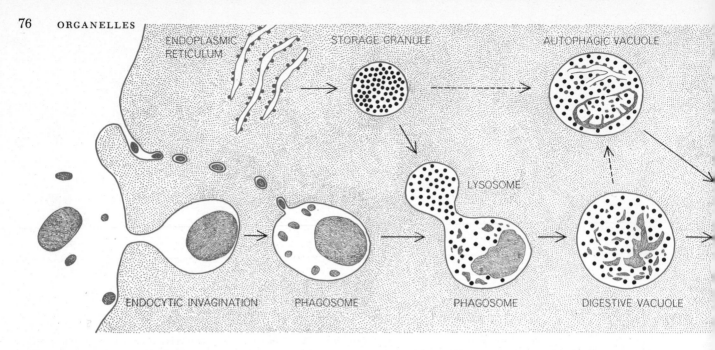

INTRACELLULAR DIGESTION involves lysosomes in various ways. It is necessary to distinguish four kinds of lysosomes: "storage granules," digestive vacuoles, residual bodies and "autophagic vacuoles." The first three are directly involved in the main digestive process. The storage granule is the original form of the lysosome; enzymes in the granule presumably are produced by the ribosomes (*small colored dots*) associated with the endoplasmic reticulum, but the origin of the lysosome membrane is unknown. When the cell ingests substances by endocytic invagination, a phagosome, or food vacuole, is formed. Several phagosomes may fuse together, forming a single vacuole. A storage granule or other lysosome fuses with the phagosome to form a digestive vacuole. Digestion products diffuse through the membrane into the cell. The digestive vacuole can continue its digestive activity, gradually

readily distinguishable in any type of cell. In this we were disappointed. The lysosomes come in a bewildering assortment of shapes and sizes, even in a single type of cell; they cannot be identified solely on the basis of their appearance. In the continuing study of lysosomes, therefore, the cell physiologist or biochemist has had to continue to provide the leads for the cell anatomist and the electron microscopist.

This polymorphism of the lysosomes is now perfectly understandable: their digestive activity causes them to be filled with a variety of substances and objects in an advanced state of disintegration, and it is their contents that determine their shape, size, density and so on. Nonetheless, the lack of any reliable visual criteria has tended to slow the progress of work in this field. The

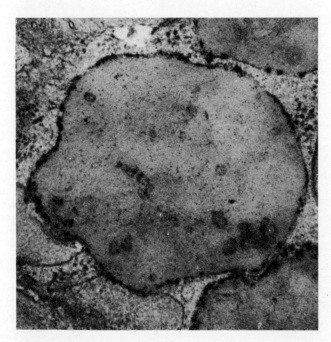

LYSOSOME is magnified 63,000 diameters. Gomori staining precipitated lead phosphate along lysosome membrane. The micrographs on these two pages, all of mouse kidney cells, were made at the Rockefeller Institute by Fritz Miller of the University of Munich.

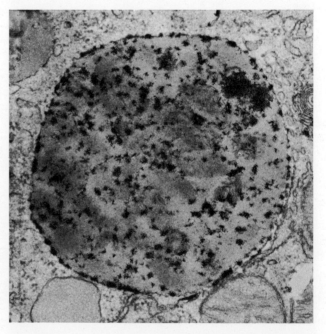

DIGESTIVE VACUOLE from kidney cell of mouse injected four hours earlier with hemoglobin is magnified 41,000 diameters. Lead phosphate appears along the membrane and in the interior. Dark gray patches are hemoglobin in the process of being digested.

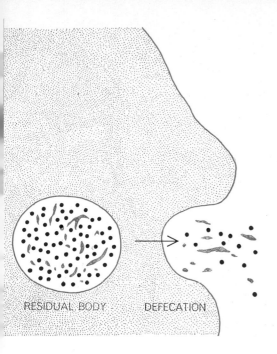

accumulating indigestible material until it becomes a residual body, which may then be eliminated by fusion with the cell membrane. The distinguishing feature of the autophagic vacuole is the material digested: parts of the cell itself, such as mitochondria and portions of the endoplasmic reticulum.

approach used on liver cells has been followed successfully in several other tissues, but it is a laborious one, usually requiring a great deal of repetitive work before one obtains fractions sufficiently pure for electron-microscope studies.

Fortunately one of the lysosomal enzymes—the same acid phosphatase that led to the discovery of the lysosomes—lends itself to visual identification. It can be stained by a method first developed by the late George Gomori of the University of Chicago. A slice of tissue is incubated with a compound susceptible to the action of the enzyme and with lead ions present in solution; at the sites where inorganic phosphate is set free by the action of the phosphatase, the phosphate precipitates in the form of an insoluble lead compound. Because lead has a high electron density the compound plainly shows up in electron micrographs; for visualization in the light microscope the compound is converted to black lead sulfide. Thus the enzyme can be localized inside the cell by means of a precipitated product of its activity. This technique, particularly in the hands of Novikoff, has greatly facilitated the study of lysosomes and their function in numerous tissues in both normal and pathological states.

Not all the substances that nurture a cell require digestion by lysosomes. In higher animals tissue cells receive most of their nutrients from the bloodstream in the form of small molecules absorbed through the cell membrane and requiring no digestion in the cell. Some materials, however, are too bulky for direct absorption and too complex chemically for immediate utilization. Objects of this kind must first be "eaten" and digested. Cells are able to engulf large molecules and even bodies as big as bacteria or other cells by a process now generally referred to as "endocytosis." A portion of the cell membrane first attaches itself to the "prey" and then appears to be sucked inward to form a small internal pocket containing the prey. The pocket pinches free from the cell membrane and drifts off into the cell interior, now forming a phagosome, as such bodies have been called by Werner Straus of the University of North Carolina.

The details of the next step vary from one type of cell to another, but they appear in all cases to involve the same fundamental mechanism. The phagosome containing the material to be digested and a lysosome containing the digestive enzymes approach each other; upon contact their membranes fuse to form a single larger vacuole. Digestion then proceeds within the membrane and the products of digestion diffuse into the cytoplasm, leaving behind only such remnants as have proved refractory to attack by the enzymes. Now that the outlines of the process are understood, lysosomes can be recognized in various cells at various stages in the performance of their function, from storage granules for newly synthesized enzymes to digestive vacuoles formed by fusion with a phagosome and finally to bodies containing the residue of previous digestive events.

In some cells, such as the amoeba and other protozoa, the residual bodies are

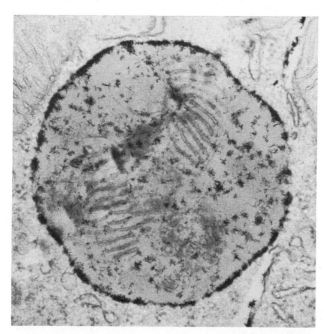

AUTOPHAGIC VACUOLE contains remnants of mitochondria from its host cell. The remnants appear as pairs of lines. "Needles" of lead phosphate were precipitated by the action of acid phosphatase, a lysosomal enzyme. The magnification is 55,000 diameters.

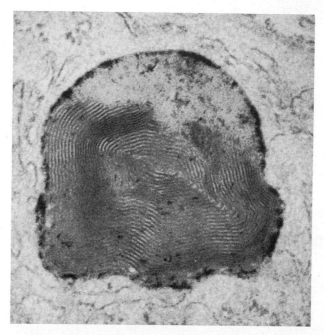

RESIDUAL BODY containing a layered collection of undigested material is enlarged 76,000 diameters. Lead phosphate is deposited mainly at membrane. The first, third and fourth micrographs on these two pages are published by permission of Academic Press.

eliminated by a kind of endocytosis in reverse, called defecation. In other cells, such as liver cells, defecation is slower or absent; the same digestive vacuoles are engaged repeatedly or continuously in digestive activity. After a time they seem to become charged with increasing amounts of residues, and this accumulation is believed to play a part in the aging of such cells.

As James G. Hirsch and Zanvil A. Cohn of the Rockefeller Institute have brought out, the cellular eating and digestive processes assume their most dramatic form in the white blood cells. These cells seem to spend most of their short life preparing for a single big burst of this activity. It has long been known that at the time the white blood cell enters the bloodstream it is filled with large granules; Hirsch and Cohn have shown that the granules are packages of digestive enzymes fitting the specifications of lysosomes. When the white cell engulfs a particle such as a bacterium, the granules can be seen to disappear one after the other, discharging their contents into the vacuole containing the ingested particle. Eventually the cells lose all their granules and are filled instead with one or more digestive pockets within which foreign particles are in process of dissolution. The cells seem not to recover from this process and eventually die.

This cycle of events in the cell matches at each point—ingestion, digestion and defecation—the process by which higher animals gain their nutrition. Digestion in both cases takes place behind a resistant envelope that protects the rest of the organism from attack by the digestive juices. In higher animals the resistant envelope forms a canal open at both ends; in most cells it surrounds a number of individual pockets. These are able to mix their contents and also to exchange matter with their environment by processes of coalescence reminiscent of the fusion of soap bubbles. Smaller pockets are also seen to pinch off from bigger ones, but the envelope always remains impermeably sealed around each pocket. One can easily imagine how a more permanent and continuous tract might, under some circumstances, evolve from such a flexible and relatively haphazard system. A primitive alimentary canal is indeed found in some single-celled organisms.

There is evidence that some cells may discharge lysosomal enzymes externally and use them to destroy surrounding structures or to open access for themselves. It is possible that the osteoclasts—bone-destroying cells that, along with bone-building osteoblasts, are responsible for the continuous remodeling of bone tissue—gnaw their way into the bone by a mechanism of this sort. They then complete their destructive action by engulfing bone fragments and digesting them in their lysosomes. It has also been suggested that in the process of fertilization spermatozoa may depend on the release of lysosomal enzymes to dissolve some of the structures that surround the egg cell. Subsequent changes in the egg seem in turn to involve the release of enzymes from the cortical granules that cover the outer surface of the cell. As a result the outer layers of the cell are broken down; a new membrane resistant to such attack is built up underneath, and the metabolism of the egg is geared toward division and development. According to Jean Brachet of the Free University of Brussels the cortical granules may belong to the lysosome family. They can also be ruptured by injury such as the prick of a needle; hence the digestive action of these bodies may have something to do with parthenogenesis: fertilization in which no sperm enters the egg.

The death of cells, even when it occurs on a large scale, is not necessarily a disastrous event in the life of a complex organism. Many of the component cells of the animal body are short-lived; they die and are replaced by newly formed cells. This is particularly true of the blood cells and of those cells that form the outer layers of the skin and of the mucous-membrane surfaces of the body. Cell death even plays a role in the early molding of the embryo and in the developmental cycle of some animals. As first shown by Rudolph Weber of the University of Berne and recently confirmed and elaborated by Yves Eeckhout in our laboratory at Louvain, when

REGRESSION OF TADPOLE TAIL, in metamorphosis of the South African frog *Xenopus laevis* into an adult, is accomplished by lysosomal digestion of cells. As metamorphosis proceeds the enzyme concentration increases (the absolute amount of enzyme remaining constant). Eventually the stub contains almost nothing but lysosomal enzymes, and it falls off. Data shown here were obtained by Rudolph Weber of the University of Berne.

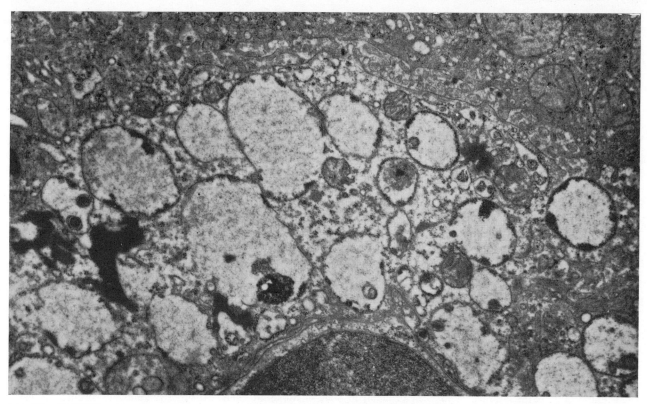

FORCED FEEDING of Kupffer cells from rat liver was achieved by injecting rats with Triton WR-1339, a detergent. The lysosomes become engorged with the detergent because they cannot digest it. Triton WR-1339 is transparent to electrons; hence the lysosomes, magnified 19,600 diameters, appear as light gray amorphous areas bounded by single membranes. The dark gray area at bottom center is a cell nucleus. The micrograph was made by Pierre Baudhuin and Robert Wattiaux of the Catholic University of Louvain.

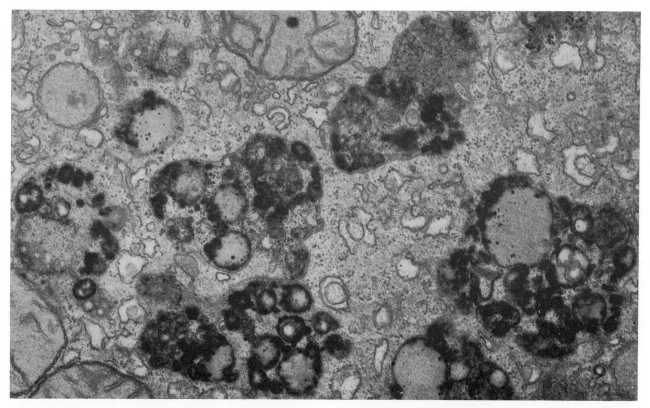

STARVATION caused a number of lysosomes in this cell from a rat liver to become autophagic vacuoles. That is, parts of the host cell (e.g., mitochondria) have found their way into the lysosomes. The mechanism that thus enables the cell to feed on its own substance without damaging itself irreparably is not known. The lysosomes are not stained; they appear as amorphous collections of objects of varying sizes, shapes and shades. The magnification of this micrograph, which was made by Beaufay, is about 38,000 diameters.

a tadpole tail has been reduced to an almost invisible stub, it still contains practically all its original complement of lysosomal enzymes and little else.

Lysosomal enzymes play their role in these processes in three different modes. In the first, white cells and other scavenger cells invariably invade the areas where cell destruction occurs: the lysosomes are there engaged through their normal digestive function inside the cell. A second mode, which has been discovered only recently, can be called cellular "autophagy": portions of a cell somehow find their way inside the cell's own lysosomes and are broken down. How the self-engulfment of the cell fragments takes place is not known. During starvation this process apparently enables the cell to use part of its own substance for fuel and for the renewal of essential constituents without doing itself irreparable damage. As in normal endocytosis, autophagy is kept localized by the limiting membrane.

The third mode of action involves the actual rupture of the lysosome membrane inside the cell and the digestion of the latter as a whole by the released enzymes. It can be described as a perforation of the cellular digestive tract. Such ruptures take place fairly quickly in dead cells, in a manner that recalls the rapid post-mortem putrefaction of the digestive mucosae in higher animals. It is obvious that once repair mechanisms are interrupted the areas most sensitive to dissolution will be those immediately adjacent to destructive enzymes. In the normal life processes of multicellular organisms lysosome rupture following death of a cell may have some value as a built-in mechanism for the self-removal of dead cells.

Of considerably greater interest is the possibility that the autodissolution of cells may occur as a pathological process. Present evidence indicates that the lysosome membranes may rupture in cells suddenly deprived of oxygen or exposed to cell poisons of certain kinds. As the enzymes are released they attack the cell itself, and they may also diffuse into the surrounding medium, damaging extracellular structures. Honor B. Fell and her co-workers at the University of Cambridge have shown that this is what happens in the cartilage and bones of animals receiving excess vitamin A. Damage by lysosomal enzymes released from the cells apparently explains the spontaneous fractures and other lesions that attend vitamin A intoxication.

Lysosomes can be involved in cell pathology in still other ways. Cells that are forced to engulf large amounts of foreign substances for the digestion of which they are not equipped will tend to accumulate such material in their lysosomes, possibly to the detriment of their general health. Plasma substitutes, such as dextran or polyvinyl pyrrolidone, have been known to cause this condition. It could also be involved in silicosis, the disease that results from the inhalation of silica dust; the particles of silica may accumulate in the lysosomes. Normal substances might accumulate in the same way if a key digestive enzyme is lacking in the lysosomes as a result, let us say, of a genetic abnormality. H. G. Hers of our department at Louvain recently discovered such a deficiency in the tissues of children who had died of a particularly severe form of glycogen-storage disease; he found that a lysosomal enzyme that attacks glycogen was missing.

If lysosomes can indeed act as "suicide bags"—and we now have good reason to believe that they can and sometimes do act in that way—the question arises as to whether or not their rupture can be influenced by means of drugs. Two possibilities come to mind. Agents acting as stabilizers of the lysosome membrane could be used to protect cells in a critical condition. Or substances that weaken the membrane could be employed to get rid of undesirable cells (for example cancer cells) if their action were sufficiently selective and specific.

So far no conscious attempt has been made to influence lysosomes in either way. But substances of both kinds are already known and some were used therapeutically before their effects on lysosomes were discovered. Vitamin A, in excess, has already been mentioned; although it is not highly specific, it appears to act preferentially on connective-tissue structures. According to recent studies performed by Lewis Thomas and Gerald Weissmann of the New York University School of Medicine, working in collaboration with the Fell group at Cambridge, cortisone and hydrocortisone appear to have a stabilizing influence on the lysosome membrane. This property may account, at least partly, for the well-known anti-inflammatory effects of these drugs. It would seem that in the individual cell, as in the multicellular organism, the digestive system occupies a pivotal position both in physiology and in pathology.

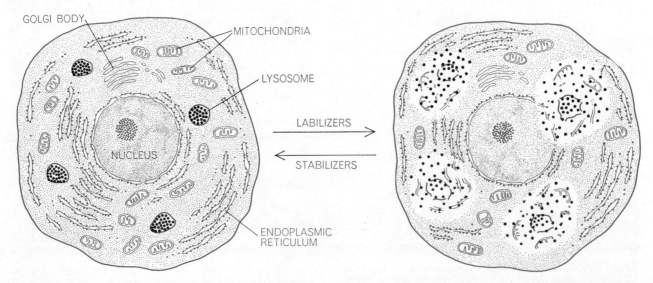

"SUICIDE BAG" is the term coined by the author to describe a lysosome that releases its complement of enzymes within a normal cell. The result is autolysis, or cell death by dissolution. It has been found that some substances affect the stability of the lysosome membrane adversely, thereby increasing the occurrence of autolysis. Other substances are known to have a stabilizing effect.

Part III

ENERGETICS

Energetics

INTRODUCTION

David E. Green's treatment, in the preceding section, of the relationship between ultrastructure and function in the mitochondrion introduces the problem of cellular energetics. The first article in this section, "How Cells Transform Energy" by Albert L. Lehninger, puts the special function of the mitochondrion into perspective by showing how the process of energy release by degradative metabolism is linked to the whole movement of energy through the living world. That cycle involves the fixation of energy in the bonds of organic molecules by photosynthesis in green plants, its release in usable form through the reactions of respiration and fermentation, and its ultimate use in a host of biological tasks—synthesis, mechanical or electrical work, light production, and so on.

Lehninger's article documents the many features that respiration and photosynthesis have in common. A decade ago biochemists often said that "photosynthesis is the reverse of respiration." This statement referred to the fact that in photosynthesis the carbon skeleton of glucose appears to be elaborated in a series of steps that trace in reverse those involved in the initial stages of respiratory disassembly. The identification of these intermediate compounds in photosynthesis is described in "The Path of Carbon in Photosynthesis" by J. A. Bassham. More recently, emphasis has been upon some features of photosynthesis and respiration that work in the *same* direction. Both processes appear fundamentally directed to the end of producing ATP, the substance that powers a variety of direct energy-requiring processes in the living cell.

The discovery that ATP can be directly generated by the photosynthetic machinery of plant cells is described by Daniel I. Arnon in "The Role of Light in Photosynthesis." Like the energy transfer systems involved in respiration, those for photosynthesis depend crucially upon the process of electron transfer. A remarkable combination of biochemical fractionations and electron microscopy has revealed that these electron transfer reactions are, in both systems, closely related to the fabric of the organelle in which they take place. The granum of the chloroplast and the "electron transport particle" of the mitochondrion are minimal elements in the series of energy transfer steps. The discovery that most of the chemistry of living energetics takes place at organized surfaces rather than in a formless soup is a remarkable achievement of modern biochemistry.

In "Biological Luminescence," William D. McElroy and Howard H. Seliger discuss one of the more unusual ways in which such energy is used. Conventional cells expend the energy obtained through photosynthesis or respiration in doing osmotic work ("active transport"), or in the synthesis of macromolecules, or—in the special cases of nerve and muscle cells—in electrical or mechanical work. A surprisingly large number of organisms, however, invest some energy in light production. Bioluminescent organisms may use such light signals for a variety of communicative purposes, which are themselves of interest; furthermore, the biochemical mechanisms underlying light production may provide some insights into the transformation of stored chemical energy into other forms.

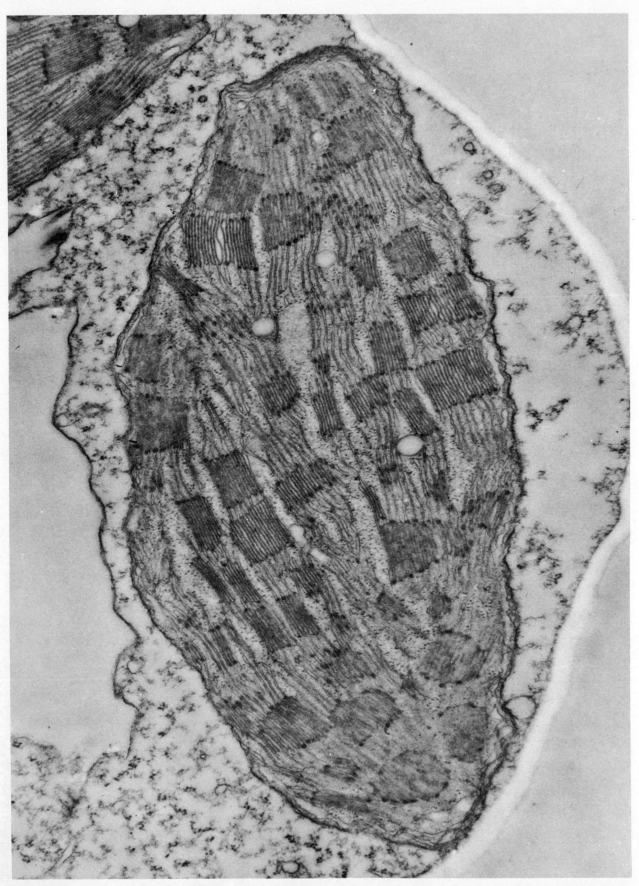

CHLOROPLAST is the site of photosynthesis, whereby light energy is transformed into chemical energy to prime the life cycle of plants and animals. A chloroplast in a maize cell is enlarged 40,000 diameters in this electron micrograph made by A. E. Vatter of Abbott Laboratories. The "light" reactions involving chlorophyll and solar energy take place within the rectangular "grana."

8
HOW CELLS TRANSFORM ENERGY

by ALBERT L. LEHNINGER   September 1961

A living cell is inherently an unstable and improbable organization; it maintains the beautifully complex and specific orderliness of its fragile structure only by the constant use of energy. When the supply of energy is cut off, the complex structure of the cell tends to degrade to a random and disorganized state. In addition to the chemical work required to preserve the integrity of their organization, different kinds of cell transform energy to do the varieties of mechanical, electrical, chemical and osmotic work that constitute the life processes of organisms.

As man has learned in recent times to use energy from inanimate sources to do his work, he has begun to comprehend the virtuosity and efficiency with which the cell manages the transformation of energy. The same laws of thermodynamics that govern the behavior of inanimate substances also govern the energy transactions of the living cell. The first law of thermodynamics says that the sum of mass and energy in any physical change always remains constant. The second law states that there are two forms of energy: "free," or useful, energy; and entropy, or useless or degraded energy. It states furthermore that in any physical change the tendency is for the free energy to decline and the entropy to increase. Living cells must have a supply of free energy.

The engineer gets most of the energy he employs from the chemical bonds in fuel. By burning the fuel he degrades the energy locked in those bonds to heat; he can then use the heat to make steam and drive a turbogenerator to produce electricity. Cells also extract free energy from the chemical bonds in fuels. The energy is stored in those bonds by the cells that manufacture the foodstuffs that serve as fuel. The cell makes use of this energy, however, in a very special way. Since the living cell functions at an essentially constant temperature, it cannot use heat energy to do work. Heat energy can do work only if it passes from one body to another body that has a lower temperature. The cell obviously cannot burn its fuel at the 900-degree-centigrade combustion temperature of coal, nor can it tolerate superheated steam or high voltage. It must therefore obtain energy and use it at a fairly constant and low temperature, in a dilute aqueous environment and within a narrow range of con-

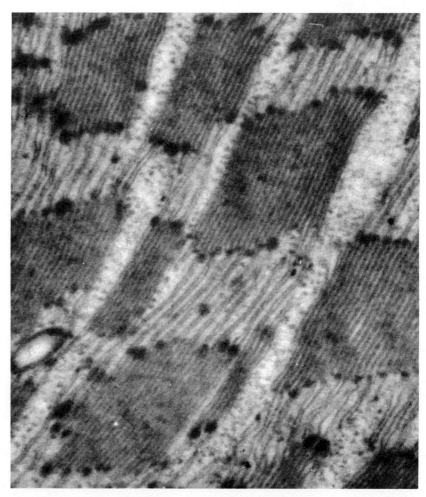

GRANA are enlarged 90,000 times in this electron micrograph by Vatter. They resemble stacks of coins in which chlorophyll is sandwiched between layers of protein and lipid. The lighter material around the grana is the stroma, in which the "dark" reactions take place.

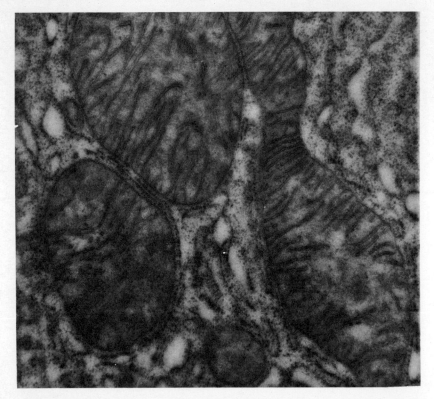

MITOCHONDRION is the site of respiration, the energy-transfer process of animal cells. Four mitochondria of a rat pancreas cell are enlarged 33,000 times in this electron micrograph by George E. Palade of the Rockefeller Institute. The inner membranes of the mitochondria's double wall are involuted to form the characteristic cristae, or folds.

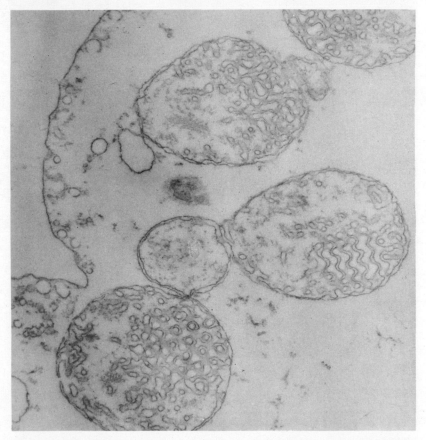

CRISTAE can be shelflike or tubular, as shown by this electron micrograph of mitochondria in a giant amoeba (*Chaos chaos*) by George D. Pappas and Philip W. Brandt of the Columbia University College of Physicians and Surgeons. Magnification is 27,000 diameters.

centration of hydrogen ions. To secure its primary energy the cell has during the eons of organic evolution perfected extraordinary molecular mechanisms that work with great efficiency under these mild conditions.

The energy-extracting mechanisms of living cells are of two kinds, and they separate all cells into two great classes. The first type of cell, called heterotrophic, includes the cells of the human body and of higher animals in general. This type of cell requires a supply of preformed, ready-made fuel of considerable chemical complexity, such as carbohydrate, protein and fat, which are themselves constituents of cells and tissues. Heterotrophic cells obtain their energy by burning or oxidizing these complex fuels, which are made by other cells, in the process called respiration, using molecular oxygen (O_2) from the atmosphere. They employ the energy so obtained to carry out their biological work, and they give up carbon dioxide to the atmosphere as the end product.

Cells in the other class get their energy from sunlight. Such cells are called autotrophic, or self-reliant. Principal among them are the cells of green plants. By the process of photosynthesis they harness the energy of sunlight for their living needs. They also use solar energy to incorporate carbon from atmospheric carbon dioxide in the elementary organic molecule of glucose. From glucose the cells of green plants and of other organisms build up the more complex molecules of which cells are made. In order to supply energy for this chemical work the cells burn some of the raw material by the mechanism of respiration. From this description of the cellular energy cycle it is clear that living things ultimately derive their energy from sunlight—plant cells directly and animal cells indirectly.

Investigations of the central questions posed here are converging on a complete description of the primary energy-extracting mechanisms of the cell. Most of the steps in the intricate cycles of respiration and photosynthesis have been worked out. Each process has been localized in a specific organ of the cell. Respiration is carried on by mitochondria, large numbers of which are found in almost all cells; photosynthesis is conducted by chloroplasts, the cytoplasmic structures that distinguish the cells of green plants. The molecular devices that make up these structures and perform their functions present the next great frontier to cell research.

From the centers of respiration

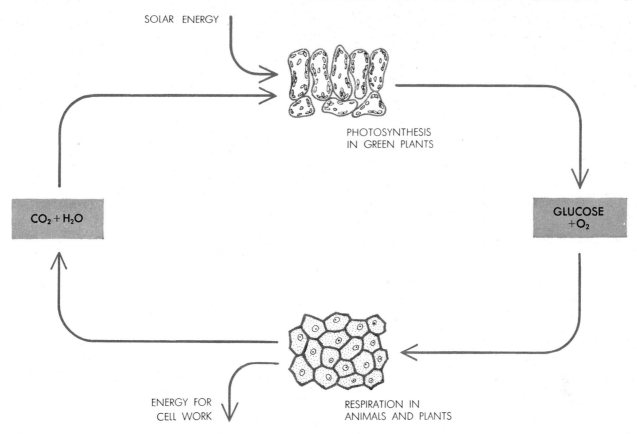

CARBON AND ENERGY CYCLE of life is based on the sun as the ultimate source of energy. Solar radiation drives photosynthesis, which builds energy-rich glucose from energy-poor carbon dioxide and water. The glucose and other fuels synthesized from it are then broken down to carbon dioxide and water by animal cells, which use the energy extracted in the process to do their work.

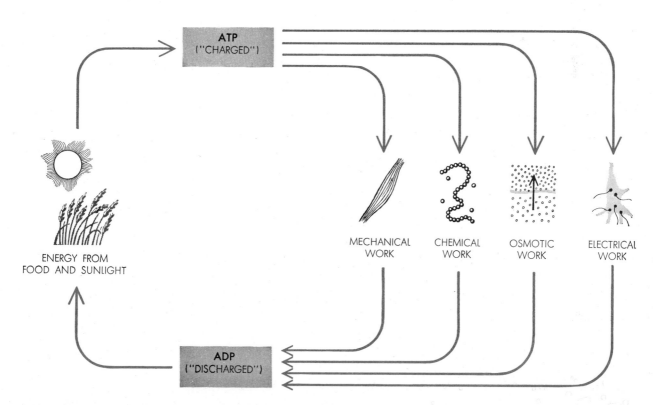

ADENOSINE TRIPHOSPHATE (ATP), the common carrier of energy in animal and plant cells, is formed in the mitochondria and chloroplasts. It supplies energy for muscle contraction, protein synthesis, absorption or secretion against an osmotic gradient and transfer of nerve impulses. "Discharged" adenosine diphosphate (ADP) thus formed is "charged" by solar or food energy.

ATP MOLECULE has one more phosphate group than ADP, attached by a high-energy bond (*wavy line at right of formula*). Solar or food energy is required to make this bond and thus to charge ATP. The chemical energy of the bond is made available again when the ATP is discharged by losing its terminal phosphate, which is transferred to an "acceptor" molecule in the cell. This

and photosynthesis the same well-defined molecule—adenosine triphosphate (ATP)—carries the free energy extracted from foodstuffs or from sunlight to all the energy-expending processes of the cell. ATP, which was first isolated from muscle by K. Lohmann of the University of Heidelberg some 30 years ago, contains three phosphate groups linked together. In the test tube the terminal group can be detached from the molecule by the drastic, one-step reaction of hydrolysis to yield adenosine diphosphate (ADP) and simple phosphate. As this reaction proceeds, the free energy of the ATP molecule appears as heat and entropy, in accordance with the second law of thermodynamics. In the cell, however, the terminal phosphate group is not merely detached by hydrolysis but is transferred to a specific acceptor molecule. The free energy of the ATP molecule is largely conserved by "phosphorylation" of the acceptor molecule, the energy content of which is now raised so that it can participate in an energy-requiring process such as biosynthesis or muscle contraction. Left over from this "coupled reaction" is ADP. In the thermodynamics of the cell ATP may be considered as the energy-rich, or "charged," form of the energy carrier and ADP as the energy-poor, or "discharged," form.

It is, of course, one or the other of the two energy-extracting mechanisms that "recharges" the carrier. In respiration in animal cells the energy of foodstuffs is released by oxidation and harnessed to regenerate ATP from ADP and phosphate. In photosynthesis in plant cells the energy of sunlight is trapped as chemical energy and harnessed to drive the recharging of ATP. Experiments employing the radioactive isotope phosphorus 32 have shown that the inorganic phosphate passes into the terminal phosphate group of ATP and out again with great rapidity. In a kidney cell the terminal phosphate group turns over so rapidly that its half-life is less than a minute, in consonance with the massive and dynamic flux of energy in the cells of this organ. It should be added that there is really no black magic associated with the action of ATP in the cell. Chemists are familiar with many similar reactions that permit the transfer of chemical energy in inanimate systems. The relatively complex structure of ATP has

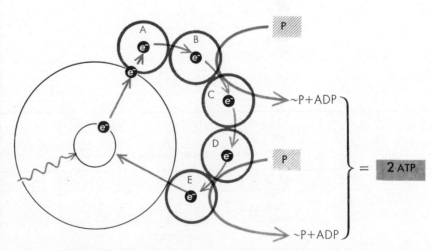

CYCLIC PHOTOPHOSPHORYLATION is the process by which an electron in chlorophyll, raised to a high-energy state by a photon of light, provides the energy to make ATP. The excited electron is captured by the first of a chain of "carriers" (*A*) and passed on around a circuit of such molecules (*B through E*), losing energy along the way. Some of the energy couples phosphate to ADP. The cycle ends as the electron returns to chlorophyll.

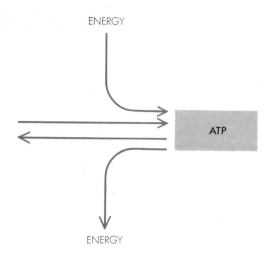

"phosphorylation" raises the energy level of the acceptor molecule. The ADP-ATP reaction is shown in schematic form at the right.

apparently evolved uniquely in the cell to produce maximum control and organization of energy-transferring chemical reactions.

The role of ATP in photosynthesis has only recently been elucidated [see "The Role of Light in Photosynthesis," by Daniel I. Arnon, page 97, for further information]. This discovery supplies a major part of the explanation of how photosynthetic cells harness the ultimate energy source of all living things, solar energy, in the synthesis of carbohydrates.

The energy of sunlight comes in packets called photons, or quanta; light of different colors or wavelengths is characterized by different energy content. When light strikes and is absorbed by certain metallic surfaces, the energy of the impinging photons is transferred to electrons of the metal. This "photoelectric" effect can be measured by the resulting flow of electric current. In the green-plant cell, solar energy of a particular range of wavelengths is absorbed by the green pigment chlorophyll. The absorbed energy raises an electron from its normal energy level to a higher level in the bond structure of this complex molecule. Such "excited" electrons tend to fall back to their normal and stable level, and when they do they give up the energy they have absorbed. In a pure preparation of chlorophyll, isolated from the cell, the absorbed energy is re-emitted in the form of visible light, as it is from other phosphorescent or fluorescent organic and inorganic compounds.

Thus chlorophyll itself in the test tube cannot store or usefully harness the energy of light; the energy escapes quickly, as though by short circuit. In the cell, however, chlorophyll is so connected spatially with other specific molecules that when it is excited by the absorption of light, the "hot," or energy-rich, electrons do not simply fall back to their normal positions. Instead these electrons are led away from the chlorophyll molecule by associated "electron carrier" molecules and handed from one to the other around a circular chain of reactions. As they traverse this external path the excited electrons give up their energy bit by bit and return to their original positions in the chlorophyll, which is now ready to absorb another photon. The energy given up by the electrons has meanwhile gone into the formation of ATP from ADP and phosphate; that is, into recharging the ATP system of the photosynthetic cell.

The electron carriers that mediate this process of "photosynthetic phosphorylation" have not yet been fully identified. One of these molecules is believed to contain riboflavin (vitamin B_2) and vitamin K. Others are tentatively identified as cytochromes: proteins containing iron atoms surrounded by porphyrin groups similar in arrangement and structure to the porphyrin of chlorophyll itself. At least two of these electron carriers are able to cause some of the energy they carry to be captured, in order to regenerate ATP from ADP [see illustration at bottom of opposite page]. This appears to be the basic scheme of the conversion of light into the phosphate-bond energy of ATP, as it has been developed by Daniel I. Arnon and his associates at the University of California and by other workers.

The complete photosynthetic process, however, involves the synthesis of carbohydrate as well as the harnessing of

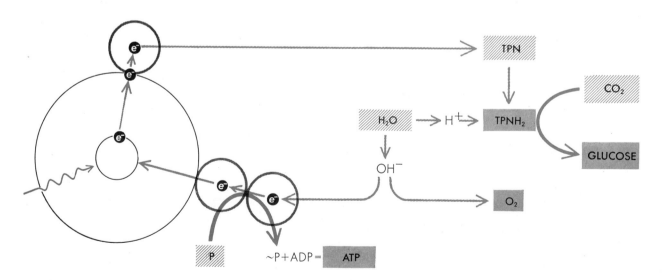

COMPLETE PHOTOSYNTHESIS requires an outside source of electrons and hydrogen ions (protons) to synthesize carbohydrate by "reducing" (adding electrons and hydrogen to) carbon dioxide. The source of electrons is chlorophyll and the source of protons is water. The reducing agent for carbon dioxide is "reduced triphosphopyridine nucleotide" ($TPNH_2$) formed by the action of protons and electrons on TPN, one of the carrier molecules. The leftover hydroxyl ions (OH^-) of water apparently lose electrons to restore the chlorophyll's supply. In this process oxygen gas, characteristic product of photosynthesis, is evolved and ATP is charged up.

solar energy. It is now believed that some of the "hot" electrons from excited chlorophyll, along with hydrogen ions derived from water, cause the reduction (that is, the addition of electrons or hydrogen atoms) of one of the electron carriers, triphosphopyridine nucleotide (TPN), which in its reduced form becomes $TPNH_2$ [see illustration at bottom of preceding page]. In a series of "dark" reactions, so named because they occur in the absence of light, $TPNH_2$ brings about the reduction of carbon dioxide to carbohydrate. Much of the energy necessary for this series of reactions is supplied by ATP [see illustration at right]. The pattern of the dark reactions was worked out largely by Melvin Calvin and his associates, also at the University of California. A by-product of the original photoreduction of TPN is the hydroxyl ion (OH^-). Although the evidence is not yet complete, it is thought that these ions donate their electrons to a cytochrome in the photosynthetic chain, releasing molecular oxygen in the process. The electrons continue down the carrier chain, contributing to the formation of ATP and finally settling—in their energy-depleted state—in the chlorophyll.

As the highly organized and sequential nature of the photosynthetic process suggests, the chlorophyll molecules are not randomly situated or merely suspended in solution inside the chloroplasts. On the contrary, the chlorophyll is arranged in orderly structures within the chloroplasts called grana, and the grana in turn are separated from one another by a network of fibers or membranes. Within the grana the flat chlorophyll molecules are stacked in piles. The chlorophyll molecules can therefore be looked on as the single plates of a battery, several plates being organized as in an electric cell, and several cells in a battery, represented by the chloroplast.

The chloroplasts also contain all the specialized electron-carrier molecules that work together with chlorophyll to extract the energy from the hot electrons and use that energy to synthesize carbohydrate. Separated from the rest of the cell, the chloroplasts can carry out the complete photosynthetic process.

The efficiency of these miniature solar

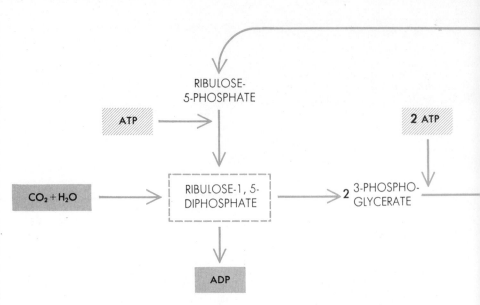

SYNTHESIS OF GLUCOSE from carbon dioxide and water is a dark reaction—that is, it involves a series of reactions that do not directly require light. But it does require two compounds made by light, ATP and $TPNH_2$, as the energy supply and reducing agent respec-

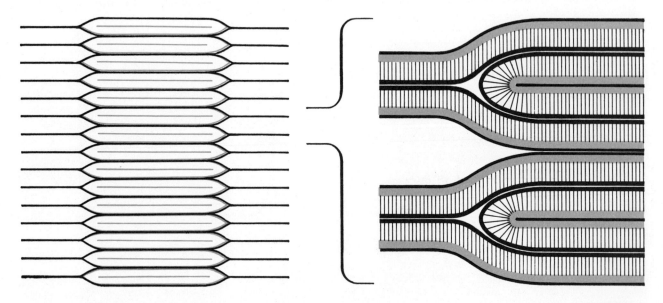

STRUCTURE OF ONE GRANUM in a chloroplast is diagrammed in successive magnifications. The chlorophyll (color) is concentrated within envelopes stacked to form the granum (left), with connecting fibers leading to adjacent grana. In the layers, two of which are magnified (second from left), the chlorophyll is sandwiched between membranes of protein, according to a hypothetical model proposed by Alan J. Hodge of the California Institute of Technology. In Hodge's model, based on electron microscopy and the "electron carrier" chemistry discussed in the text, the individual chlorophyll molecules are oriented (third from left) be-

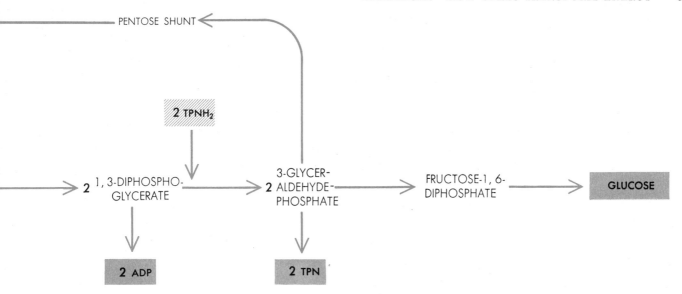

tively. In this complex cycle, shown here only in outline, the key intermediate is ribulose diphosphate, which picks up the carbon dioxide and makes two molecules of phosphoglycerate. This is reduced by TPNH$_2$ and rearranged in steps, ultimately to become glucose. Meanwhile the ribulose diphosphate is regenerated in a series of reactions abbreviated here as "pentose shunt."

power plants is impressive. Though the exact figures are subject to controversy, it can be demonstrated under special laboratory conditions that the photosynthetic process converts as much as 75 per cent of the light that impinges on the chlorophyll molecule into chemical energy. On the other hand, the efficiency of energy recovery of a field of corn, given the random and uneven exposure of the leaves to sunlight and other conditions of nature, is considerably lower: on the order of only a few per cent.

The molecule of glucose, as the end product of photosynthesis, can therefore be visualized as having a considerable amount of solar energy locked in its molecular configuration. In the process of respiration heterotrophic cells extract this energy by carefully taking apart the glucose molecule step by step, conserving its energy of configuration in the phosphate-bond energy of ATP.

There are different kinds of heterotrophic cell. Some, such as certain marine microorganisms, can live without

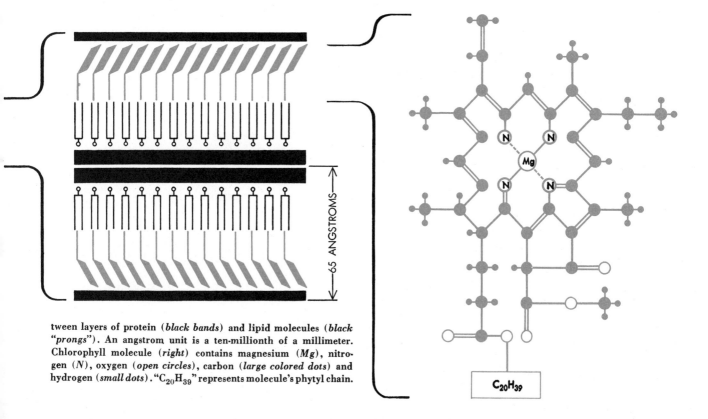

tween layers of protein (*black bands*) and lipid molecules (*black "prongs"*). An angstrom unit is a ten-millionth of a millimeter. Chlorophyll molecule (*right*) contains magnesium (*Mg*), nitrogen (*N*), oxygen (*open circles*), carbon (*large colored dots*) and hydrogen (*small dots*). "$C_{20}H_{39}$" represents molecule's phytyl chain.

92 ENERGETICS

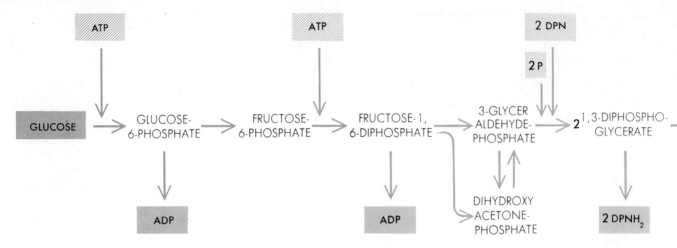

GLYCOLYSIS is the first step in energy recovery from glucose. As comparison of this diagram with the upper one on pages 90 and 91 will show, many of the steps are the reverse of those in the dark synthesis of glucose by plants. Six-carbon glucose is broken down

oxygen; some, such as brain cells, absolutely require oxygen; some, such as muscle cells, are more versatile, being able to function either aerobically or anaerobically. Furthermore, although most cells prefer glucose as the major fuel; some can live exclusively on amino acids or fatty acids synthesized from glucose as the basic raw material. The disassembly of the glucose molecule by the liver cell may be taken, however, as typical of the process by which most known aerobic heterotrophs obtain energy.

The total amount of energy available in a molecule of glucose may be quite simply determined. By burning a sample in the laboratory it can be shown that the oxidation of the glucose yields six molecules of water and six molecules of carbon dioxide, with the evolution of

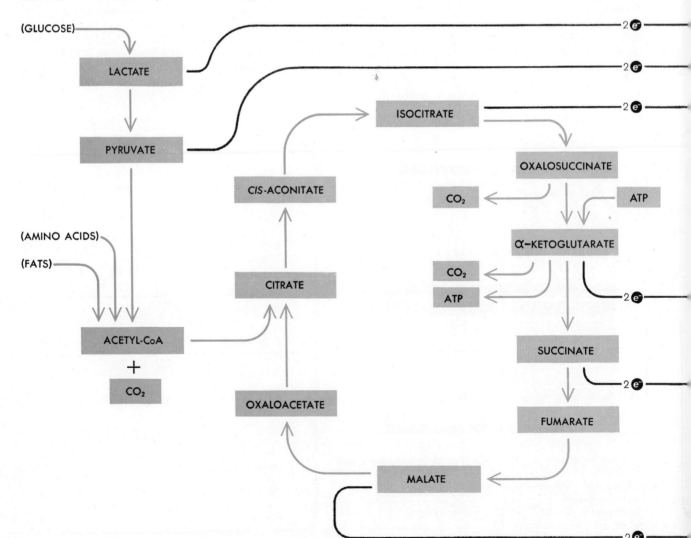

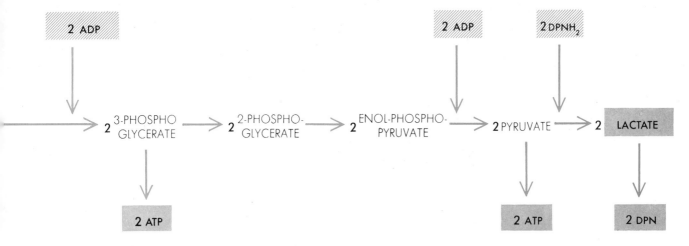

into two molecules of three-carbon lactate (or lactic acid; the ionic rather than the acid form of the intermediate compounds is shown in the diagrams). Two molecules of ATP are used up in glycolysis but four are formed, for a net gain of two molecules of ATP.

some 690,000 calories of energy per gram molecular weight (that is, per 180 grams of glucose) in the form of heat. Energy in the form of heat is, of course, useless to the cell, which functions under essentially constant temperature conditions. The step-by-step oxidation of glucose achieved by the mechanism of respiration occurs in such a way, however, that much of the free energy of the glucose molecule is conserved in a form that is useful to the cell. In the end more than 50 per cent of the available energy is recovered in the form of phosphate-bond energy. This recovery compares most favorably with the standard of the engineer, who rarely converts more than a third of the heat of combustion into useful mechanical or electrical energy.

The oxidation of glucose in the cell proceeds in two major phases. The first, or preparatory, phase, called glycolysis, brings about the splitting of the six-carbon glucose molecule into two three-carbon molecules of lactic acid. This seemingly simple process occurs not in one step but at least 11 steps, each catalyzed by a specific enzyme. If the complexity of this operation seems to contradict the Newtonian maxim *Natura enim simplex est*, then it must be borne in mind that the function of the reaction is to extract chemical energy from the glucose molecule and not merely to split it in two. Each of the intermediate products contains phosphate groups, and a net of two molecules of ADP and two phosphates are used up in the reaction.

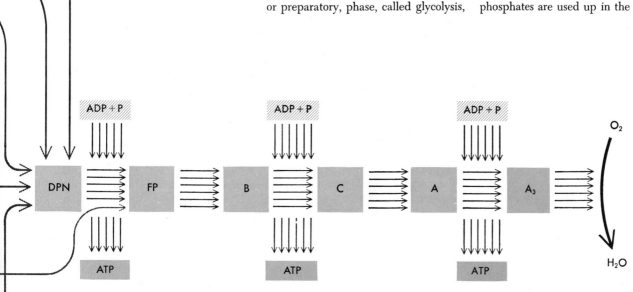

KREBS "CITRIC ACID" CYCLE finally oxidizes the products of glycolysis to carbon dioxide and water. Lactate is first converted to pyruvate, which in turns goes to acetyl coenzyme A. (Here fat and protein join carbohydrate in the metabolic process.) There follows a cycle of reactions, involving the regeneration of oxaloacetate, in which carbon compounds are broken down to carbon dioxide. Electrons removed at various stages are passed down a "respiratory chain" of electron carriers: diphosphopyridine nucleotide (DPN), a flavoprotein enzyme (FP) and a series of iron-containing enzymes: cytochromes B, C, A and A_3. As the electrons pass down the chain, ultimately to reduce oxygen to water, they drive the phosphorylations in which ATP is formed. Each molecule of lactate contributes six pairs of electrons; five of these charge up three ATP molecules each and the sixth makes two ATP's. One more ATP is formed in the citric acid "mill" itself, so a total of 36 molecules of ATP is produced by the two molecules of lactate that were formed from the original glucose molecule.

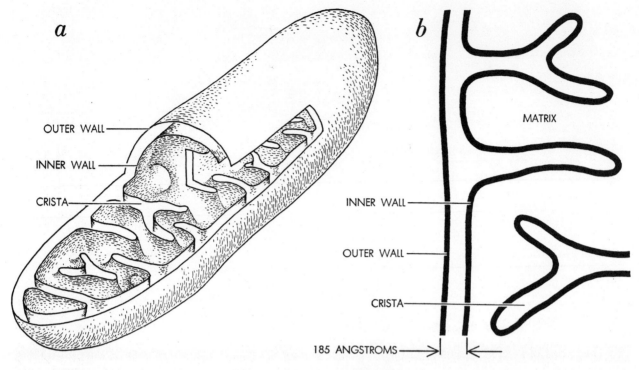

STRUCTURE OF MITOCHONDRION is basically that of a fluid-filled vessel with an involuted wall (*a*). The wall consists of a double membrane (*b*), with infoldings of the inner one forming cristae. Each membrane is apparently constructed of a layer of pro-

Ultimately the splitting of glucose not only yields two molecules of lactic acid but also generates two new molecules of ATP [*see illustration at top of preceding two pages*].

What does this mean in terms of energy? Thermodynamic equations show that the splitting of a gram molecule of glucose to lactic acid makes a total of 56,000 calories available. Since the charging of each gram molecule of ATP captures about 10,000 calories of energy, the yield at this stage is about 36 per cent, a respectable figure by engineering standards. The conversion of 20,000 calories represents, however, a small fraction—only 3 per cent—of the total of 690,000 calories bound in the glucose. Yet many cells, such as anaerobic cells or muscle cells in exercise (which are unable to conduct the process of respiration), function on this small yield.

With glucose now broken down to lactic acid, aerobic cells proceed to extract a major portion of the remaining energy by the process of respiration, in which the three-carbon lactic acid molecule is broken down to single-carbon molecules of carbon dioxide. The lactic acid, or rather its oxidized form pyruvic acid, undergoes an even more complex series of reactions, each step again being catalyzed by a specific enzyme system [*see illustration at bottom of preceding two pages*]. First the three-carbon compound is broken down to an activated form of acetic acid—acetyl coenzyme A—and carbon dioxide. The two-carbon acetic acid compound then combines with a four-carbon compound, oxaloacetic acid, to make the six-carbon citric acid. This is degraded again to oxaloacetic acid by a series of reactions, and the three-carbon atoms of pyruvic acid that were fed into this cyclic mechanism at last appear as carbon dioxide. This "mill," which oxidizes not only glucose but also fat and amino acid molecules previously broken down to acetic acid, is known as the Krebs citric acid cycle. It was first postulated by Sir Hans Krebs in 1937 in one of the great landmarks of modern biochemistry and honored by a Nobel prize in 1953.

Although the Krebs cycle accounts for the oxidation of lactic acid to carbon dioxide, it alone does not explain how the large amount of energy remaining in these molecules is extracted in useful form. The process of energy recovery that accompanies the action of the Krebs cycle has been an intensely active field of investigation in recent years. While the over-all picture can be described with some assurance, there are many details yet to be solved. In the course of the cycle, it appears, electrons are extracted from the intermediates by enzymes and fed into a series of electron-carrier molecules, collectively called the respiratory chain. This chain of enzyme molecules is the final common pathway of all electrons removed from foodstuff molecules during biological oxidation. At the last link in the chain, the electrons combine with oxygen to form water. The breakdown of foodstuffs by respiration therefore in essence reverses the process of photosynthesis in which electrons are removed from water to form oxygen. Moreover, it is striking that the electron carriers in the respiratory chain bear many chemical similarities to those of the corresponding chain in photosynthesis. They contain, for example, riboflavin and cytochrome structures similar to those in the chloroplast. The Newtonian simplicity of nature is thereby affirmed.

As in photosynthesis, the energy of the electrons passing along the chain to oxygen is tapped off and used to drive the coupled synthesis of ATP from ADP and phosphate. Actually this respiratory-chain phosphorylation, or oxidative phosphorylation, is better understood than the more recently discovered photosynthetic phosphorylation. One thing known with certainty is that there are three points along the chain at which ATP is recharged. For each pair of elec-

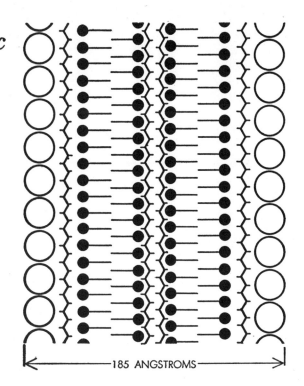
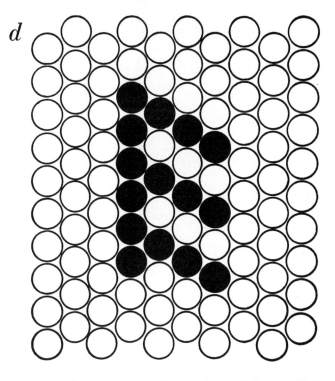

tein molecules (*spheres at "c"*) lined by a double layer of lipid molecules (*black-tailed spheres*). The respiratory-chain electron carriers and enzymes appear to be regularly spaced elements (*black spheres at "d"*) of the protein monolayers. The "matrix" is fluid.

trons split from lactic acid in the course of its oxidation in the Krebs cycle, therefore, an average of three molecules of ATP are formed.

From the total yield of ATP molecules it' is now possible to calculate the thermodynamic efficiency with which the cell extracts the energy made available by the oxidation of glucose. The preliminary splitting of glucose to two molecules of lactic acid yields two molecules of ATP. Each molecule of lactic acid in turn delivers ultimately six pairs of electrons to the respiratory chain. Since three molecules of ADP are "charged up" to ATP for each pair of electrons traversing the chain, 36 molecules of ATP are formed in the respiratory process proper. On the rough estimate of 10,000 calories each, the 38 molecules of ATP incorporate in their phosphate bonds some 380,000 of the 690,000 calories contained in the original gram molecule of glucose. The efficiency of the combined processes of glycolysis and respiration can therefore be estimated as a minimum of 55 per cent.

The intricacy of the respiratory process in particular suggests again that the enzymatic machinery involved could not do its work if its component parts were randomly mixed together in solution. Just as the molecular devices of photosynthesis appear to be spatially oriented to one another in the chloroplast, so the organ of respiration in the cell, the mitochondrion, presents the same picture of structured order. There may be anywhere from 50 to 5,000 mitochondria in a cell, depending on its type and function. A single liver cell of the rat contains about 1,000 mitochondria. They are large enough (three to four microns long) to be seen in the cytoplasm with a light microscope. But their ultrastructure requires the electron microscope in order to be seen.

In electron micrographs it can be seen that the mitochondrion has two membranes, the inner one occurring in folds in the body of the structure. Recent research on mitochondria isolated from cells of the liver has shown that the Krebs-cycle enzyme molecules are located in the matrix, or soluble portion of the inner contents, but the respiratory-chain enzymes, in the form of molecular "assemblies," are located in the membranes [see "Energy Transformation in the Cell," by Albert L. Lehninger, Offprint #69, for more information]. The membranes consist of alternating layers of protein and lipid (fatty) molecules, just as do the membranes of the grana of the chloroplasts. Indeed, there is a remarkable similarity in the structure of these two fundamental power plants of all cellular life, one capable of capturing solar energy in ATP and the other of transforming the energy of foodstuffs to ATP energy.

Modern chemistry and physics have recently been able to specify the three-dimensional structure of certain large molecules, such as those of proteins and of DNA, the molecules that carry genetic information. The next great step in cell research is to find out how the large enzyme molecules, themselves proteins, are arranged together in the mitochondrial membranes, together with the lipids, so that each catalyst molecule is properly oriented and therefore able to react with the next one in the working assembly. The "wiring diagram" of the mitochondrion is already clear!

If the classical engineering science of energy transformation is humbled by what is now known about the power plants of the cell, so are the newer and more glamorous branches of engineering. The technology of electronics has achieved amazing success in packaging and miniaturizing the components of a computer. But these advances still fall far short of accomplishing the unbelievable miniaturization of complex energy-transducing components that has been perfected by organic evolution in each living cell.

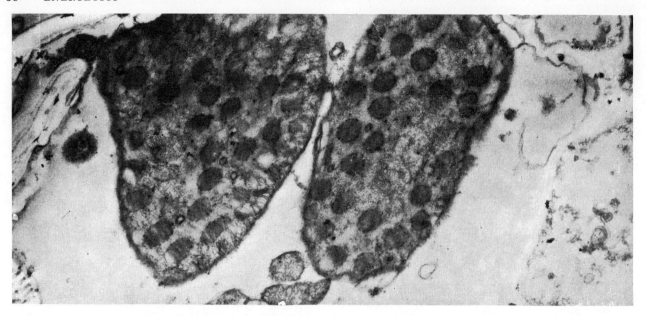

TWO CHLOROPLASTS within a maize cell are magnified 12,000 diameters in this electron micrograph made by A. E. Vatter. A single cell of a green leaf usually contains from 20 to 100 of these structures. They carry out the complete photosynthetic process.

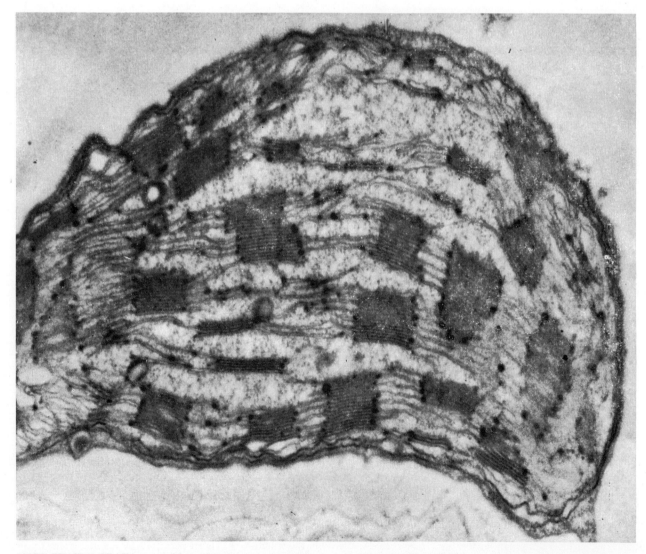

MORE THAN 20 GRANA are visible in this electron micrograph, made by A. E. Vatter, which magnifies one maize chloroplast 33,000 diameters. Grana resemble stacks of coins. Since they contain all of the chlorophyll in green plants, they are the receptors that trap the energy of light. The material around the grana, called stroma, performs the "dark" reactions of photosynthesis.

THE ROLE OF LIGHT IN PHOTOSYNTHESIS

by DANIEL I. ARNON   November 1960

When the subject of photosynthesis was last surveyed in SCIENTIFIC AMERICAN, Eugene I. Rabinowitch wrote: "The photosynthetic process, like certain other groups of reactions in living cells, seems to be bound to the structure of the cell; it cannot be repeated outside that structure."

At that time (November, 1953), my co-workers at the University of California and I were in our fifth year of a vain effort to separate photosynthesis from other life processes in the cell. The next year our luck turned. We learned how to remove the chlorophyll-containing particles (chloroplasts) from spinach leaves by a technique that preserved in the isolated particles the ability to carry out the complete process of green-plant photosynthesis: the conversion of carbon dioxide and water to carbohydrates and oxygen, with no outside supply of energy except visible light.

Previously it had been known that isolated chloroplasts can evolve oxygen when exposed to light. This was called the Hill reaction, after R. Hill of the University of Cambridge, who discovered it in 1937. But isolated chloroplasts could not be made to assimilate carbon dioxide with the techniques then in use. Thus until 1954 carbon dioxide assimilation was believed to be a property of the intact cell.

Once the whole photosynthetic sequence had been dissected out, so to speak, the way to a biochemical attack on the mechanism of photosynthesis became much clearer. When the biochemist studies a physiological process, his goal is to take the process apart, separate its successive reactions, find out where and how each one takes place and identify the biological catalysts (enzymes) that make it possible. In the intact cell such an analysis is extremely difficult. An individual step of one physiological process may be hidden or modified by a similar step of another. Or the step may take place so rapidly that it escapes detection altogether.

The difficulty was especially acute in the study of photosynthesis because of the process of respiration. At the same time that a living cell is building up carbohydrate out of carbon dioxide and water, and throwing off oxygen, it is respiring: absorbing oxygen, with which it breaks carbohydrate down to carbon dioxide and water again. Isolated chloroplasts provided a photosynthetic system that does not respire.

The past six years have been busy ones in my own and other laboratories. This article will summarize what has been learned, with particular emphasis on the question that has always intrigued me most: What is the role of light in photosynthesis?

Early Investigations

The story will be easier to follow if we go back to its beginning. In 1772 Joseph Priestley made his epochal discovery that green plants, "instead of affecting the air in the same manner with animal respiration, reverse the effect of breathing." Over the next 30-odd years the gross facts of photosynthesis were uncovered. Its raw materials and end products were identified. Its dependence on light was discovered in 1779 by the Dutch physician Jan Ingenhousz, who later offered a plausible suggestion as to the role that light energy plays. Sunshine, he said, splits apart the carbon dioxide (CO_2) that a plant has absorbed from the air, the plant "throwing out at that time the oxygen alone and keeping the carbon to itself as nourishment."

Later other investigators proposed that the "C" from the broken CO_2 combines with water to form a product with the empirical formula (CH_2O). (By enclosing a formula in parentheses the chemist indicates that the symbols inside do not necessarily represent an individual molecule, but merely a grouping that may be part of some larger structure.) The idea made eminently good sense because it is in exactly this proportion that carbon, hydrogen and oxygen occur in carbohydrates, the chief product of photosynthesis in green plants. And so, for the next 100 years and more, the role of light was regarded as settled.

The hypothesis became a fixed principle of biology. All living organisms were divided into two groups: (1) green plants, which were thought to be the only organisms capable of assimilating carbon dioxide (and only in the presence of light); (2) all other forms of life, which do not have this power and must

---

AUTHOR'S NOTE

The author wishes to acknowledge the contribution of his associates and graduate students to the work in his laboratory reported in this article. In chronological order of their association with the investigations they are: Frederick R. Whatley, Mary Belle Allen, Lois J. Durham, John B. Capindale, Lawson L. Rosenberg, Harry Y. Tsujimoto, Achim V. Trebst, Hans R. Müller, Joseph and Colette Bové, Manuel Losada, Shoitsu Ogata, Mitsuhiro Nozaki and David O. Hall.

98 ENERGETICS

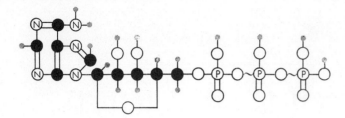

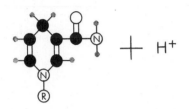

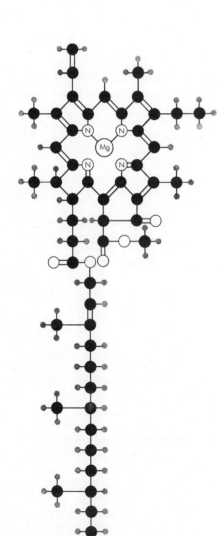

- ● CARBON
- ○ OXYGEN
- ⓡ RADICAL
- ⒡ IRON
- ⓜ MAGNESIUM
- • HYDROGEN
- ⓝ NITROGEN
- ⓢ SULFUR
- ⓟ PHOSPHORUS

THREE KEY MOLECULES in photosynthesis are adenosine triphosphate (ATP), at top; reduced pyridine nucleotide (P-NH$_2$), in middle; and chlorophyll, at bottom. Energy from light, trapped by chlorophyll, serves to bind third phosphate group to adenosine diphosphate to make ATP. TPNH$_2$ differs from DPNH$_2$ in having a third phosphate group in the part of the molecule shown here as R (*radical*).

therefore subsist by consuming the organic products of the photosynthetic group. So firmly was the scheme entrenched that in several European languages the term "carbon dioxide assimilation" also meant what is now called photosynthesis.

This seemingly logical structure was undermined in the 1880's by two developments, although no one realized it at the time. First, Sergei Winogradsky, then at the University of Strasbourg, discovered the so-called chemosynthetic bacteria: cells which contain no chlorophyll, but which can make organic material by assimilating carbon dioxide in the dark. Second, Theodor Wilhelm Engelmann at the University of Utrecht found that purple bacteria which metabolize sulfur compounds perform a type of photosynthesis apparently without giving off oxygen.

As so often happens when an experimental result contradicts the prevailing fundamental notions, the implications of Winogradsky's discovery were simply ignored by most of his contemporaries. A notable exception was the Russian microbiologist A. F. Lebedev. With remarkable insight, Lebedev insisted that carbon dioxide assimilation could no longer be considered peculiar to photosynthetic cells. He suggested, in fact, that all cells possess this ability. What set apart the photosynthetic group was its use of light as the source of energy for the process. This view was too far ahead of its time, and had little impact.

Engelmann's discovery had still less. The then indigestible facts of bacterial photosynthesis were either denied or explained away by ingenious hypotheses. Photosynthesis without the evolution of oxygen seemed a contradiction in terms. Even Lebedev, while arguing for the assimilation of carbon dioxide in the dark, was convinced that it took place by splitting of the carbon dioxide molecule, with oxygen as an inevitable by-product.

Van Niel's Hypothesis

It took 40 years and many more experiments to establish the significance of chemosynthetic and photosynthetic bacteria. In the 1930's, as we shall see shortly, a variety of cells were found to assimilate carbon dioxide and manufacture carbohydrates, without light. At that time, moreover, C. B. van Niel of Stanford University conclusively demonstrated that bacteria can carry on photosynthesis without evolving oxygen. The splitting of carbon dioxide by light was no longer a tenable hypothesis.

Van Niel supplied a new hypothesis. He proposed that light splits not carbon dioxide but water. According to this idea, which applied both to bacteria and green plants, light acts on water to make (H) and (OH) radicals. The (H) supplies the hydrogen to convert carbon dioxide to carbohydrate. In plants the (OH) reacts to form oxygen. Bacteria lack the enzymes to catalyze this reaction. Instead, they combine (OH) with hydrogen from an outside source to form water again. The model accounted for everything then known, including the fact that bacterial photosynthesis, unlike that in green plants, does require an outside hydrogen donor (such as hydrogen gas, sulfur compounds or certain organic acids).

Van Niel's hypothesis was widely accepted. Now it became necessary to trace the path of the (H) that is presumably formed in light. However, the splitting of water by radiation with energy as low as that of visible light is a reaction unknown in chemistry. Therefore there was no clue as to the properties of the initial (H) material, or as to how it is able to convert carbon dioxide to carbohydrate.

In time the thinking of workers in photosynthesis crystallized into two points of view. One considered that the (H) is an "active" species of hydrogen atom that can force itself on the carbon dioxide molecule. Such a reaction is also unknown elsewhere in biochemistry. But the adherents of the idea saw in it the very uniqueness of photosynthesis, and buttressed their hypothesis with a number of theoretical arguments. Without going into the theory, it can be said that no convincing experimental evidence has so far appeared to support it, and there is much evidence against it.

On the second viewpoint the conversion of carbon dioxide and water to carbohydrates proceeds by a path that is not peculiar to photosynthesis. In fact, the path is considered to be essentially a reversal of the route by which carbohydrates are decomposed to carbon dioxide and water during respiration. If so, then the role of the hypothetical (H) would be to form compounds that can drive the "dark" reactions of respiration backward, in the direction of synthesis.

Energy Transformations

To appreciate what is implied in this proposal we must consider the process of respiration itself. The biological "burning" of carbohydrate in oxygen has been intensively studied since the turn of the century. The intermediate steps through which comparatively complicated starch and sugar molecules pass on their way to carbon dioxide and water have been isolated. The enzyme systems that catalyze the successive reactions have been identified. Perhaps most important, the disposition of the energy released during oxidation is now known.

The energy is transferred to two compounds that are literally the power supply of life. One of them, familiar to readers of SCIENTIFIC AMERICAN, is adenosine triphosphate, or ATP [see "Energy Transformation in the Cell," by Lehninger, Offprint #69, for more information]. ATP has been called the universal energy currency of living cells. Every sort of vital process, from the contraction of a muscle to the synthesis of a hormone, draws on ATP. As its name implies, ATP contains three phosphate groups [see illustration on opposite page]. It is made in cellular bodies known as mitochondria, by the addition of a third phosphate to adenosine diphosphate (ADP). The bond of ADP to the third phosphate is where new energy is stored. It is as though the bond were a coil spring which is compressed when the phosphate is attached. When the phosphate group is removed, the spring extends, thus releasing the stored energy. In respiration the energy that compresses the spring comes from the oxidation of carbohydrate. Hence the formation of ATP in the mitochondria of a respiring cell is called oxidative phosphorylation.

The second of the key compounds is reduced pyridine nucleotide. There are two kinds: triphosphopyridine nucleotide ($TPNH_2$) and diphosphopyridine nucleotide ($DPNH_2$); here they are collectively abbreviated PNH_2. Each is a powerful biological "reductant," that is, it can readily force its hydrogen atoms on other molecules. PNH_2 participates in many oxidation-reduction reactions in all living cells. One of these reactions provides energy for phosphorylation itself. Thus some PNH_2 is oxidized to make ATP.

At this point it may be worth while to consider the terms "oxidation" and "reduction," on which our entire story turns. Essentially, oxidation means removing an electron from a molecule, and reduction means adding an electron. Whenever an electron is exchanged between two substances, one is oxidized and the other reduced. Moreover, almost every such exchange is accompanied by the release or absorption of energy. It makes no difference whether we

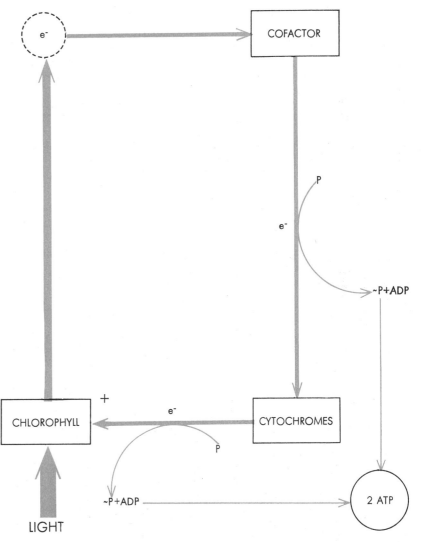

CYCLIC PHOTOPHOSPHORYLATION is represented by this diagram. The chlorophyll molecule absorbs a photon of light. This sends an electron to a high-energy state (arrow at left), where it is captured by vitamin K or another cofactor. The electron then moves to the cytochrome enzyme-system, losing energy (arrow at right) to a phosphorylating enzyme-system which employs the energy to couple a phosphate group onto ADP. As the electron passes back to chlorophyll from cytochromes, it again gives up energy to form another molecule of ATP. The chlorophyll acquires a positive charge when it ejects the electron and is thus able to pull the returning electron from the cytochromes. No outside electron donor is used, and the system functions as a self-contained cyclic mechanism.

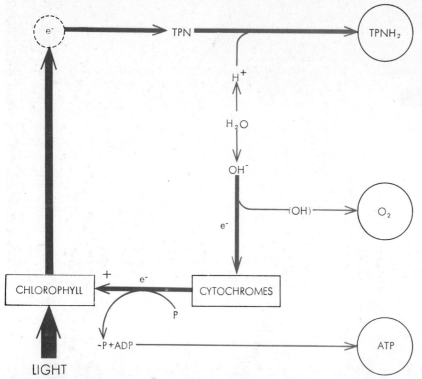

NONCYCLIC PHOTOPHOSPHORYLATION in chloroplasts differs from the cyclic system in that the electrons raised to a high energy by chlorophyll end up in $TPNH_2$. Some water molecules are normally dissociated into H^+ and OH^- ions, and it is from among these that the system takes the electrons and protons (H^+) it needs to make ATP and $TPNH_2$. The (OH) radicals that are left over combine with each other to produce water plus the oxygen gas that is characteristically evolved by green plants.

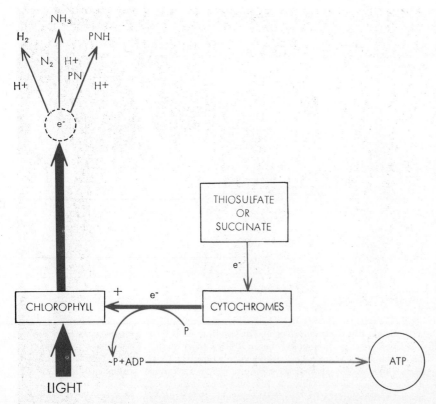

NONCYCLIC ELECTRON-TRANSPORT MECHANISM in bacterium *Chromatium* takes electrons from thiosulfate and succinate. Energized in chlorophyll, they can reduce pyridine nucleotide, join nitrogen to make ammonia or make hydrogen gas. ATP is also formed.

think of the energy as arising out of the pull exerted on the electron by "oxidizing power" or the push exerted by "reducing power." These terms have meaning only in relation to a specific pair of substances, which always interacts in the same way.

Often, though not invariably, an electron travels in company with a proton; in short, as part of a hydrogen atom. In that case, oxidation means removing hydrogen, and reduction means adding hydrogen. Thus pyridine nucleotide (PN) is "reduced" to PNH_2, and carbon dioxide to carbohydrate, by the addition of hydrogen atoms.

To return now to photosynthesis, the idea that it must involve a special way of converting carbon dioxide to carbohydrate was dealt a severe blow by modern studies of cellular metabolism that have utilized radioactive tracers. Investigators have found that all kinds of cells devoid of chlorophyll—for example, liver cells—can synthesize carbohydrates from carbon dioxide, if they are furnished the necessary energy in the form of ATP and PNH_2. Apparently the breakdown of carbohydrates can be reversed. And if in liver, why not in plants?

There grew a strong suspicion that there is no special, photosynthetic, way for assimilating carbon dioxide, and that all cells, whether they contain chlorophyll or not, may accomplish it essentially by reversing respiration. Soon the idea drew support from the work of Melvin Calvin, A. A. Benson and their co-workers in the Radiation Laboratory of the University of California, who traced the path of carbon in photosynthesis from carbon dioxide to carbohydrate. They identified many intermediate products identical with those formed when carbohydrate is burned in respiration. Once the "photosynthetic carbon-cycle" had been established, each of its features was discovered in various cells that assimilated carbon dioxide in the dark. In fact, P. A. Trudinger at the University of Sheffield in England, and J. P. Aubert and his colleagues at the Pasteur Institute in Paris, have now demonstrated the complete cycle in the nonphotosynthetic bacterium *Thiobacillus denitrificans*.

In all these dark reactions energy was provided by ATP and PNH_2. This implied that the function of light must be to manufacture ATP and PNH_2. Then in 1951 workers in three laboratories discovered that isolated chloroplasts do make PNH_2 in the light. The exact mechanism remained to be identified, but the whole problem of the role of

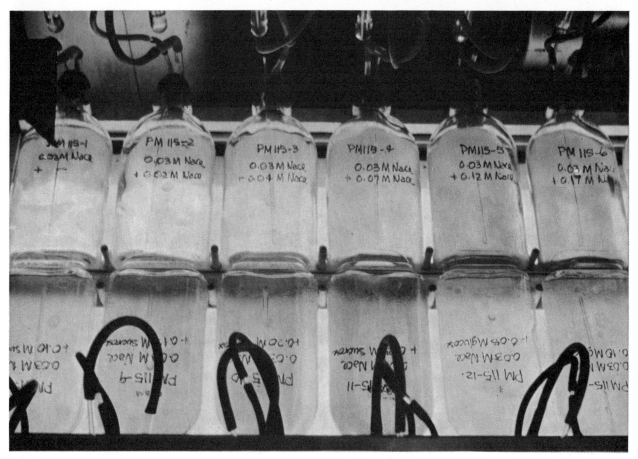

GREEN ALGA *Platymonas* is grown in bottles in the laboratory of the author of this article at the University of California. The bottles are illuminated from below by fluorescent lights. Lettering on bottles records the different experimental nutrient solutions.

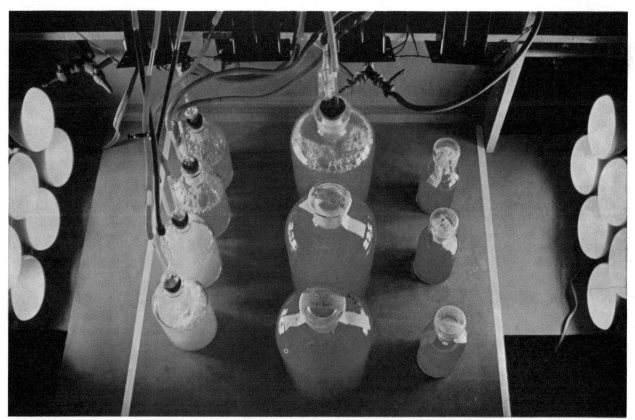

RED BACTERIUM *Chromatium* is similarly grown. Bottles at left receive nitrogen gas; top two bottles contain molybdenum. Top bottle in center receives hydrogen gas. Middle bottle in center obtains hydrogen from malic acid; bottom bottle, from thiosulfate.

ISOLATION OF CHLOROPLASTS in laboratory of author is shown in sequence of photographs on this and following pages. Here spinach leaves are prepared for grinding with sand and water.

SLURRY IS POURED into filter after hand-grinding with mortar and pestle. The ice in the bowl is used to keep the flask cool to preserve the enzymes that will mediate the photosynthetic reactions.

light seemed very near solution. As has been mentioned, PNH_2 is used to make ATP by oxidative phosphorylation in mitochondria. Presumably the mitochondria in photosynthetic cells could obtain PNH_2 from chloroplasts and with it manufacture ATP. The model was reasonable, and widely accepted. At that time I almost abandoned further work on the role of light in photosynthesis in the belief that the problem was essentially solved.

But the model posed one major difficulty: The most specialized photosynthetic cells in leaves have very few mitochondria, being almost filled with chloroplasts. How could these few bodies produce enough ATP to support the vigorous photosynthetic activity of all the chloroplasts? Because this question seemed to have no satisfactory answer, I continued my attempts to find out if chloroplasts produced anything other than PNH_2.

In 1954 we found that chloroplasts alone carry out complete photosynthesis. If they converted carbon dioxide into carbohydrates by a reversal of the breakdown reactions, they must have been able to make ATP.

Isolated Chloroplasts

Our first task was to be sure that photosynthesis in our chloroplast preparation was really the same process that occurs in living cells. In the next few years we subdivided chloroplasts into various parts and identified in them, or isolated from them, the individual enzyme systems that catalyze the step-by-step transformation of carbon dioxide into carbohydrates. The products of assimilation—identified with radioactive tracers, chromatography and radioautography—proved to be the same as in photosynthesis by whole cells.

The experiments demonstrated that the assimilation of carbon dioxide in isolated chloroplasts is indeed a reversal of the carbohydrate breakdown reactions. Further, the energy for the process is provided jointly by ATP and PNH_2. When experimental conditions were arranged so that only one of the pair was formed in light, no carbohydrates were made.

We were now virtually certain that the role of light in photosynthesis is to supply ATP and PNH_2. Final evidence was forthcoming when we separated the light and dark phases of photosynthesis in chloroplasts. We illuminated a chloroplast preparation, but did not supply any carbon dioxide, so there was no raw material for manufacturing carbohydrates. Instead, we supplied large amounts of ADP, inorganic phosphate and TPN (triphosphopyridine nucleotide). The result was the evolution of oxygen and an accumulation of ATP and $TPNH_2$.

Now we extracted the enzymes for carbon dioxide assimilation, discarding the green part of the chloroplasts and with it the light-absorbing chlorophyll. Using ATP and $TPNH_2$ made when the light was on, the enzymes proceeded to assimilate carbon dioxide in the dark and to produce the same carbohydrates that whole chloroplasts and intact green leaves manufacture. In a further experiment PNH_2 and ATP taken from animal cells were supplied to the enzymes in total darkness; again the extracts assimilated carbon dioxide and made the familiar compounds.

At this point the objectives of our research narrowed to the problem of identifying the reactions through which light energy forms ATP and PNH_2. At first we looked for a mechanism within the framework of Van Niel's water-splitting scheme, which had guided our research for many years. We tried to envisage a process by which the hypothetical (H), produced when light splits water, could form ATP and PNH_2. As time went on, however, new experimental facts showed that this approach was inadequate. We had to abandon it to look for different guideposts.

Photosynthetic Phosphorylation

Since the manufacture of PNH_2 by chloroplasts in light had been observed before, we decided to concentrate on the completely unknown process by which ATP is synthesized. Very early we established that isolated chloroplasts could apply absorbed light purely to the formation of ATP. We discovered this by depriving illuminated chloroplasts of carbon dioxide and pyridine nucleotide while giving them large amounts of ADP and inorganic phosphate. With no raw

FILTRATE IS POURED INTO TEST TUBES which will be placed in a refrigerated centrifuge. Spinning the material at high speed separates the chloroplasts from the other portions of the cells.

ISOLATED CHLOROPLASTS in stoppered cylinder in bowl at left are placed in vessels having long vents (*foreground*) along with the reagents that contain radiocarbon or radiophosphorus.

material for making carbohydrate or even PNH_2, the chloroplasts used light energy to force the third phosphate group on ADP to form ATP, which accumulated in substantial quantities by the end of the experiment. We called this process photosynthetic phosphorylation—photophosphorylation for short—to distinguish it from the oxidative phosphorylation by mitochondria. To biochemists, imbued with the idea that the energy of ATP comes from burning carbohydrates, it was as if we had suddenly learned how to get electric power directly from coal or oil, without burning them in a generating plant.

In our early experiments photosynthetic phosphorylation showed a puzzling dependence on oxygen. Chloroplasts would manufacture ATP only in the presence of oxygen, although none of the gas was consumed. Apparently the oxygen acted as a catalyst. Further probing soon revealed two other catalysts in the process: vitamin K and flavin mononucleotide (FMN), a component of the vitamin B complex. Both are constituents of green leaves, and vitamin K had long ago been found to be localized in the chloroplasts. Because its location suggested a connection with photosynthesis, earlier workers had tried adding vitamin K in experiments on whole cells. Paradoxically, it inhibited photosynthesis. But when we added it to our isolated chloroplasts, it increased the rate of ATP production almost twentyfold. Adding FMN had the same result. And, when enough catalyst was added, the reaction no longer required the presence of oxygen. Thus oxygen is not essential to the process of photophosphorylation. The process is anaerobic.

Soon after we discovered photosynthetic phosphorylation in chloroplasts, Albert W. Frenkel, then at Harvard University, found that cell-free extracts of the photosynthetic bacterium *Rhodospirillum rubrum* also make ATP anaerobically, in the presence of light. (Bacteria do not have chloroplasts. Their chlorophyll is contained in structures called chromatophores, and it was these that Frenkel had extracted.) Subsequent investigations in various laboratories, on chloroplasts and chromatophores from several different plants and bacteria, have established anaerobic photophosphorylation as a general process, common to all photosynthetic organisms.

In trying to understand how photophosphorylation works, it was natural to compare it with the oxidative phosphorylation of respiration, about which so much had already been learned. In the respiratory process the energy required to force the third phosphate on ADP is obtained when an electron (attached to a hydrogen atom) drops from a higher to a lower energy-level while moving from an electron donor to an electron acceptor. The drop is accomplished in a series of steps that divides the total energy available in the electron into portions of the required size. Each chemical step may be thought of as a sort of water wheel that is turned by the falling electron and uses its power to attach phosphate to ADP (or to drive some other necessary reaction in the cycle). The original electron donor is sugar or starch; the ultimate acceptor is oxygen. When oxygen receives the electron (with its hydrogen ion), it is converted to water.

It seemed reasonable to suppose that "falling" electrons also power photosynthetic phosphorylation. The problem was to account for the electron donor and the electron acceptor. They cannot come from the outside; photophosphorylation consumes neither chemical fuel nor oxygen, only light.

About two years ago my associates and I constructed a theoretical model that seems to fit the experimental facts of photosynthetic phosphorylation. It was suggested by another photochemical reaction whose details were worked out in 1942 by Gilbert N. Lewis and David Lipkin of the University of California. We proposed that, in the primary photochemical act, a photon (quantum unit) of light strikes a chlorophyll molecule, exciting one of the electrons to an energy sufficient to remove it from the molecule. Having lost an electron, the molecule is now in a position to act as an electron acceptor. If it took back the electron directly, it would merely re-emit the light energy that had just been absorbed. (Under the proper conditions light does cause pure chlorophyll to fluoresce in just this way.) The reaction that makes phosphorylation possible is the capture of the excited electrons by a molecule such as vitamin K or FMN. Now the

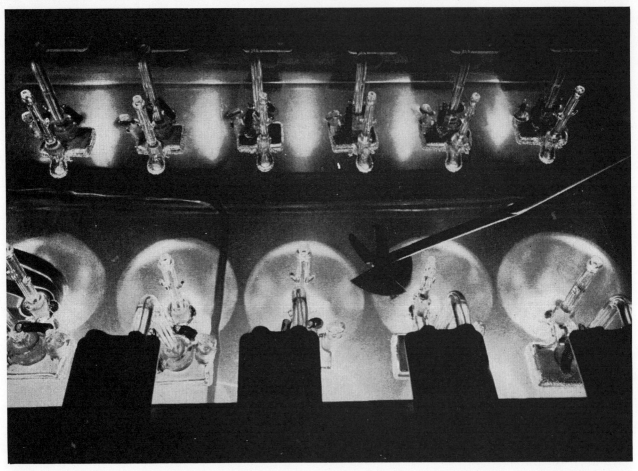

PHOTOSYNTHESIS TAKES PLACE in Warburg apparatus. The 11 small vessels seen in two rows contain the chloroplasts and the reagents. The vessels are in a constant-temperature water bath that is stirred by the propeller. Light for photosynthesis comes from reflector lamps under glass bottom of tank. Apparatus gently shakes the containers in order to facilitate the chemical reactions.

electrons are forced to return to chlorophyll in a series of graded steps, resembling those in respiration.

The downhill path we have traced takes electrons from vitamin K or FMN through a number of cytochromes—iron-containing pigments that catalyze many biological oxidations—and finally back to chlorophyll. The "water wheels" that drive the synthesis of ATP are thought to be linked with the cytochrome chain.

The electron transport and its coupled phosphorylation reactions are analogous to, and in some ways possibly identical with, their counterparts in oxidative phosphorylation. Only the light-induced production of a high-energy electron and its ultimate acceptor are peculiar to photosynthesis. Because the electron donor and acceptor are the same substance—chlorophyll—and because no outside donors are involved, we have named the process cyclic photophosphorylation.

There seems little doubt that photophosphorylation is a primary and critical reaction of photosynthesis. In most organisms, however, it is not the only one.

Light energy is also required to make PNH_2, which furnishes the hydrogen to reduce carbon dioxide to carbohydrate. Yet the fact that chloroplasts do, under certain conditions, use light solely for the manufacture of ATP raises an interesting question: Are there any photosynthetic cells in which phosphorylation is really the only function of light, and which manufacture their reducing substance by a dark reaction? If so, they would exhibit the simplest, and therefore perhaps the most primitive, form of photosynthesis.

Recently we have found that the anaerobic red sulfur bacterium *Chromatium* is just such an organism. As has long been known Chromatium grows only in the light. The bacterium can use either carbon dioxide or acetate as the source of carbon for its cellular substance. To assimilate carbon dioxide it requires a supply of hydrogen gas, with which it can form the necessary PNH_2 in the dark. Acetate, however, is assimilated by a somewhat different chemical pathway, which does not require an external supply of hydrogen. When supplied with hydrogen gas and carbon dioxide, or with acetate, the only thing that Chromatium makes under the influence of light is ATP. Chromatium grows anaerobically and thus cannot make ATP by oxidative phosphorylation. To test whether ATP is indeed the sole product of the light reaction in the assimilation of CO_2 or acetate, we placed a cell-free preparation of Chromatium in the dark and added ATP. This preparation synthesized organic carbon compounds in the dark just as Chromatium does in the light, when it is not supplied with ATP but has to make it from ADP and inorganic phosphate. Light and ATP were entirely equivalent!

Under certain circumstances higher plants may possibly act like Chromatium and go on making ATP even when carbon dioxide assimilation stops. In leaves, for example, the tiny pores that admit carbon dioxide are sometimes seen to close in the middle of the day. Usually they close when the leaves have accumulated an abundance of starch, and water is scant. It seems at least possible that the chloroplasts would continue to make ATP, which could be used to convert carbohydrate reserves into other compounds such as proteins and fat. Certainly plants conduct such reactions, and they need energy from ATP to do so.

This conjecture has recently received experimental support from the work of G. A. Maclachlan and Helen K. Porter at the University of London. They discovered that leaf tissue uses light energy to synthesize starch from glucose under conditions when carbon dioxide is excluded but photosynthetic phosphorylation can occur.

In any case, in Chromatium supplied either with hydrogen gas and CO_2 or with acetate, the role of light is limited to making ATP by cyclic photophosphorylation. As was mentioned in connection with Van Niel's work, however, photosynthetic bacteria can also make use of other hydrogen donors: inorganic materials such as thiosulfate and organic acids such as succinate. But the hydrogen in these substances does not have enough reducing capacity to convert PN to PNH_2 in the dark. Additional energy is required.

Noncyclic Photophosphorylation

In the photosynthetic mode of life of these bacteria, which grow without oxygen, the energy must come from light. How do they use light to reduce PN with thiosulfate or succinate? Recently we have found that our picture of electrons excited by light applies here too.

A number of investigators had shown that the cytochromes in photosynthetic cells become oxidized (that is, lose electrons) when the cells are illuminated. Our theory suggests that electrons are transferred from cytochrome to chlorophyll, replacing the ones expelled from chlorophyll by the action of light. Now we find that the oxidized cytochromes are in turn reduced by thiosulfate or succinate. The result is that electrons donated by thiosulfate or succinate are transferred via cytochromes to chlorophyll and are there raised at the expense of light energy to a reducing potential sufficient to make PNH_2.

The fate of these activated electrons is different, however, from that in the cyclic route. They do not return to chlorophyll but are eventually transferred to external acceptors. Three of these have now been identified: nitrogen gas, which is converted to ammonia (NH_3); PN, which is converted to PNH_2; and protons (hydrogen ions), which become hydrogen gas. Here light energy is being used either to produce PNH_2 or to fix atmospheric nitrogen. It further appears that, in traveling this noncyclic route, the electrons also give up some of their energy to the formation of ATP.

The complete picture of noncyclic electron flow has been uncovered very recently. It was foreshadowed, however, by earlier experiments in other laboratories. Howard Gest and Martin D. Kamen, then at Washington University in St. Louis, showed that in light photosynthetic bacteria fix nitrogen and evolve hydrogen gas. Frenkel, working at the University of Minnesota, and Leo P. Vernon of Brigham Young University demonstrated that the same organisms reduce PN with succinate in light. Hydrogen gas seems to be evolved when the reducing electrons donated by thiosulfate or succinate are a surplus because they are not consumed in metabolic reactions: fixing nitrogen or reducing PN.

Do green plants also have an open-ended route of electron transport for making PNH_2 as well as the closed path of cyclic phosphorylation? The answer turns out to be yes. The essential difference between their noncyclic electron-transport mechanism and that of bacteria resides in the fact that chloroplasts derive their external electron supply from water. The electrons from water have an even smaller reducing capacity than those from thiosulfate or succinate. Therefore green plants depend unconditionally on light for "raising" the electrons from water to a reducing potential sufficient to form PNH_2.

As we visualize the noncyclic process in plants, it involves a cytochrome-catalyzed reaction, thus far hypothetical, which is found in green plants but not in bacteria. The plant cytochrome is presumably so highly oxidized in light that it can take from hydroxyl ions (OH^-) the electrons to be fed to chlorophyll. There they are raised to the energy necessary for reducing pyridine nucleotide.

The special cytochrome reaction may possibly resemble a reaction, familiar to

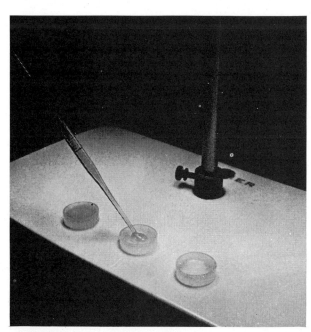

AFTER PHOTOSYNTHESIS a sample is taken in a pipette from each container and placed upon a planchet. There the material is allowed to dry preparatory to placing it into a radiation counter.

RADIOACTIVITY IS MEASURED by placing planchet in apparatus at left. Machine at right registers counts, showing the amount of radiocarbon or phosphorus absorbed in photosynthesis.

106   ENERGETICS

RADIOAUTOGRAPH OF A CHROMATOGRAM is used to analyze products of photosynthesis by isolated chloroplasts. A sample of the reacting mixture is placed on moist chromatogram paper at spot in lower left corner. The various compounds then travel at characteristic rates along paper, thereby separating from each other. Dried paper is placed against photographic film. Radioactivity exposes the film, making spots seen here. Phosphorylated glucose and other compounds were identified in this chromatogram.

every high-school chemistry student, which takes place in the electrolysis of water. There the positive electrode, using energy supplied by a battery, takes electrons from hydroxyl ions. These then change to (OH) radicals, which combine to yield oxygen and water. This type of reaction would account for the evolution of oxygen by green plants. The whole sequence appears thermodynamically feasible, but it cannot be considered proved until the postulated cytochrome reaction is found. It is possible that the transfer of electrons from hydroxyl ions to cytochromes requires an additional input of energy.

In the course of our studies of PN reduction by chloroplasts we found that it is accompanied by the formation of ATP. We named this reaction noncyclic photophosphorylation. Compared with the "classical" type of oxidative phosphorylation it is a remarkable process indeed. It is accompanied by the evolution rather than the absorption of oxygen. And the formation of ATP results not in the ultimate oxidation, but in the reduction, of pyridine nucleotide. As with cyclic photophosphorylation, the noncyclic type has now been confirmed and elucidated in a number of laboratories.

The noncyclic mechanism, however, cannot by itself supply all the ATP needed for assimilating carbon dioxide in the manufacture of carbohydrates. Additional ATP must be supplied by cyclic photophosphorylation.

There remains one further phenomenon to account for in the chloroplast experiments: the catalytic role of oxygen in photophosphorylation when the concentration of vitamin K and FMN is low. We have worked out a scheme to accommodate oxygen in the general picture developed above. But the details need not be considered here.

Our theoretical model of photosynthesis is reasonably complete, although several features remain to be confirmed by experiment. The role of light in the primary photochemical act is simply to raise the energy of the electrons in chlorophyll. Thereafter cellular chemistry takes over, shunting the excited electrons into different downhill paths, where their energy is converted into chemical energy and is harnessed to drive several possible reactions. Thus the essence of photosynthesis is the conversion of light into chemical energy, which can be used by the cell in various ways. Most commonly, to be sure, it is applied to the assimilation of carbon dioxide, but this is by no means its only possible use.

In all photosyntheses, bacterial or plant, the common denominator of the energy conversion process is cyclic photophosphorylation: the manufacture of the universal biological energy currency —ATP—at the expense of light energy alone, and with no consumption of material substances. All photosynthetic organisms, and only photosynthetic organisms, perform this feat. The assimilation of carbon dioxide and the evolution of oxygen are processes that the cell may, but need not, carry out while performing photosynthesis. Indeed, as our recent bacterial studies show, the list of processes that are driven by trapped light energy should be extended to include nitrogen fixation and the evolution of hydrogen gas.

Photosynthesis and Evolution

Finally it is worth noting that our model seems to fit nicely into present ideas about early biochemical evolution. There is a good deal of evidence that when life appeared on earth, the atmosphere contained little oxygen but did contain free hydrogen gas. It is reasonable to suppose that as soon as the chlorophyll molecule evolved, cells were able to harness the energy in the visible spectrum of sunlight for the production of ATP, by the same method still observed in Chromatium. This was a momentous event; it provided cells, living in an anaerobic environment, with a mechanism for making ATP that is much more efficient than fermentation, the only process they could have used earlier. It is particularly interesting from an evolutionary point of view that the aerobic plants of today have retained the ancient capacity for cyclic photophosphorylation even though they have acquired the ability to make ATP through respiration as well.

The next step up the ladder of evolution was probably the noncyclic process by which bacteria reduce pyridine nucleotide without molecular hydrogen, using electron donors such as thiosulfate and succinate. Finally, in the most advanced type of photosynthesis observed in green plants, water became the electron donor. By developing a mechanism for obtaining their reducing electrons from this ubiquitous substance, green plants were able to live virtually everywhere and were no longer restricted to areas where special electron donors could be found. At this point also, oxygen production became an inseparable part of photosynthesis. As plants spread over the surface of the earth, they released to the atmosphere the oxygen locked in the water molecule, opening the gateway to the biochemical evolution of higher organisms that depend on molecular oxygen.

THREE CULTURE TUBES are part of the continuous-culture apparatus used by the Bio-Organic Chemistry Group at the University of California to grow green algae under constant conditions in an aqueous medium. Two tubes are empty; the third contains algae.

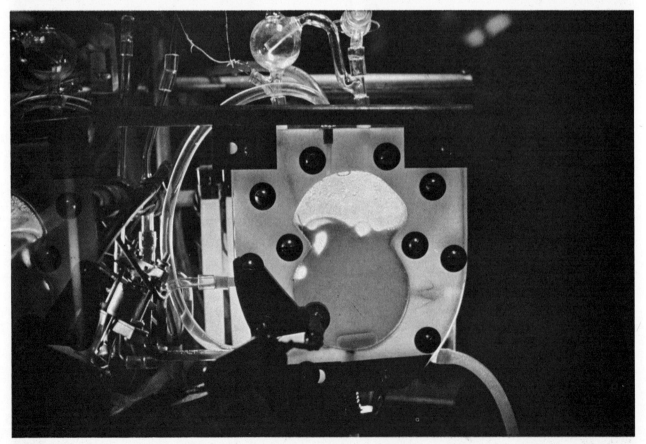

"LOLLIPOP" is a thin, transparent vessel to which algae are transferred from culture tubes. Carbon dioxide containing radioactive carbon is bubbled through the algae suspension in experiments to determine the path taken by carbon in the photosynthetic process.

THE PATH OF CARBON IN PHOTOSYNTHESIS

by J. A. BASSHAM   June 1962

The processes of life consist ultimately of the synthesis and breakdown of carbon compounds. Because a carbon atom can bind four other atoms to itself at a time and is thereby able to link up with other atoms—especially other carbon atoms—in chains and rings, carbon lends itself to the construction of a virtually endless variety of molecules. These molecules derive their physical characteristics and chemical activity not only from their composition but also from their size and intricacy of structure. The rich variety of life suggests in turn that living cells have gone far in the elaboration of such compounds and the processes that make and unmake them. All these processes depend in the end on a first one. This is the process of photosynthesis, which takes carbon and several other common elements from the environment and builds them into the substances of life.

The plant finds most of these elements already bonded to oxygen in oxides such as carbon dioxide (CO_2), water (H_2O), nitrate (NO_3^-) and sulfate (SO_4^{--}). Before the plant can bind the elements other than oxygen together as organic compounds, it must remove some of the excess oxygen as oxygen gas (O_2), and this accomplishment takes a large amount of energy.

In the simplest terms photosynthesis is the process by which green plants trap the energy of sunlight by using that energy to break strong bonds between oxygen and other elements, while forming weaker bonds between the other elements and forcing oxygen atoms to pair as oxygen gas. For example, to make the sugar glucose ($C_6H_{12}O_6$) the plant must split out six molecules of oxygen in order to combine the carbon and half the oxygen of six carbon dioxide molecules with the hydrogen of six water molecules.

The glucose and other organic compounds taken up in the chemical machinery of the plants and the animals that live on plants serve both as fuel and as the raw materials for the synthesis of higher organic compounds. That considerable solar energy is bound by photosynthesis becomes apparent when wood or coal is burned. In living cells the controlled combustion of respiration extracts this energy to power the other processes of life. Both kinds of combustion take oxygen from the air and break down organic compounds to carbon dioxide and water again. In its end result photosynthesis can be defined as the opposite of respiration. Together these complementary processes drive the cyclic flow of matter and the noncyclic flow of energy through the living world [see illustration below].

From such generalizations about the effect and function of photosynthesis in nature it is a long step to the explanation of how photosynthesis works. Yet much of the explanation is now complete. The work has been greatly facilitated by the earlier and more nearly complete resolution of the chemistry of respiration. The two processes, it turns out, are in some ways complementary on the molecular scale, just as they are on the grand scale in the biosphere. Each involves some 20 to 30 discrete reactions and as many intermediate compounds; half a dozen of these reactions and their intermediates are common to both photosynthesis and respiration. Only the first few steps in photosynthesis are driven directly by light. The energy of light is trapped in the bonds of a few specific compounds. These energy carriers deliver the energy in discrete units to the steps of synthesis that follow. The same or closely similar carriers perform corresponding operations in respiration, picking up energy from the stepwise dismemberment of the fuel molecule and

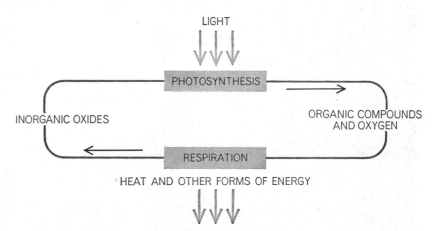

PHOTOSYNTHESIS AND RESPIRATION are the complementary processes that drive the cyclic flow of matter and the noncyclic flow of energy through the biosphere. Photosynthesis uses light energy to convert inorganic oxides to oxygen and organic compounds such as glucose. In respiration of plants and animals oxygen reacts with these compounds to produce the inorganic oxides carbon dioxide and water as well as biologically useful energy.

110 ENERGETICS

delivering it to the energy-consuming processes of the cell. Although the first, energy-trapping stage in photosynthesis remains to be clarified, it is now possible to trace the path of carbon from the very first step in which a single atom of carbon is captured in the bonds of an evanescent intermediate compound.

The term "carbohydrate" recalls the deduction of early 19th-century investigators that photosynthesis made glucose directly by combining atoms of carbon with molecules of water, as the formula for glucose suggests. In line with this idea it was thought that the oxygen transpired by green leaves came from the splitting of carbon dioxide. The progress of chemistry, however, failed to disclose any processes that would accomplish these results so simply. Accumulating evidence to the contrary became convincing some 30 years ago, when C. B. van Niel of Stanford University discovered that certain bacteria produce organic compounds by a process of photosynthesis similar to that in plants but with one important difference. These bacteria use hydrogen sulfide (H_2S) instead of water and liberate elemental sulfur instead of gaseous oxygen. The

PATH OF RADIOACTIVE CARBON (color) was determined from experiments described in the text. Five molecules of PGA, the first stable intermediate product to appear, are reduced (1) by cofactors ATP and TPNH to five triose phosphate molecules. A circled P represents a phosphate group ($-HPO_3^-$). Two of these are converted to a different type of triose phosphate (2); the subsequent condensation of one of each kind of triose into hexose diphosphate (3) is mediated by the enzyme aldolase. Hexose diphosphate then loses a phosphate group (4). Transketolase, another enzyme, removes two carbons from the hexose and adds them to a triose

otherwise complete similarity of the two processes strongly suggested that the oxygen evolved by green plants must come from the splitting of water.

The Capture of Light

Photosynthesis could now be described in terms of familiar chemistry. The splitting of water would be accomplished by the process of oxidation (which means the removal of hydrogen atoms from a molecule), with oxygen gas as the product of the reaction. The free hydrogen atoms would then be available to carry through the equally familiar and opposite process of reduction (which means the addition of hydrogen atoms to a molecule). By the addition of hydrogen atoms (or electrons plus hydrogen ions) the carbon dioxide would be reduced to an organic compound.

It is during the first, energy-converting stage of photosynthesis that the water molecule is split. Initially the energy of light impinging on the plant cell is transformed into the chemical potential energy of electrons excited from their normal orbits in molecules of the green pigment chlorophyll and other plant pig-

phosphate (5), making one tetrose and one pentose phosphate. The tetrose condenses with a triose (6) to form a heptose diphosphate, which then loses a phosphate group (7). Transketolase removes two carbons from the heptose and adds them to a triose (8), making two more pentose phosphates for a total of three; these are converted to ribulose-5-phosphate (9), then to ribulose diphosphate (10). Addition of three carbon dioxide molecules (11) produces three unstable compounds (*brackets indicate unknown structure*) that begin the cycle again. Addition of three water molecules (12) results in six PGA molecules for a net gain of one in the cycle.

ments. A large part of this energy eventually goes into the splitting of water as electrons and hydrogen ions are transferred from water to the substance triphosphopyridine nucleotide (TPN^+), which is thereupon reduced to the form designated TPNH. The TPNH thus becomes not only a carrier of energy but also the bearer of electrons for the subsequent reduction of carbon dioxide. Along with the movement of electrons from water to TPNH, some energy goes to charging the energy-carrying molecule adenosine triphosphate (ATP), specifically by promoting the attachment of a third phosphate group ($-OPO_3^{--}$) to adenosine diphosphate (ADP), the discharged form of the carrier. Both ATP and TPNH belong to the family of compounds known as cofactors or coenzymes, which work with enzymes in the catalysis of chemical reactions. ATP, the universal currency of energy transactions in the cell, plays a significant role in respiration as well as in photosynthesis.

Needless to say, the manufacture of each of these cofactors involves an intricate cycle of reactions [see "The Role of Light in Photosynthesis," by Daniel I. Arnon, page 97, for further information]. Although the cycles are not yet fully understood, it is enough for the purpose of the present discussion to know that ATP and TPNH, or closely similar compounds, furnish the energy for the second stage of photosynthesis, during which the carbon atom of carbon dioxide is reduced and joined to a hydrogen atom and a carbon atom in place of an oxygen.

The process of reduction goes forward in small steps. Each reaction brings about some change in a carbon compound until the starting material is at last transformed to the final product. For each reaction there is therefore an intermediate compound. Since every life process involves a more or less extended series of intermediates, cells typically contain a large number of intermediates. Many of them turn up in two or more pathways leading to different end products. The tracing of the path of carbon in photosynthesis required first of all a technique for identifying the intermediates proper to it and for establishing their sequence along the path.

Samuel Ruben and Martin D. Kamen, then at the University of California, met this need some 20 years ago by their discovery of the radioactive isotope of carbon with a mass number of 14. This isotope has a conveniently long half life of more than 5,000 years; over the time period of an experiment, therefore, carbon 14 has an effectively constant radioactivity. Ruben and his colleagues recognized at once the potential usefulness of this isotope as a label for the identification of compounds in biological processes. They prepared carbon dioxide in which the carbon atoms were carbon 14. When they exposed green plants to an atmosphere containing this gas instead of normal carbon dioxide ($C^{12}O_2$), the plants took up the $C^{14}O_2$ and made compounds from it. The presence of the carbon 14 in these compounds could be detected by various devices, such as the Geiger-Müller counter, and by radioautography on X-ray film. Unfortunately this work was cut short by the war and by Ruben's death in a laboratory accident.

In 1946 Melvin Calvin organized a new group at the Lawrence Radiation Laboratory of the University of California with the primary objective of tracing the path of carbon in photosynthesis with $C^{14}O_2$ as one of its principal tools. Starting as a graduate student in 1947, I had the good fortune to participate in this work with Calvin, Andrew A. Benson and others.

The early experiments were quite simply contrived. We used leafy plants and often just the leaves of plants. After allowing a leaf to photosynthesize for a given length of time in an atmosphere of $C^{14}O_2$ in a closed chamber, we would bring biochemical activity to a halt by immersing the leaf in alcohol. With the enzymes inactivated, the reactions converting one intermediate compound into another would stop, and the pattern of labeling would be "frozen" at that point. We soon discovered, however, that photosynthesis proceeds too rapidly for completely reliable observation by such a procedure. With a few seconds' delay in the penetration of alcohol into the cell, for example, the labeling pattern would be disarrayed and no longer representative of the stage at which we tried to halt the photosynthesis.

Since rapid and precisely timed killing of the plant is important, we adopted single-celled algae—*Chlorella pyrenoidosa* and *Scenedesmus obliquus*—as the subject for many of our experiments. In both species the plant consists of a single cell so small that it can be seen only with a microscope. Alcohol can quickly penetrate the cell wall and deactivate the enzymes. The algae offer another advantage: they can be grown in continuous cultures, assuring us a supply of material with highly constant properties.

An experimental sample is taken from the culture in a thin-walled, transparent closed vessel. Illuminated through the walls of the vessel and supplied with a stream of ordinary carbon dioxide, which is bubbled through the suspension, the algae photosynthesize at the normal rate. We shut off the supply of carbon dioxide and inject a solution of

REDUCTION OF PGA to triose phosphate requires both ATP and TPNH. At top ATP gives up its terminal phosphate group to PGA to produce phosphoryl-3-PGA. At bottom TPNH donates a hydrogen atom and an electron (*broken circle and arrow*), thereby displacing a phosphate group and forming triose phosphate. The second step is in reality more complex than shown here and involves other cofactors in addition to TPNH.

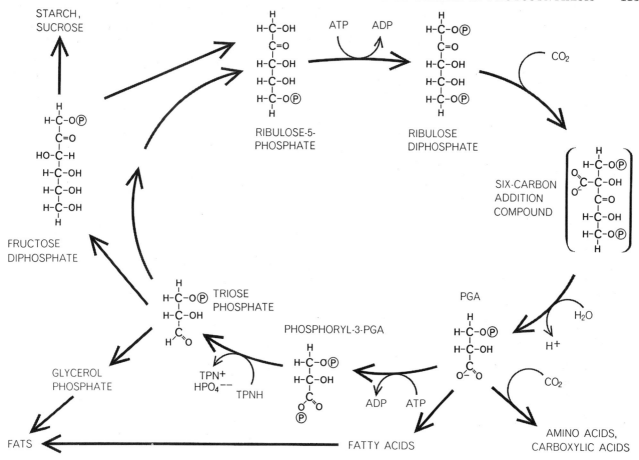

END PRODUCTS OF PHOTOSYNTHESIS are not limited to carbohydrates (e.g., sucrose and starch), as first thought, but include, among other things, fatty acids, fats, carboxylic acids and amino acids. Carbon cycle shown here is highly simplified; it involves at least 12 discrete reactions. Moreover, the steps from PGA to fatty acids and to amino and carboxylic acids have not been indicated.

radioactive bicarbonate ion (carbon dioxide dissolved in our algae culture medium is mostly converted into bicarbonate ion). After a few seconds or minutes the cells are killed. We then extract the soluble radioactive compounds from the plant material and analyze them.

The Reduction of CO_2

Calvin and his colleagues soon found that the carbon 14 label was distributed among several classes of biochemical compounds, including not only sugars but also amino acids: the subunits of proteins. As the exposure time was reduced to a few seconds, the first stable intermediate product of photosynthesis was found to be the three-carbon compound 3-phosphoglyceric acid (PGA).

The next step was to determine which of the three carbon atoms in the first generation of PGA molecules synthesized in the presence of radioactive carbon dioxide bears the carbon 14 label. We first removed from PGA the phosphate group [see illustration on opposite page] and then diluted the free glyceric acid with glyceric acid containing the stable carbon 12 isotope in order to have enough material for analysis by ordinary chemical methods. Treatment with reagents that severed the bonds between the carbons produced three different products, one from each carbon atom. By measuring the radioactivity of each of the products we were able to identify the labeled carbon.

In PGA from plants that had been exposed to labeled carbon dioxide for only five seconds we found that virtually all the carbon 14 was located in the carboxyl atom, the carbon at one end of the chain that is bound to two oxygens. This was not surprising because the carboxyl group most nearly resembles carbon dioxide. The carbon is bound to the oxygens by three bonds, however, instead of four; the fourth bond now ties it to the middle carbon in the PGA chain. The transfer of this bond from one of the oxygens to a carbon constitutes the first step in the reduction of the carbon dioxide. This was evidence also that the reduction is accomplished by some sort of carboxylation reaction, a reaction in which carbon dioxide is added to some organic compound. Ultimately, of course, the two other carbons of PGA must come from carbon dioxide. But it was some time before investigation disclosed the specific compound that picks up the carbon dioxide and the cyclic pathway that makes this carbon dioxide acceptor from PGA.

The discovery of the pathway intermediates was made much easier by a then comparatively new technique: two-dimensional paper chromatography, developed by the British chemists A. J. P. Martin and R. L. M. Synge. Closely similar compounds can be distinguished in this procedure by slight differences in their relative solubility in an organic solvent and in water. The extract from the plant is dropped on a sheet of filter paper near one corner. An edge of the paper adjacent to the corner is immersed in a trough containing an organic solvent; the paper is held taut by a weight and the whole assembly is placed in a water-saturated atmosphere in a vapor-tight box. The solvent traveling through the paper by capillarity dissolves the compounds and carries them along with it. As they move along in the solvent, however, the compounds tend to distribute themselves between the solvent and the water absorbed by the fibers of the paper. In general the more soluble the compound is in water compared with the organic solvent, the slower it travels. If the compound is also absorbed to some extent by the cellulose fibers, its movement will be even slower. As a result the compounds are distributed in a row in one dimension. Depending on the solubility of the compounds and the nature of the solvent used, some compounds may still overlap one another. Repetition of the procedure, with a different solvent traveling at right angles

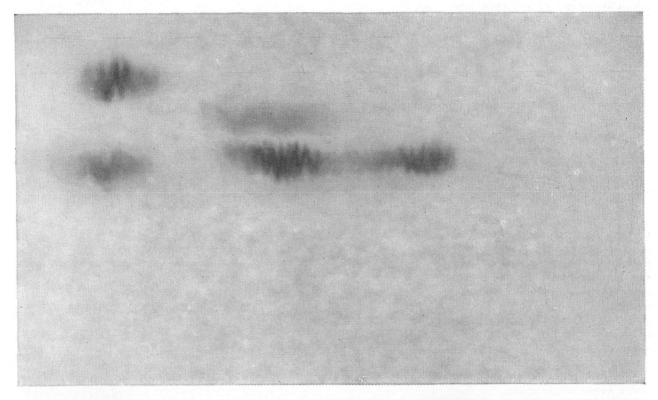

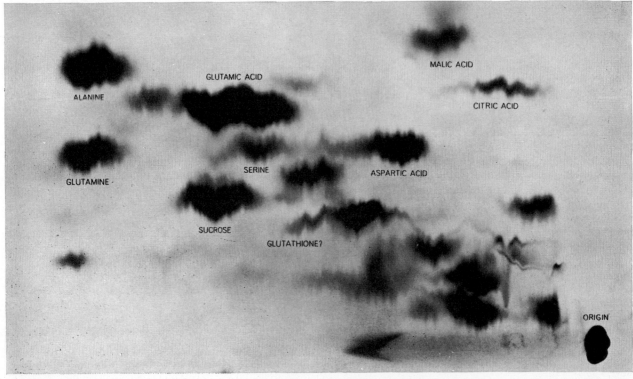

CHROMATOGRAM AND RADIOAUTOGRAPH used to corroborate the identity of amino acids produced by photosynthesizing algae appear at top and bottom respectively. The method of identifying such substances is described in the text. Areas of the radioautograph corresponding to colored areas in the chromatogram are alanine, glutamine, glutamic acid, serine and aspartic acid.

to the direction of the first run, will usually separate these compounds in a second dimension.

Since most of the compounds are colorless, special techniques are needed to locate them on the paper. Those that are radioactive will locate themselves, however, if the chromatographic paper is placed in contact with a sheet of X-ray film for a few days. The resulting radioautograph will show as many as 20 or 30 radioactive compounds in the substances extracted from algae exposed to carbon 14 for only 30 seconds [*see illustration at right*]. Clearly the synthetic apparatus of the plant works rapidly.

Chromatographs and Radioautographs

In order to identify these compounds we prepared a chromatographic map by running samples of many known compounds through the same chromatographic system and recording the locations at which we found them on the paper. The locations can be made visible in these cases by spraying the paper with a mist of some chemical that is known to react with the compound to produce a colored spot. Comparison of the radioautograph of an unknown compound with the map yields a first clue to its identity. This can be corroborated by washing the radioactive compound out of the paper with water and mixing it with a larger sample of the suspected authentic substance. The mixture is applied to a new piece of filter paper and chromatographed. With enough of the authentic material to yield a colored spot, comparison with a radioautograph of the same paper now shows whether or not the radioactive and the authentic material really coincide. The possibility that the authentic material and the radioactive material are still not the same can be tested by using different solvent systems in the preparation of the chromatograph and by other means.

Over the years these procedures have established the identity of a great many of the intermediate and end products of photosynthesis. Some of the sugar phosphates labeled by carbon 14 proved to be well-known derivatives of triose (three-carbon) and hexose (six-carbon) sugars. Others were discovered for the first time among the intermediates produced by our algae. Benson showed that among these are a seven-carbon sugar phosphate and also five-carbon phosphates, including in particular ribulose-1,5-diphosphate.

The rapid building of carbon 14 into the more familiar triose and hexose phosphates suggested certain biochemi-

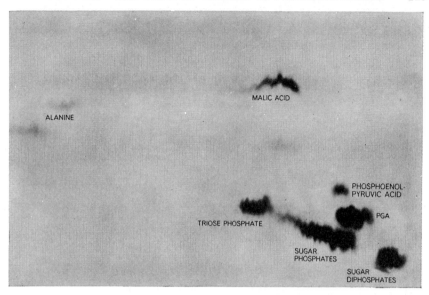

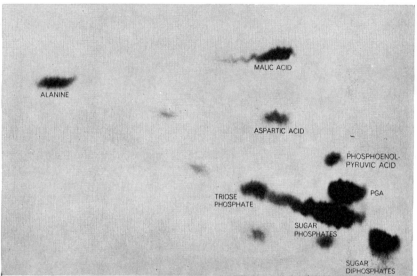

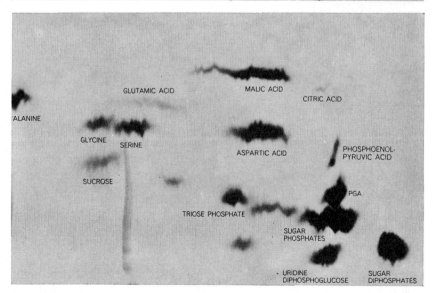

THREE RADIOAUTOGRAPHS reveal the compounds containing radioactive carbon that were produced by *Chlorella* algae during five (*top*), 10 (*middle*) and 30 seconds (*bottom*) of photosynthesis. Alanine, the first amino acid to appear in the process, shows up very faintly at first (*top*); glycine, serine, glutamic acid and aspartic acid appear as photosynthesis progresses. These radioautographs were made at the author's laboratory by exposing X-ray-sensitive film to chromatograms of compounds extracted from three samples of algae.

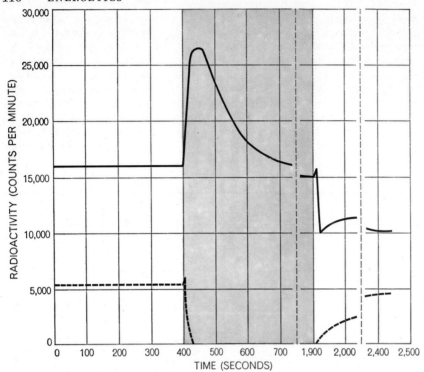

EFFECT OF SUDDEN DARKNESS on PGA (*solid curve*) and ribulose diphosphate (*broken curve*) is shown here. The conversion of ribulose diphosphate to PGA continues after the light is turned off (*colored area*), so that the ribulose concentration drops to zero. The concentration of PGA, which is no longer reduced to triose phosphate by ATP and TPNH, increases momentarily before it is used up in the production of other compounds.

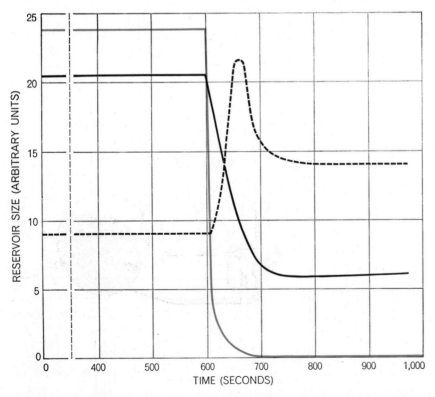

SUDDEN DEPLETION OF CARBON DIOXIDE (*colored curve*) slows the carboxylation of ribulose diphosphate to PGA. Because the light remains on after depletion, ribulose diphosphate (*broken curve*) continues to be formed and its concentration rises. PGA (*black curve*) is still reduced to triose phosphate, so that its concentration drops. "Reservoir size" refers to the average size of the "pool" of any one compound per unit quantity of algae.

cal pathways already established in studies of respiration. It seemed likely that PGA might be linked to these phosphates by the reverse of a sequence of respiratory reactions first mapped many years ago by the German chemists Otto Meyerhof, Gustav Embden and Jakob Parnas. In the respiratory pathway hexose phosphate is split into two molecules of triose phosphate, with the split occurring between the two carbon atoms in the middle of the chain. The triose phosphate is then oxidized to give PGA. The electrons from this energy-yielding operation are picked up by diphosphopyridine nucleotide (DPN^+), which is thereupon reduced to DPNH. The DPN^+ is a close relative of the TPN^+ that turns up in photosynthesis. In addition this oxidation yields enough energy to make a molecule of ATP from ADP and phosphate ion.

In the reverse pathway of these reactions in photosynthesis, Calvin concluded, the plant uses the cofactors ATP and TPNH, made earlier by the transformation of the energy of light, to bring about the reduction of PGA to triose phosphate. In the first step the terminal phosphate group of ATP is transferred to the carboxyl group of PGA to form a "carboxyl phosphate" (really an acyl phosphate). Some of the chemical potential energy that was stored in ATP is now stored in the acyl phosphate, making the new intermediate compound highly reactive. It is now ready for reduction by TPNH. This reducing agent donates two electrons to the reactive intermediate. One carbon-oxygen bond is thereby severed and the oxygen atom, carried off with the phosphate group, is replaced by a hydrogen atom. In this way the carboxyl carbon atom is reduced to an aldehyde carbon atom; that is, it now has two bonds to oxygen instead of three and one bond each to carbon and hydrogen [*see illustration on page 112*] This is the point in the cycle at which the major portion of the solar energy captured in the first stage of photosynthesis is applied to the reduction of carbon.

The Unstable Intermediate

The next development in the plotting of the carbon pathway came from a series of experiments first performed in our laboratory by Peter Massini. He hoped to see which intermediates would be most strongly increased or decreased in concentration by turning off the light and allowing the synthetic process to go on for a while in the dark. In order to establish the concentration of the various intermediates when the reaction proceeds in the light, he bubbled radioactive carbon dioxide through the culture for more than half an hour. At the end of this period every intermediate was as highly radioactive as the incoming carbon dioxide. The radioactivity from each compound therefore gave a measure of the concentration of the compound. He then turned off the light and after a few seconds took another sample of algae in which he measured the relative concentration of compounds by the same technique. Comparison with the compounds sampled in the light showed that the concentration of PGA was greatly increased. This finding could be readily explained: turning off the light stopped the production of the ATP and TPNH required to reduce PGA to triose phosphate.

Of the sugar phosphates present, only one, the five-carbon ribulose diphosphate, was found to have changed significantly; its concentration dropped to zero. Because the PGA had simultaneously increased in concentration, it was apparent that ribulose diphosphate was consumed in the production of PGA. This finding was of great significance because it indicated for the first time that

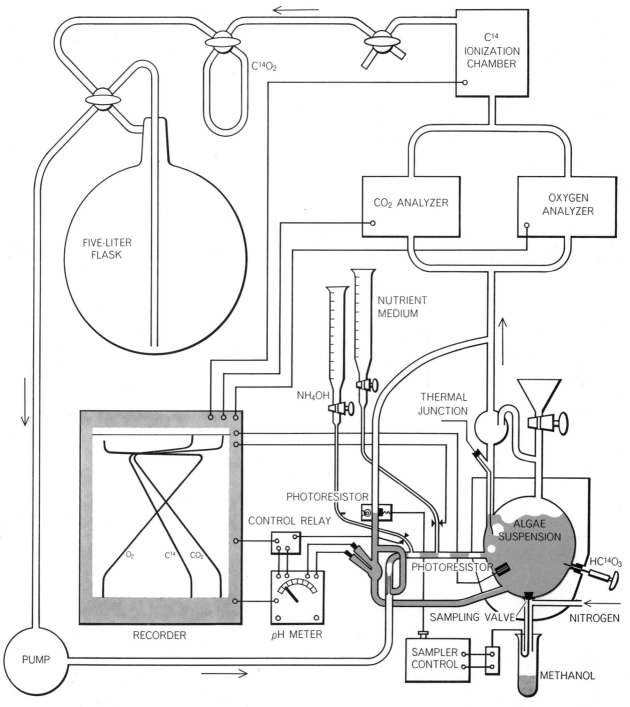

STEADY-STATE APPARATUS permits experimental control and study of photosynthesis. The algae are suspended in nutrient in a transparent vessel (*lower right*). A gas pump circulates a mixture of air, ordinary carbon dioxide and labeled carbon dioxide (when needed) to the vessel, where it bubbles through the suspensions. Labeled carbon can also be added in the form of bicarbonate ($HC^{14}O_3$). Measurements of the oxygen, carbon dioxide and labeled carbon levels in the gas are recorded continuously. The pH is maintained at a constant value by means of the pH meter. The sampler control allows removal of samples into the test tube.

ribulose diphosphate is the intermediate to which carbon dioxide is attached by the carboxylation reaction.

For this reaction ribulose diphosphate is prepared by an earlier reaction that goes on in the light and in which ATP donates its terminal phosphate group to ribulose monophosphate. The more reactive diphosphate molecule now adds one molecule of carbon dioxide by carboxylation. The details of this reaction remain obscure because the resulting six-carbon intermediate is so unstable that we have not been able to detect it by our methods of analysis. As its first stable product this sequence of events yields two three-carbon PGA molecules.

Massini's experimental results were confirmed by a parallel experiment devised by Alex Wilson, then a graduate student in our laboratory. Instead of turning out the light Wilson shut off the supply of carbon dioxide. In this situation one might expect to find an increase in the concentration of the compound that is consumed in the carboxylation reaction; ribulose diphosphate showed such an increase. Correspondingly, one would look for a decrease in the concentration of the product of this reaction; PGA did in fact decrease in concentration.

The first steps along the path were thus established. The photosynthesizing plant starts with ribulose monophosphate and converts it to ribulose diphosphate, using chemical potential energy trapped from the light in the terminal phosphate bond of ATP. Carbon dioxide is joined to this compound, and the resulting six-carbon intermediate splits to two molecules of PGA. With energy and

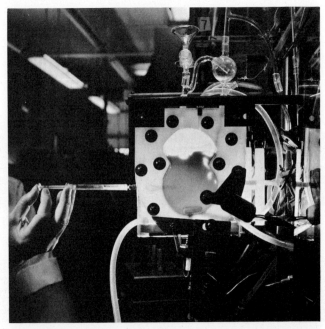
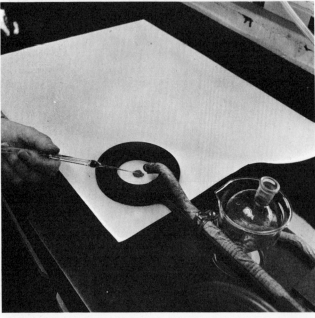

EXPERIMENT to determine the path of carbon in photosynthesis is outlined. After removal of an algae sample from its culture tube, the sample is placed in a transparent vessel (*top left*). At start of experiment labeled bicarbonate is injected into the vessel (*second from top left*). A sample is then removed by pressing a button on the sampler control (*third from top left*); alcohol in the test tube kills the algae. The sample is concentrated by evaporation in a special flask to which a vacuum has been applied (*top right*)

electrons supplied by ATP and TPNH, PGA is reduced to triose phosphate. In the next step, it was apparent, two triose phosphates must be joined end to end in the reverse of a familiar respiratory pathway to form a hexose phosphate. The pathway from hexose to pentose phosphate remained to be uncovered.

We continued the carbon-by-carbon dissection and analysis of these chains by the methods that had earlier shown the carbon 14 in PGA to be located first in the carboxyl carbon. In the hexose molecules we had found the labeled carbon concentrated in the two middle carbons, just where it should be if two triose molecules made from PGA were linked together by their labeled ends. We also took apart the seven-carbon and five-carbon sugar phosphates to establish the position of the carbon 14 atoms in their chains. As the result of these degradations we were able to show that the overall economy of the photosynthetic process starts with five three-carbon PGA's, variously transforms them through three-, six-, four- and seven-carbon phosphate intermediates and returns three five-carbon ribulose diphosphates to the starting point [*see illustrations on pages 110 and 111*]. From the carboxylation of these three chains and their immediate bisection, the cycle at last yields six PGA molecules. The net result, therefore, is the conversion of three carbon dioxide molecules to one PGA molecule.

The Calvin Cycle

With these steps filled in, the carbon reduction cycle in photosynthesis, called the Calvin cycle, was complete. The in-

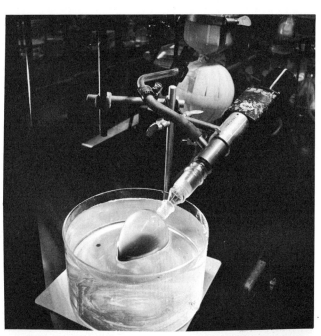

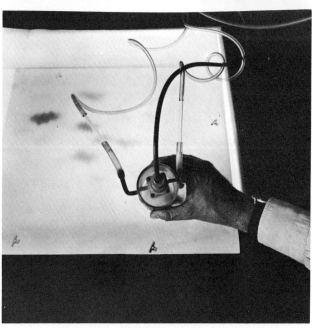

and the extract applied to chromatogram paper (*bottom left*). The paper is placed in a trough of chromatographic solvent (*second from bottom left*), which diffuses through the paper; after eight hours the paper is turned at right angles and the process is repeated. A radioautograph is made by exposing X-ray-sensitive film to the chromatogram (*third from bottom left*). The radioactivity of the compounds in the chromatogram is then measured (*bottom right*), using the radioautograph as a guide to their location.

termediates formed in the cycle depart from it on various pathways to be converted to the end products of photosynthesis. From triose phosphate, for example, one sequence of reactions leads to the six-carbon sugar glucose and the large family of carbohydrates.

Because the cycle had been established primarily by experiments with algae and the leaves of a few higher plants, it was important to see whether or not the cycle prevailed throughout the plant kingdom. Calvin and Louisa and Richard Norris carried out experiments with a wide variety of photosynthetic organisms. In every case, although they found variation in the amounts of particular intermediates formed, the pattern was qualitatively the same.

It also had to be shown that the pathway we had traced out is quantitatively the most important route of carbon reduction in photosynthesis. To this end Martha Kirk and I undertook an intensive study of the kinetics of the flow of carbon in photosynthesis. Our study has helped to solve other general problems, particularly the question of how carbon enters into the pathways leading to the synthesis of proteins and fats. The biological materials for this work are supplied by an algae culture system under automatic feedback control. In this apparatus we are able to maintain the photosynthetic process in a steady state, with nutrients supplied at a constant rate and with temperature, density, salinity and acidity held within narrow limits.

At the start of a run we inject radioactive bicarbonate ion into the culture medium along with radioactive carbon dioxide gas and so bring the ratio of carbon 14 to carbon 12 immediately to its final level in both the gas and the liquid phase. An automatic recorder measures the rate at which carbon is absorbed by the photosynthesizing cells. We take samples every few seconds and kill the cells immediately by immersing them in alcohol. After we have chromatographed the photosynthetic intermediates and measured their radioactivity we then plot the appearance of labeled carbon in each of these compounds as a function of time.

By the end of three to five minutes, our records show, all the stable intermediates of the cycle are saturated with carbon 14. Taking the total amount of carbon thus fixed in compounds and comparing it with the rate of uptake of carbon in the culture, we found that the cycle accounts for more than 70 per cent of the total carbon fixed by the algae. A small but significant amount is also taken up by the addition of carbon dioxide to a three-carbon compound, phosphoenolpyruvic acid, to give four-carbon compounds.

From the earliest work with carbon 14 in our laboratory, it had been apparent that carbon dioxide finds its way rather quickly into products other than carbohydrates in the photosynthesizing plant. This was at variance with traditional ideas about photosynthesis that regarded carbohydrates as the sole organic products of the process. It was important to ask, therefore, whether fats and amino acids could be formed directly from the cycle as products of its intermediates or whether these noncarbohydrates were synthesized only from the carbohydrate end products of photosynthesis. Our kinetic studies show that certain amino acids must indeed be formed from the intermediates and must therefore be regarded as true products of photosynthesis. The amino acid alanine, for example, shows up labeled by carbon 14 at least as rapidly as any carbohydrate; it would be labeled with carbon 14 much more slowly if it were made from carbohydrate, since the carbohydrate would have to be labeled first. We have been able to show that more than 30 per cent of the carbon taken up by the algae in our steady-state system is incorporated directly into amino acids. There is some evidence that fats may also be formed as products of the cycle.

The discovery that plants make these other compounds as direct products of photosynthesis lends new interest and importance to the chloroplast, the subcellular compartment of green cells that contains pigments and the rest of the photosynthetic apparatus. It has been known for some time that this highly structured organelle is responsible for the absorption of light, the splitting of water and the formation of the cofactors for carbon reduction. More recent studies have shown that it is the site of the entire carbon-reduction cycle. Now the chloroplast emerges as a complete photosynthetic factory for the production of just about everything necessary to the plant's growth and function.

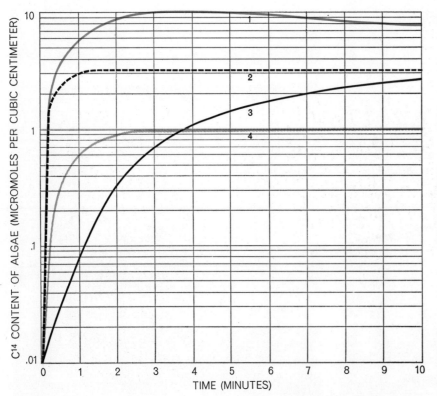

CALVIN CYCLE (*see illustration on pages 110 and 111*) was shown to be the most important route of carbon reduction in photosynthesis in studies by the author and Martha Kirk. As seen here, all stable intermediates of the cycle become saturated with labeled carbon within three to five minutes. Comparison with the rate of carbon uptake in the algae culture showed that the cycle accounts for more than 70 per cent of the carbon fixed in compounds.

1 HEXOSE AND HEPTOSE MONOPHOSPHATES
2 PGA
3 SUCROSE
4 HEXOSE AND HEPTOSE DIPHOSPHATES

11

BIOLOGICAL LUMINESCENCE

by WILLIAM D. McELROY and HOWARD H. SELIGER
December 1962

The light of fireflies and other luminescent organisms has always charmed human observers. What benefit does the ability to produce light confer on an organism? Is the light completely "cold," that is, is its production 100 per cent efficient? Exactly how is the light produced? In spite of much study, dating back more than a century, these questions cannot be answered completely. But the main steps in the process have been established, its efficiency has been measured and the principal substances involved in it have been identified. With this knowledge in hand it is possible to make a reasonable guess as to how biological luminescence, or bioluminescence, arose in the course of evolution.

Bioluminescence is not only of interest in itself but also provides a sharp tool for studying other biological processes. This has resulted in part from the development of highly sensitive and rapid recording devices for the measurement of light. Because the emission of light by an organism is a chemical reaction catalyzed by an enzyme, the intensity of the light provides direct evidence of the rate of a kind of reaction that is common to all life processes. Consequently light emission by cells or cell extracts can be studied under various conditions and can serve as a valuable quantitative tool for biochemical and biophysical investigations.

One of the most striking features of bioluminescence is the sheer diversity of organisms that have developed the ability to emit light. They include certain bacteria, fungi, radiolarians, sponges, corals, flagellates, hydroids, nemerteans (vividly colored marine worms), ctenophores (small jellyfish-like animals), crustaceans, clams, snails, squids, centipedes, millepedes and insects. Among the last are of course the insects familiarly known as fireflies and glowworms. Many fishes are also luminous, but there are no self-luminous forms among amphibians, reptiles, birds or mammals. None of the higher plants is luminous. With the possible exception of a few strains of luminous bacteria, no freshwater organism is luminous, even though many of them are closely related to light-emitters that live in the sea.

Light emission can occur whenever a physical system undergoes a discrete change in free energy. The source of the original excitation of the system can be thermal, as in an incandescent lamp; electrical, as in a flash of lightning; mechanical, as in the scintillation that attends the breaking of a sugar crystal; or chemical, as in the glow of phosphorus. Bioluminescence is chemical luminescence, or chemiluminescence.

In a secluded bay in Jamaica certain protozoa that glow when they are disturbed are so abundant that they brightly illuminate the fish swimming in their midst. In Thailand one species of firefly congregates on certain trees, and all the fireflies flash on and off in unison like Christmas-tree lights. In Brazil the "railroad worm," the larva of a large beetle, bears green lights along its sides and a red light on its head. In the waters off Bermuda the female of a species of marine worm comes to the surface three days following a full moon and secretes a glowing circle of luminous material. The male, emitting puffs of light, heads straight for the circle of light, and both eggs and sperm are discharged into the water. A deep-sea angler fish carries a luminous organ at the tip of a retractable rodlike appendage with which it lures victims into its jaws. One deep-sea member of the squid family, quite unlike its ink-emitting cousins, spurts a luminous cloud when it wants to hide. In this symphony of living light most chords are blue. The greens and yellows of the fireflies and the green and red of the railroad worm are grace notes ornamenting the azure theme of the luminous bacteria and larger marine organisms.

The Luminous Clams

Probably the best known of the luminous mollusks is the boring clam *Pholas dactylus*, which men have regarded as a delicacy since antiquity. In Greek *pholas* means lurking in a hole, which describes the mollusk's habit of boring into soft rock and hiding there with only its siphon exposed. In 1887 the French physiologist Raphaël Dubois used *Pholas* in his pioneering studies of the substances involved in bioluminescence. Dubois demonstrated that a cold-water extract of *Pholas* would continue to emit light for several minutes. He found that after the light emission had ceased it could be restored by adding a second extract obtained by washing a fresh clam in hot water and cooling the juice. Dubois concluded that there was some substance in the hot-water extract that was essential for light emission and that it was not affected by heating. He called this material luciferin, a name he coined from Lucifer, meaning light-bearer. The substance in the cold-water extract he called luciferase, indicating by the suffix "-ase" that it had the properties of an enzyme. Enzymes are biological catalysts, and like most enzymes luciferase is heat-sensitive. Dubois reasoned that both luciferin and luciferase were extracted by water—hot or cold—but that hot water inactivated the luciferase, leaving only the luciferin active.

The other pioneer in the field of bioluminescence was the late E. Newton

GLOWING TOADSTOOLS of the genus *Mycena* were photographed by self-emitted light. The light is given off principally from the gills beneath the caps of the fungus. Luminous mold is often seen in the vegetative state on rotting logs, but it is rare for luminescence to continue into the fruiting state shown here. The photograph was made by Yata Haneda of the Yokosuka City Museum.

124   ENERGETICS

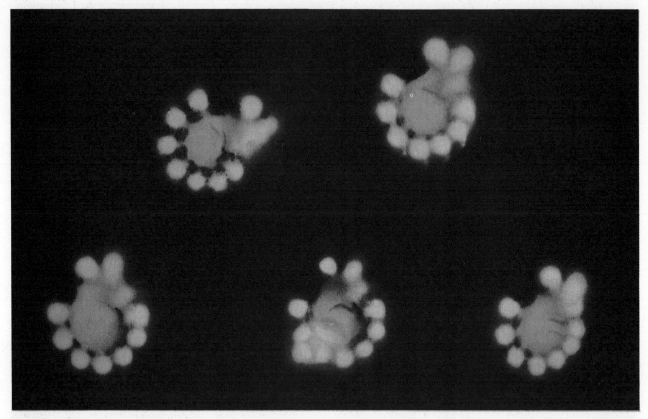

RAILROAD WORM is the larval form of a South American beetle. F. W. Goro made this unusual photograph by pressing a sheet of Kodachrome directly against a single larva several times to obtain multiple images. The picture is consequently a true autograph, made without the intervention of lens or camera. It is reproduced through the courtesy of *Life*. Copyright 1945 by Time Inc.

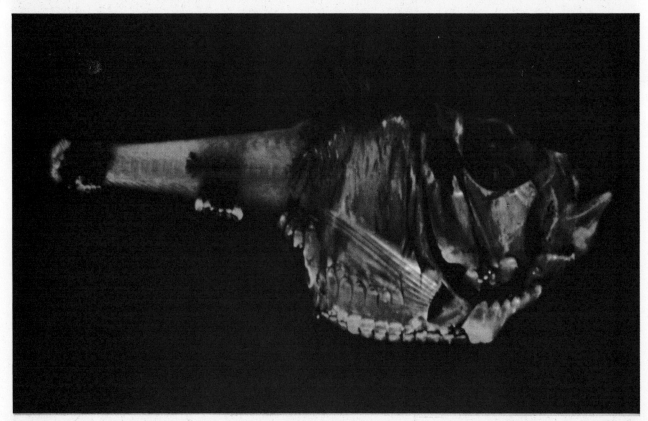

HATCHET FISH, *Argyropelecus hemigymnus,* is one of the more than 100 species of light-emitting fish. The photograph is by Haneda. In classifying many thousands of fish netted below 400 meters, the deep-sea investigator William Beebe found that more than 95 per cent were luminous. It seems likely that the light patterns found in various species play a role in hunting and recognition.

Harvey of Princeton University. Following Dubois's observations Harvey did much to show clearly that light emission in organisms is an enzymatic process. He described luciferin-luciferase reactions in a variety of organisms and demonstrated that they were of several different types. On a field trip to Japan he found a crustacean (*Cypridina hilgendorfii*) that, when dried, provided a convenient source of both luciferin and luciferase. Harvey and his students used many pounds of the material in their studies. *Cypridina* is a small crustacean with two hinged valves covering its body. It is found in both fresh and salt water, but only the marine forms are luminous. During World War II, Japanese soldiers used dried *Cypridina* as a source of low-intensity light when they did not want to run the risk of using a flashlight. A small quantity of *Cypridina* powder placed in the palm of the hand and moistened provided enough light for reading a map or a message.

Cypridina live in the sea bottom near the shore and come out to feed at night. The organism is not itself luminous; it excretes luciferin and luciferase into the surrounding water, and the interaction of the two substances produces a blue light. The luciferin is apparently synthesized in one gland and the luciferase in another. Japanese biochemists have recently purified the luciferin from *Cypridina* and have published a tentative description of its molecular structure. This appears to resemble the structure of firefly luciferin, about which we will have more to say.

Fireworms of the Sea

There are a large number of luminous forms among the annelid worms, which range in length from a fraction of an inch to several inches. The luminescence is particularly striking during the mating period of the "fireworms," annelids of the order Polychaeta. It seems likely that Columbus saw fireworms on his first voyage to the New World. He wrote of seeing lights in the water resembling moving candles as he approached the Bahamas. The relation between luminescence, the phase of the moon and periodicity in the breeding of these marine organisms is beautifully illustrated by the Bermuda annelid *Odontosyllis enopla*. The worms begin to swarm two or three days after the full moon, the females appearing first. Each swims in a small circle at the surface, emitting a greenish light. Invariably the performance reaches a peak between 55 and 56 minutes after sunset. The circles of light evidently attract the male worms, which normally stay well below the surface. As the males swim toward the females, traveling 15 to 20 feet with remarkable accuracy, they emit short flashes of light. Commonly several males will converge on a single female; the whole group then rotates in a tight, glowing circle as its members discharge eggs and sperm into the water. The eggs are accompanied by a secretion that leaves a luminous cloud in the wake of the female. The females, which range up to 35 millimeters in length, are often twice as long as the males. The body of the female glows strongly and almost continuously. The male continues to glow with sharp intermittent flashes.

After the mating process has begun the males exhibit an additional positive response to light. For example, if a flashlight is aimed into the water, males will start swimming toward its beam. There is no evidence to indicate that the females will respond positively to the light, although they are obviously stimulated to release their eggs by the presence of the males. Recently we have been able to obtain from *Odontosyllis* extracts of luciferin and luciferase that give off light when the two are mixed together. We do not yet have enough of the two materials, however, to study the chemistry of the bioluminescent reaction in detail.

Marine Dinoflagellates

The "burning of the sea" presented a mystery to fishermen and other observers for centuries. The "burning" refers to the glow sometimes seen in the wake of a ship as it moves through tropical waters. The glow is due to the presence of large numbers of dinoflagellates that luminesce when they are disturbed. These one-celled organisms often develop in large quantities at favorable seasons of the year. In secluded bays a permanent heavy culture can develop; the waters of the bay can become so thick with dinoflagellates that the water itself is colored. Such luminescent bays have become famous tourist attractions. One of the most spectacular is Oyster Bay, near Falmouth on the northern coast of Jamaica; another is on the southern coast of Puerto Rico near Parguera. The two bays are inhabited chiefly by the luminous dinoflagellate *Pyrodinium bahamense*. If one travels across one of the bays at night, looking down from the bow of a moving boat, one can see fish sharply silhouetted against the glowing water as they dart out of the way. The movement of the fish triggers the luminescence, and every wave looks as if it were aflame.

The discovery that this luminescence comes always from living things was only slowly appreciated because most of the dinoflagellates are invisible to the naked eye. The mystery of the burning sea was not definitely settled until about 1830. In recent years dinoflagellates have been grown in the laboratory and their bioluminescence has been studied in great detail.

The "red tides" of the sea are due in most cases to dinoflagellates, which are also capable of forming patches of brown and yellow. The color of their nighttime luminescence, however, is always blue. On occasion the daytime red patches are due to the flagellate *Noctiluca,* which is large enough to be seen without a microscope. Along the Pacific coast of the U.S. it is not unusual to find patches of *Gonyaulax polyhedra*, a quite luminous dinoflagellate. The red tides reported along the Gulf Coast of Florida in recent years are produced by an organism (*Gymnodinium brevis*) that is unrelated to the flagellates and is nonluminous.

By growing cultures of *Gonyaulax* in the laboratory it has been found that they stop producing light at dawn and luminesce again in the evening. In addition to being luminescent *Gonyaulax* is a photosynthetic organism requiring light for growth. Under laboratory conditions one can readily obtain cultures of 10,000 to 20,000 cells per liter. When the culture vessel is shaken, the cells emit bright flashes of light lasting less than a tenth of a second. If the organisms are illuminated continuously with a dim light, so that they are no longer exposed to a normal day-night cycle, an interesting phenomenon takes place. When one shakes the culture to measure the maximum light output, one finds that the maximum output continues to occur each night at about 1 a.m. and decreases to a minimum some 12 hours later. In other words, the organism's normal day-night rhythm will continue unbroken for weeks under a steady weak light sufficient to supply energy. This remarkable biological clock can be altered, however, by subjecting the cells to an artificial light-dark cycle. For example, if *Gonyaulax* cells are exposed to eight hours of darkness followed by eight hours of light, they adopt a new rhythm in which they can emit light during the eight hours of darkness and are nonluminous for the eight hours of light. When the cells are removed from this artificial 16-hour cycle and are again placed under a continuous light of low intensity, the original

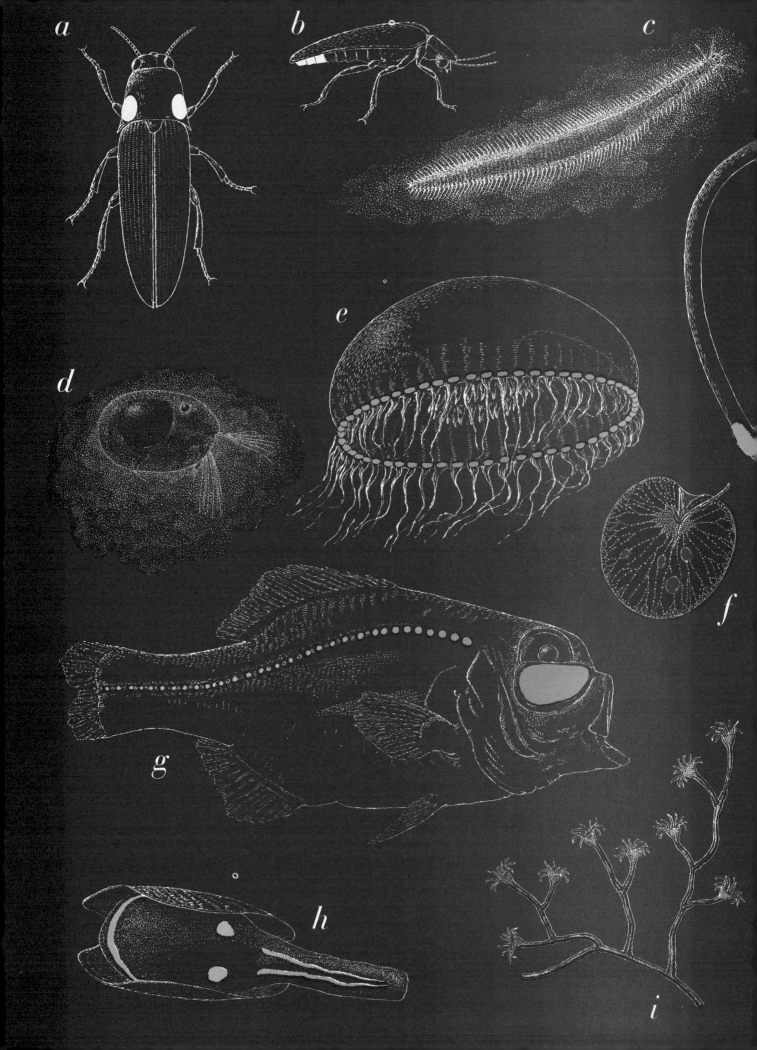

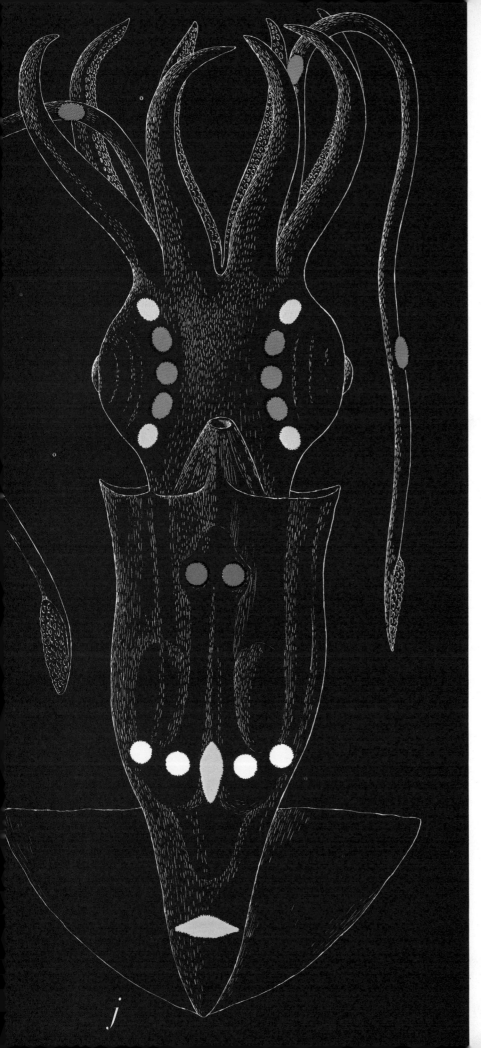

24-hour rhythm resumes. The mechanism underlying this behavior has not been discovered.

Recently we have looked for rhythmic behavior in dinoflagellates under natural conditions and have found that some species show a night-day rhythm and that others do not. In *Gonyaulax* both luciferin and luciferase, the necessary components for light production, are found in greater amounts in cell-free extracts prepared during the night hours than during the day, showing that the rhythm of luminescence reflects rhythmic biochemical processes. It would appear that the luminescent system is not the clock but rather that it is controlled by a master clock that regulates other physiological processes.

Luminous Fungi and Bacteria

If you should chance to stumble over a rotten log in the woods at night, you might be surprised to find that freshly exposed parts of the log were glowing brilliantly. Luminescence of this sort is caused by fungi. The phenomenon was known to Aristotle, and it was studied by such illustrious figures as Francis Bacon and Robert Boyle. Not until early in the 19th century, however, was the role of the fungus properly appreciated.

One of the best known luminous fungi is *Panus stipticus*, which exists in two varieties: a North American form that is luminous and a European form that is not. The threadlike mycelia of the two varieties are able to fuse, and it can be shown by this mating technique that luminescence is under genetic control. Evidently the European variety lacks one or more genes needed to produce enzymes required for bioluminescence.

GALLERY OF ANIMALS at left suggests the diversity of bioluminescent organisms. Roughly speaking, bioluminescent organisms exist in about a third of the 33 phyla and a third of the 80 classes given in the official American classification of the animal kingdom. The 10 luminous animals at left are: *a*, a click beetle (*Pyrophorus noctilucus*); *b*, a common North American firefly (*Photuris pennsylvanica*); *c*, the Bermuda fireworm (*Odontosyllis enopla*); *d*, a Japanese crustacean (*Cypridina hilgendorfii*); *e*, a jellyfish (*Aequorea aequorea*); *f*, a protozoan (*Noctiluca miliaris*); *g*, a fish (*Photoblepharon*) in which the light is supplied by symbiotic bacteria; *h*, an edible clam (*Pholas dactylus*); *i*, one of the luminous hydroids (*Campanularia flexuosa*); *j*, deep-sea squid (*Thaumatolampas diadema*).

128　ENERGETICS

At least one of these enzymes is luciferase, which is found in the North American *Panus* but not in the European.

Before electric refrigerators came into general use there were often reports in the newspapers about "mystery meat" that gave off light. There should have been no mystery about the light; it has been known for a long time that luminous bacteria—all quite harmless—readily grow on meat and dead fish. Boyle experimented with such bacteria and in 1668 demonstrated that they need air if they are to emit light. Subsequently luminescent bacteria found in salt water became a favorite subject for studying bioluminescence. Most of these forms will grow easily on ordinary nutrient agar containing 3 per cent sodium chloride (the salinity of sea water) and

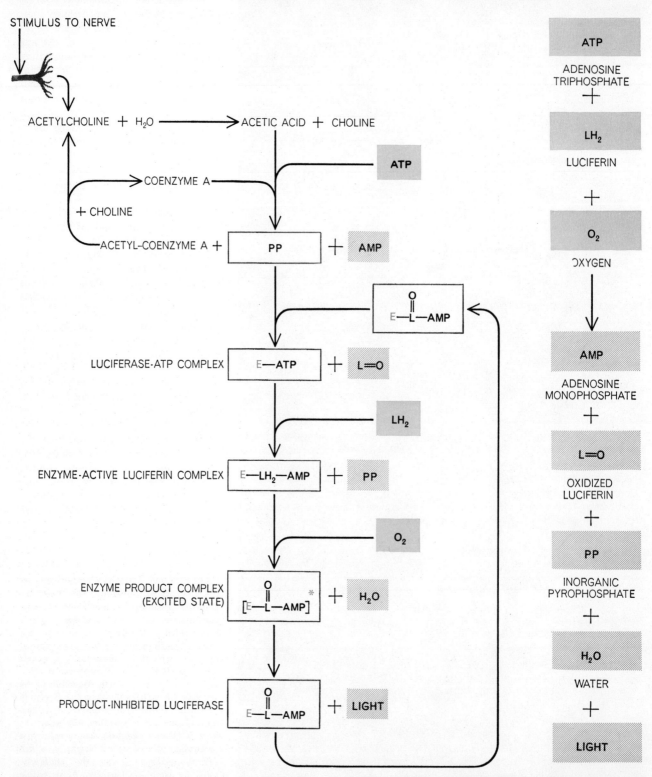

FIREFLY FLASH is probably triggered by a nerve impulse delivered to the luminous gland. A sequence of chemical reactions then produces light. The substances consumed in the reaction, as shown in the summary at right, are adenosine triphosphate (ATP), luciferin (LH_2) and oxygen (O_2). The products are oxidized luciferin ($L=O$), two phosphate compounds, water and light. The reaction is catalyzed by the enzyme luciferase, represented by E. One quantum of light is produced for each molecule of luciferin oxidized.

glucose or glycerol. Among the best sources of salt-water luminous bacteria are dead fish or squids that have not been washed in fresh water. If such material is incubated overnight at 15 or 20 degrees centigrade, it is usually covered with colonies of luminous bacteria by morning. The bacteria can then be transferred to agar plates and readily developed into pure cultures that emit a strong blue or blue-green light. This culture technique was exploited by Dubois, who wrote: "In 1900, at the Palace of Optics, at the International Exposition in Paris, I have been able to illuminate, as from the clearest light of the moon, a vast chamber using large glass flasks of 25-liter capacity... containing very brilliant photobacteria... In the evening as soon as one entered the chamber one could read and see all the people in the room." The light emitted by luminous bacteria is usually a broad band in the blue or blue-green region of the spectrum (wavelengths between 480 and 500 millimicrons).

Some of the most interesting luminous bacteria live in symbiosis with other organisms, frequently squids and fishes. The host often has a complicated luminous organ in which the light is supplied by bacteria. Although the bacteria emit light continuously, the fish or squid may develop a special device, such as a movable screen, that serves to turn the light on and off. One of the most striking instances of bacterial symbiosis occurs in the Indonesian fish *Photoblepharon*. This fish has under each eye an oval white spot, richly supplied with blood vessels, in which the luminous bacteria grow. To turn off the light there is a black fold of skin that can be drawn over the luminous spot like an eyelid [*see illustration on pages 126 and 127*].

The physiology and biochemistry of bacterial luminescence have been studied in great detail. Although we do not know the exact mechanism for creating the luminescent state, we are reasonably certain of the compounds involved. It is now clear that the light-emitting reaction is intimately related to the oxidative, or electron-transport, processes of the bacterial cell. The top illustration on the next page outlines the current hypothesis, in which the light-emitting reaction is a side branch of the general electron-transport process by which the cell extracts energy from food. The requirements for luminescence are a reduced form of riboflavin, an aldehyde, oxygen and an enzyme.

Luminous bacteria have been favored organisms for studying the action of drugs and other inhibitors of cell respiration because the effects are observable externally by means of a photoelectric cell. It is also possible to obtain mutant strains of luminescent bacteria that are nonluminous or only weakly luminous. One can then examine the ability of various chemicals to restore luminescence. The illustration on page 11 shows how the dim light emitted by a suspension of certain mutant bacteria can be increased by the addition of a long-chain aldehyde such as dodecanal. To determine the rate at which the aldehyde penetrates the cell membrane one simply uses a photocell to measure the rate at which the light intensity increases.

Fireflies and Glowworms

Among the insects true instances of self-luminescence are to be found in the springtails, lantern flies, click beetles, the larvae of certain flies and, of course, in the fireflies and their larvae, called glowworms. It is a spectacular sight to see the glowworms that live in caves in New Zealand, the most famous being at Waitomo, about 200 miles north of Wellington. The ceilings of these caves are covered with thousands of glowing larvae, and from each is suspended a long luminescent thread that apparently serves to catch food particles or small insects. If one talks loudly, or if the wall of the cave is tapped sharply, the larvae turn off their lights virtually as one. After a brief period the lights come on again, tentatively at first and then more boldly, until the whole ceiling is once again ablaze.

The true fireflies, or lightning bugs, are found in many parts of the world and provide perhaps the most familiar example of bioluminescence. (Curiously, fireflies are almost unknown in England.) The scientific literature on this group of insects far exceeds that of any other luminous organism. The old hypothesis that the light of the firefly is a mating device to attract the sexes is now universally accepted. Nothing could be simpler than a flashing light to advertise the whereabouts of a flying male to a responsively flashing female waiting in the grass.

Each species of firefly has a characteristic flash that the female of the species can recognize. The signaling system of one common American species of firefly, *Photinus pyralis*, is fairly typical of the mating behavior of a number of species. At dusk the male and female emerge separately from the grass. The male flies about two feet above the ground and emits a single short flash at

STRUCTURE OF LUCIFERIN in the firefly has been established by the authors and their associates at Johns Hopkins University. In the light-producing reaction it combines with one molecule of oxygen to form oxidized luciferin and water. Other luciferins are known.

130 ENERGETICS

```
FOOD → DIPHOSPHOPYRIDINE NUCLEOTIDE → FLAVINS → COENZYME Q → CYTOCHROMES → OXYGEN
                                        ↓
                                   FLAVIN MONONUCLEOTIDE
                                        ↓  + LUCIFERASE
                                           + ALDEHYDE
                                           + OXYGEN
                                      LIGHT
```

SOURCE OF BACTERIAL LIGHT is a side branch of the oxidation-reduction reactions that extract energy from nutrients. In this sequence hydrogen atoms (or their equivalent electrons) are removed from nutrient and passed along (*colored arrows*) to a series of compounds. The final hydrogen-acceptor is oxygen and the final product is water. At certain steps in the sequence energy is removed from the reactants and stored in the form of ATP (*not shown*). Light is emitted when one of the reduced flavins (flavin mononucleotide) reacts with luciferase and oxygen in the presence of an aldehyde. In this reaction flavin takes the role of luciferin.

regular intervals. The female climbs some slight eminence, such as a blade of grass, and waits. Ordinarily she does not fly at all, and she never flashes spontaneously. If a male flashes within three or four yards of her, she will usually wait a decorous interval, then flash a short response. At this the male turns in her direction and glows again. The female responds once more with a flash, and the exchange of signals is repeated—usually not more than five or 10 times—until the male reaches the female, waiting in the grass, and the two mate.

Recognition apparently depends on the time interval between the male flash and that of the female. This interval in certain species is approximately two seconds at 25 degrees centigrade (77 degrees Fahrenheit) and varies with temperature. A flash of artificial light of about a second's duration, simulating the delayed response of a female firefly, will usually induce a male to fly toward it.

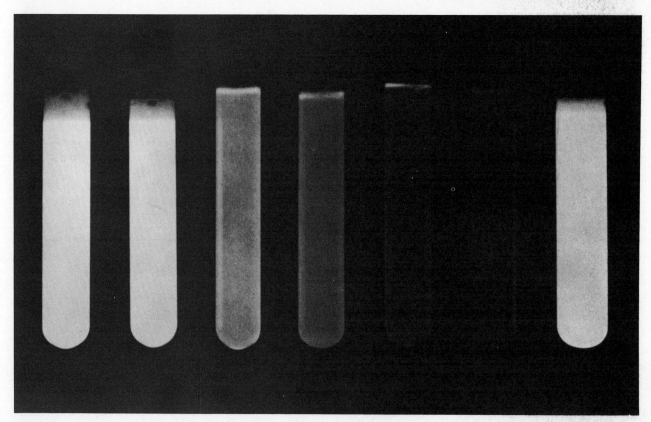

ROLE OF OXYGEN in bioluminescence can be nicely demonstrated with suspensions of luminous bacteria. The first two tubes at left had been aerated continuously prior to the making of the photograph. The next four tubes had been standing unaerated for two, three, four and five minutes respectively. With the passage of time their light emission declined. The last tube at the right, which had been standing undisturbed for 10 minutes, was shaken vigorously to introduce fresh oxygen just before the photograph was made. For a brief period it glows even more brightly than the two tubes that had been supplied with oxygen continuously.

Other species of fireflies have other systems and types of flashes. Synchronous flashing of a number of males to one female has been observed, but it is rare in North American species. Among tropical fireflies, however, it is fairly common. In Burma and Thailand, for example, all the fireflies on one tree may flash simultaneously, whereas those on another tree some distance away may also flash in unison but out of step with those on the first tree. It is conceivable that all the fireflies on one tree are males and those flashing out of phase nearby are all females, but this has not been established.

The eggs of American fireflies are laid on or near the ground and hatch in about three weeks. The larvae differ considerably in habit. They live mostly in damp places among fallen leaves, becoming active at night and feeding on slugs, snails and the larvae of smaller insects. The firefly larvae usually winter under stones or a short distance underground, often in specially constructed chambers. The larvae metamorphose into pupae near the surface.

The first indication of the formation of the light organ takes place about 15 days after egg development begins. After about 22 days of development the light organ has become functional and appears as two bright spots of light. The larvae emerge on about the 26th day of incubation and become glowworms, with the two small lights at one end. In about two years they reach maturity as pupae. During pupation additional light organs develop, which are to become the light organs of the adult firefly. The light organs of both the larva and the adult develop out of fatty bodies that differentiate into specialized luminescent and reflector layers.

The light emission of fireflies depends on a rich supply of oxygen. The light organs are supplied with blood through an extensive capillary system and with oxygen through an extensive system of tracheal tubes. Unfortunately it is difficult to trace the air-supplying tracheae into the photogenic tissue; it is equally difficult to trace the nerve fibers that must control the flashing of the firefly. Investigators have been able to isolate an individual nerve fiber going to the luminous gland and have been able to stimulate light emission by applying an electric current to the nerve. Probably the best indication of nervous control of the flash is to be observed, however, when the animal is decapitated. Flashing ceases immediately. Subsequently the light organ may glow dimly, with random scintillations, for a long time. The

DARK MUTANT BACTERIA, in the cylinder at left, are barely luminous because they cannot make a long-chain aldehyde (dodecanal) essential for high luminosity. When this aldehyde is added to a suspension of the mutant organisms, they glow brightly (*right*).

132   ENERGETICS

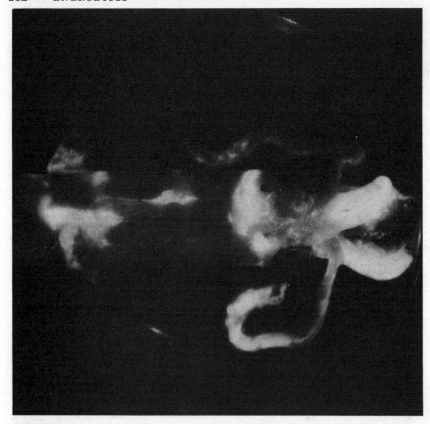

LUMINOUS BACTERIA will usually develop on the surface of a salt-water squid kept overnight in a warm place. Only salt-water varieties of bacteria are luminous. The photograph, made by the authors, required a 15-second exposure at f/4.7 with Polaroid 3,000-speed film.

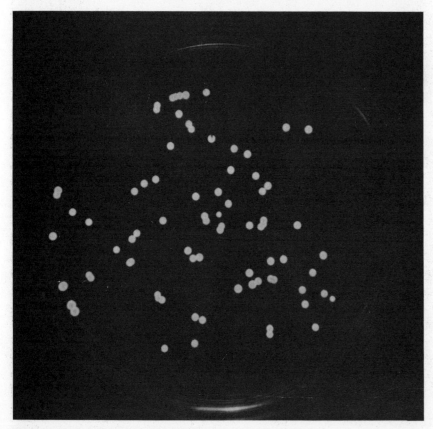

COLONIES OF LUMINOUS BACTERIA can readily be produced by removing bacteria from a decaying sea animal, such as the squid shown at the top of the page, and transferring them to a saline nutrient agar. Robert Boyle experimented with such bacteria in 1668.

exact mechanism of nervous control remains unknown. According to one hypothesis the nerve impulse simply liberates oxygen into the luminous gland, thereby stimulating luminescence. A second hypothesis, which we favor, proposes a series of steps triggered by the release of acetylcholine at a nerve ending in the luminous organ [see illustration on page 128].

The chemistry of the firefly light has been worked out in considerable detail since Harvey first established in 1916 that the glow of the firefly results from the same luciferin-luciferase reaction that Dubois had found in the luminous clam. We now know that firefly luminescence requires, in addition to oxygen, the ubiquitous energy-supplying substance adenosine triphosphate (ATP). If a cold-water extract obtained from firefly lanterns is allowed to stand until the light disappears, the light can be restored to more than its original intensity with the addition of ATP.

Within the past few years we have isolated firefly luciferin in our laboratory at Johns Hopkins University. We have established its chemical structure and have confirmed its validity by synthesizing the compound and showing that under the appropriate conditions it luminesces. We have also isolated and obtained in pure form the light-stimulating enzyme of the firefly, luciferase. It appears to contain about 1,000 amino acid subunits and is therefore larger than any of the proteins whose structure has so far been established.

The peak wavelength of the light emitted by the firefly *Photinus pyralis* is 562 millimicrons, in the yellow-green part of the spectrum. We have found that extracts of firefly lanterns emit light at the same wavelength when the acid-alkaline balance of the solution is neutral. If the solution is made acid, or if high concentrations of inorganic phosphate are added, the light shifts to red, with a peak emission at 614 millimicrons. Presumably shifts of this sort can explain the slight differences in the color of the light emitted by various fireflies.

The availability of luciferin in pure form has also enabled us to determine the efficiency of the light-emitting process. To do this we compare the number of luciferin molecules oxidized with the number of light quanta produced. It turns out that for each molecule of luciferin consumed exactly one light quantum is emitted. It has been fashionable for many years to describe bioluminescence as "cold light" to distinguish it from thermal luminescence. The finding that the quantum efficiency

"INSTANT" BIOCHEMICAL LIGHT is produced simply by adding water to the powder obtained by drying and pulverizing a small marine crustacean called *Cypridina*. Rich in luciferin and luciferase, dried *Cypridina* was sometimes used as a light source by Japanese soldiers during World War II when the use of a flashlight under battle conditions might have revealed their position.

SENSITIVE ASSAY FOR ATP uses the lanterns from four or five fireflies as the indicating agent. A small sample containing an unknown amount of ATP is added to a suspension of the pulverized lanterns. The more ATP present, the more intense the light emitted. The photograph shows the light produced by .1-milliliter samples containing various microgram amounts of ATP.

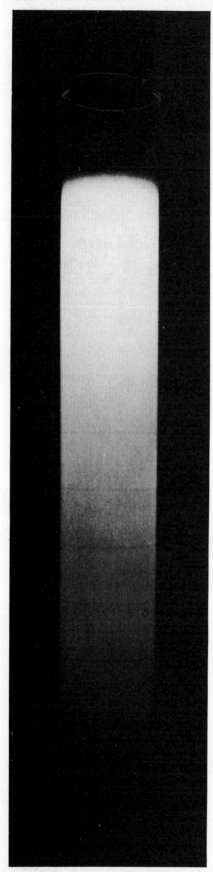

DIMMING OF BACTERIA is observed when a well-aerated suspension is allowed to stand undisturbed. The dimming begins at the bottom, as oxygen is depleted, and works upward. Rising air bubbles postpone the dimming in the upper part of the cylinder.

of firefly light production is indeed 100 per cent makes the term "cold light" strictly accurate.

One of the few creatures to luminesce in two colors is the Central and South American beetle *Phrixothrix*. The larva of these insects is decorated with 11 pairs of luminous green spots that form two parallel rows running along the sides of the body; on the head of the larva are two luminous spots that glow a bright red. At night, when only the red spots are shining, the animal looks like a glowing cigarette. When the animal is disturbed and crawling, however, the green lights flash on, so that it rather resembles a railroad train with red head lamps. Not surprisingly, *Phrixothrix* is commonly called the railroad worm.

Other luminescent insects are found among the click beetles, the Elateridae. In some ways they look much like ordinary fireflies. Most of them, however, are decorated with two oval greenish spots, one on each side of the front part of the body. Because these luminous spots have the appearance of automobile headlights the insects are sometimes called "automobile bugs." In addition the click beetle usually has on its first abdominal segment a heart-shaped spot that glows orange and that is visible only when the beetle is in flight.

Luminescence in Evolution

Among the more advanced multicellular organisms, light emission has been adapted to fulfill very definite functions: as a mating signal for the fireworms and the fireflies, as a lure for the deep-sea angler fish and as a protective screen for certain squids and other marine animals. What function, if any, light emission has in the lower organisms such as the bacteria, the fungi and the dinoflagellates is not immediately obvious. The wide distribution of this large variety of different luminous organisms with entirely different chemical reactions for light emission would indicate that at some time this mechanism must have had some selective advantage.

Even though the luciferins from various luminous organisms are different, we are reasonably certain that all are associated either directly or indirectly with the energy-liberating reactions of the cell. In all cases where the detailed chemistry of the reactions leading to light emission has been examined, oxygen is an essential ingredient. For example, in the luminous bacteria the light-emitting reaction is a branch of the electron-transport system that is essential for growth and reproduction. It seems reasonable to expect that the origin of the light-emitting processes was in some way closely associated with the early evolution of life on earth. Furthermore, it is our belief that various "practical" adaptations of bioluminescence in the more advanced organisms came late in evolution.

We propose that bioluminescence was originally an incidental concomitant of the chemical reactions that were most efficient in removing oxygen from living systems. It is generally believed that the earliest forms of life on earth developed in the absence of oxygen. The first organisms, therefore, were anaerobes. When in the course of the millenniums free oxygen slowly appeared—as a result of solar decomposition of water vapor, augmented, perhaps, by primitive photosynthesis—it would have been highly toxic to anaerobic organisms that could not quickly get rid of it. Chemically the most efficient way to remove oxygen is to reduce it to form water. In the forms of life then present, the most likely reducing agents would have been those organic compounds that were already part of the hydrogen-transport system of the primitive anerobes. When oxygen is converted to water by such compounds, enough energy is liberated in single packets, or quanta, to excite organic molecules to emit light. Low-energy packets will not do. Thus all the successful oxygen-removing organisms would have been potentially luminescent.

During subsequent evolution anaerobic organisms evolved that could use oxygen directly in their metabolic machinery. Then the oxygen-removing light reaction was no longer a selective advantage. But since it had evolved with the primitive electron-transport process, it was not easily lost. In most cases where it has been studied carefully bioluminescence is produced by a nonessential enzyme system. It is possible, for example, to grow luminous bacteria and luminous fungi under conditions that inhibit light emission without affecting growth. And it is possible to obtain mutant strains of luminous fungi and bacteria that are fully vigorous although nonluminous. We find additional support for our hypothesis in the observation that all luminescent reactions can detect and use oxygen at extremely low concentrations. Bacteria can easily produce measurable light when the oxygen concentration is as low as one part in 100 million. Thus we argue that bioluminescence is a vestigial system in organic evolution and that through the secondary processes of adaptation the system has been preserved in various and unrelated species.

Part IV

SYNTHESIS

Synthesis

INTRODUCTION

A dozen years ago, the introduction for a section with this title would have dealt with discoveries about how various building-block molecules were linked together in the construction of the various kinds of macromolecules. At that time, we were learning of the ways in which these components, such as the acetate molecules used in making long-chain lipids, were "activated" biochemically so that, with the assistance of various enzymes, they could be coupled into chains. The concern, in short, was with the basic chemistry of biological polymerization.

To be sure, this concern is important, and it is with us still. But it has become subordinate to a new theme, the theme of specification. Whereas once we wondered *how* monomers were joined together, our attention is now focused on the mechanism that generates a particular sequence. The new biology of synthesis emphasizes two classes of polymers: the nucleic acids, which are the carriers of genetic information, and the proteins, which—through their role as enzymes—are the agents of biochemical specificity. The underlying assumptions are simple: that each of the thousand-plus protein enzymes in a cell has a unique structure enabling it to act as catalyst in a particular reaction; that this unique structure is directly related to the sequence of amino acids in the protein; that this amino acid sequence is specified by the chemistry of the gene. The mechanism underlying the last of these has been revealed largely during the past few years; its exposure has surely been the most compelling and dramatic triumph of twentieth-century biology.

In "How Cells Make Molecules," Vincent G. Allfrey and Alfred E. Mirsky supply the background for these discoveries and recount all but the most recent advances. At the time this article was written, the molecular structure of deoxyribonucleic acid (DNA) was clear, and it was known that the base sequence could be employed, with the aid of appropriate enzymes, in replicating a new strand of DNA or "transcribing" its specificity to a strand of ribonucleic acid (RNA). RNA was known to constitute half the weight of the ribosome, the cytoplasmic particle involved in protein synthesis, and to exist also in soluble forms, as "messenger" and "transfer" RNA. From the general scheme described by Allfrey and Mirsky, activated amino acids, coupled to their own specific transfer RNA molecules, could be arrayed in proper order on the ribosome along a template constructed of messenger RNA sent out from the nucleus. Formal proof of the postulates concerning the role of the various forms of RNA was lacking, however, and there was no information about the nature of the code according to which nucleic acids specified the amino acid sequence in the protein.

"The Genetic Code: II," written by Marshall W. Nirenberg just eighteen months later, describes the sudden breakthrough that has brought the second problem near a solution. The four bases of DNA (and secondarily, those of the complementary messenger RNA for which DNA acts as template) can, in combinations of three, specify 64 different code words—more than enough to handle the amino acid "dictionary" of 22 words. Nirenberg describes the system in which synthetic polynucleotides can substitute for messenger RNA in directing the incorporation of specific amino acids into proteins. Although these experiments have not been successful in defining the specific order of "letters" in many of the words, and though we still need to know something about the punctuation system, the mechanism of hereditary specificity now seems nearly worked out.

The nature of the cytoplasmic machinery that assembles protein according to this code still raises problems. In "Polyribosomes," Alexander Rich describes advances that may clarify these. Electron micrographs reveal "strings" of ribosomes, held together by threads whose dimensions are consistent with the configuration of messenger RNA. Although final evidence is still lacking, the theory that several ribosomes may be "reading" a given strand of messenger RNA at the same point, proceeding always from one particular end to the other, is a plausible one, and it is consistent with the remarkable micrographs made in Rich's laboratory.

These exciting new advances in molecular genetics bring us to the level of the synthesis of a polypeptide chain, an elongate strand containing the specific sequence of amino acids that make up a particular protein molecule. In other words, they account for the *primary structure* of the protein. "The Three-dimensional Structure of a Protein Molecule" by John C. Kendrew dramatically testifies to the layers of organization that overlie this simple structure. The chain folds back on itself, twists, and rolls up into a shape of exquisite complexity. Kendrew describes the exhaustive and difficult effort that produced a complete description of this "native" structure of one of the simpler proteins, myoglobin. At first glance, the result might not seem commensurate with the labor; but in fact, the three-dimensional structure embodies everything functionally significant to the protein molecule. In most enzymes, a particular region of the surface functions in the chemical associations so crucial to activity. Such an "active site" is formed of folds from very different regions of the polypeptide chain, so that the immediate neighbor-relationships derived from chemical studies are not very meaningful. Indeed, recent genetic analyses have revealed that amino acid "switches" at distant points along the chain (arising through mutations) have related effects upon activity; when these analyses are ultimately combined with structural data like Kendrew's, the correlation between structure and function will have been carried to the finest detail yet.

The final article, "Collagen" by Jerome Gross, serves as a reminder that not all important proteins are enzymes. Collagen is perhaps the most important single chemical factor in the structural integrity of the vertebrate body. Other fibrous proteins also play important mechanical roles, sometimes passive, sometimes active. One of the most tantalizing problems in biology is how organization at a certain level grows out of the properties of units at the level below. For this reason, collagen is an excellent subject for case study; relatively large fibers with perfectly normal properties can be reconstituted from solutions of the protein, and interconverted with aberrant forms if the conditions of the experiment are changed. Appropriately, the story ends with a developmental question: how are the natural fibers oriented during development into the highly structured arrays one sees in connective tissue?

HOW CELLS MAKE MOLECULES

by VINCENT G. ALLFREY and ALFRED E. MIRSKY
September 1961

In our laboratory we have the portraits of Gregor Mendel and Friedrich Miescher side by side. Mendel in 1866 set forth evidence, from his observation of inheritance in the pea plant, for the idea that genetic information is carried in discrete units from one generation to the next. Miescher in 1869 isolated from the nucleus of cells a substance that he called nuclein and that is known today as deoxyribonucleic acid (DNA). He knew that he had in his hands a novel substance containing nitrogen and phosphorus, and he was well aware of its location within the nucleus. But he could have had no idea of what DNA does in the nucleus, because the role of the nucleus in heredity was at that time unknown, even to Mendel. It was well over half a century, long after the death of both men, before their work could be fused. The fusion required much more nucleic-acid and protein chemistry than there was in Miescher's time and a vast amount of new biology, including the unearthing of Mendel's work in 1900.

In the last three decades of the 19th century, work in biology, led by August Weismann, demonstrated the continuity of the germ plasm and showed that the nucleus of the cell plays a central role in heredity. Attention soon focused on the chromosomes. Since sperm and egg nuclei provide equal complements of chromosomes (except for the sex chromosomes) and since there is precisely equal cleavage and distribution of chromosomes at cell division, it seemed clear that chromosomes are concerned with the continuity that is essential in heredity. With the rediscovery of Mendel and the growth of genetics, biologists talked less about the germ plasm and more about the genes as discrete units of the germinal material. It became increasingly clear that each gene is derived from a pre-existing gene. The genes were located in the chromosomes, which now far more conclusively than before were shown to contain the materials determining heredity.

Meanwhile the chemistry of nucleic acids was making progress. But there was little contact between the two movements. The now standard color test for DNA was first demonstrated in 1914 by the German chemist Robert Feulgen in a test tube. Not until 10 years later did Feulgen use the test to stain cells and show that the chromosomes are the locus of DNA concentration in the nucleus. Yet it cannot be said that this experience led to the idea that DNA is the essential gene material.

The evidence that genetic information is carried by DNA came in the late 1940's from a number of sources. André Boivin and Roger and Colette Vendrely of the University of Strasbourg and Alfred E. Mirsky and Hans Ris of the Rockefeller Institute measured the DNA content of nuclei in germ cells and various somatic cells and found that in a given organism the DNA content is constant per set of chromosomes. This constancy pointed to DNA as the essential material of the genes. In experiments on pneumococci Oswald T. Avery, Colin M. MacLeod and Maclyn McCarty of the Rockefeller Institute showed that hereditary traits can be transmitted from one strain of bacteria to another by transferring to cells of the latter DNA extracted from the former. Their experiments conclusively established DNA as the carrier of genetic information.

This development made a great impression on geneticists, for it went a long way toward answering one of their outstanding questions: What is the nature of the gene? The impact on biochemists was far greater; it revealed to them what the problems of biochemistry could be. The principle of genetic continuity, the rule that each gene comes from a pre-existing gene, was now transformed to the biochemical question: How does the molecule of DNA replicate itself? There was also the problem of the passage of genetic information from DNA in the chromosome to the fabric of the cell. As the carrier of genetic information, it was plain, DNA does much more than replicate itself. It plays an active role, directing the life of the cell. Biochemists could now study the molecular basis of gene action, tracing the effect of DNA in discrete observable events in the synthesis of the molecules of which the cell is made.

The fusion of biochemistry and cell biology has brought an accelerating growth of understanding over the past 20 years. It is now possible to answer at the molecular level some of the fundamental questions of genetics and cell biology that go back to the era of Weismann. Some of the most significant knowledge of DNA activity has come in recent months. It is difficult to believe that the discovery of this substance goes back to 1868!

When the chromosomes are coiled up tightly in stumpy rods, the DNA they contain is the repository of genetic information, but it is inert. When DNA is actively communicating its information to the cell, the chromosomes have an entirely different appearance. Then, in their "lampbrush" configuration, the chromosomes uncoil into delicate filaments forming a lacelike structure in which DNA and other components are readily accessible for interaction with the surrounding medium [*see illustration on page 146*].

Chemists have shown that the molecule of DNA consists of a long, un-

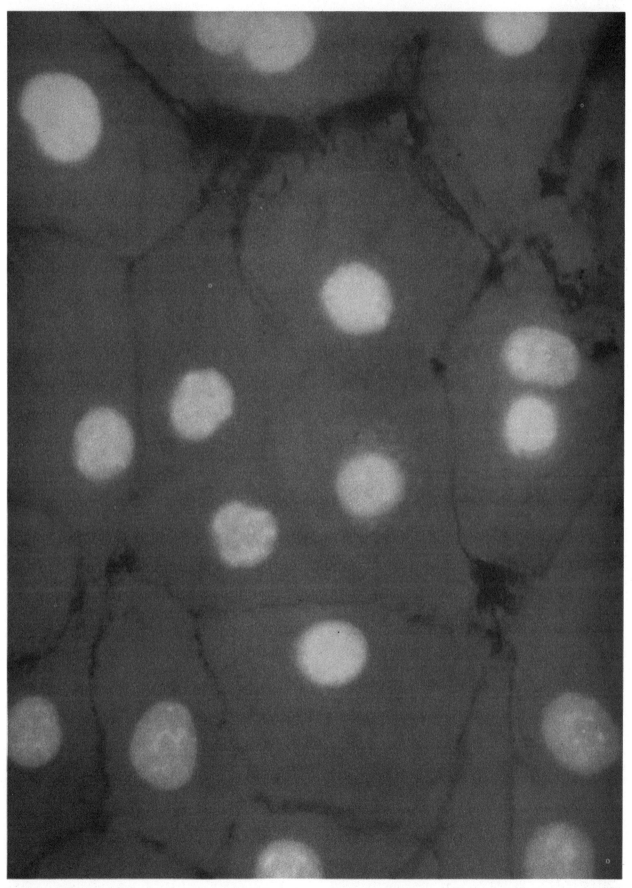

NUCLEIC ACIDS, the "blueprints" and "templates" of protein synthesis, are visualized in this photomicrograph of human amnion cells made by Suydam Osterhout at the Rockefeller Institute. Stained with acridine orange, the deoxyribonucleic acid (DNA) fluoresces yellow-green and is seen to be localized in the nuclei. Ribonucleic acid (RNA) fluoresces orange-red and is seen throughout the cytoplasm. There is RNA in the nucleus also, but it is obscured by the DNA. The magnification is 3,500 diameters.

branched chain, the backbone of which is made up of alternate five-carbon sugar (deoxyribose) and phosphate groups. To each sugar is attached a nitrogenous base; in most DNA's there are four such bases: adenine, guanine, thymine and cytosine. The unit in the chain, consisting of phosphate-sugar-base, is called a nucleotide. From X-ray crystallography, particularly the work of F. H. C. Crick and J. D. Watson, came the understanding that a DNA molecule consists not of a single polynucleotide chain but of two, twined around each other in a double helix and held together by hydrogen bonds between the bases. The companion bases are never identical but are always specifically complementary, with adenine joined to thymine and guanine to cytosine. This is demonstrated by experiments with simple, synthetic polynucleotides. Thus a synthetic DNA, or polynucleotide, made up exclusively of thymine bases (polythymidylic acid) binds a complementary chain made up exclusively of adenine bases (polyadenylic acid). In chains made up of all four bases, as in the natural molecule, the sequence of bases in one chain governs the sequence in the other; if the actual order of bases in one chain of a DNA were known, one could write down the order of its complementary chain. To complete the picture, Crick and Watson proposed that the genetic information carried by the molecule is encoded in the sequence of its bases.

The usefulness of this work to the classical concerns of genetics began to be demonstrated in 1957 when Arthur L. Kornberg, then at Washington University in St. Louis, brought about the synthesis of DNA in a cell-free system containing the four nucleotides, the enzyme polymerase and DNA. In the presence of polymerase the nucleotides linked together to form long chains of DNA. More important, Kornberg found that the polymerization of the four nucleotides will proceed only if a small amount of DNA is present to "prime" the reaction. He soon found that the over-all proportions of the bases in the product DNA paralleled the base composition of the primer used. Early this year Kornberg made the point more strongly. As primers he used a number of different DNA's prepared from virus, bacterial and animal sources, each DNA having its own characteristic sequence of nucleotides. Using a statistical method (known as nearest-neighbor nucleotide frequencies) to analyze the products of these reactions, he found that each primer DNA directs the polymerization so that the sequence of the nucleotides in the enzymatically synthesized DNA is the same as its own.

According to the Watson-Crick model, when DNA primes the making of more DNA, the double helix uncoils. Then along each chain a complementary chain is formed. In Kornberg's experiments the pattern of nearest-neighbor frequencies in every case showed the pairing of adenine to thymine and guanine to cytosine, just as in the model. A geneticist can now see how it is that each gene is derived from a pre-existing gene.

It was of course the more general activity of the genes—by which they bring about the expression of hereditary traits in the organism—that led to their discovery. The way to an understanding

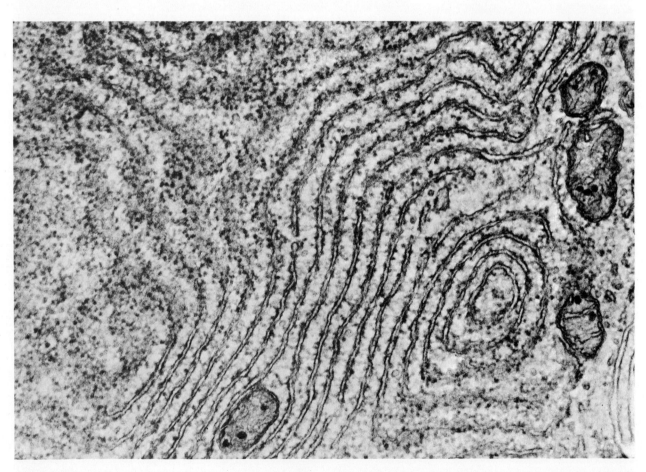

RIBOSOMES, the sites of protein synthesis, are shown in this electron micrograph made by Bernard Tandler of the Sloan-Kettering Institute. The cytoplasmic membrane system of a human submaxillary gland cell is enlarged 70,000 times. The ribosomes are the small, dark particles, each about .000015 millimeter in diameter, lining the membranes. The large oblong bodies are mitochondria.

of this activity at the molecular level—how DNA governs the biosynthetic processes of the cell—has become clear only in the last 10 years.

The heritable changes that geneticists first studied were necessarily those most readily observed. Many of these were changes in color. One of the early Mendelians was the English physician Archibald Garrod, who made a penetrating study of a rare hereditary disease in man characterized by the appearance of a black pigment, alcapton, in the urine. The pigment is formed because there is a derangement in the metabolism of the amino acid tyrosine; one particular reaction that normally occurs fails to occur. This failure, Garrod perceived as far back as 1909, is due to the absence of an enzyme that is normally present. The role of the normal gene, therefore, is to determine the production of a particular enzyme, and this is what the abnormal gene fails to do.

The idea that the action of a gene is concerned with the formation of a particular enzyme was ignored by most geneticists for some 30 years. It was revived by George W. Beadle and Edward L. Tatum when they showed the principle at work in heritable metabolic derangements of the red bread mold *Neurospora*. This time the idea made a deep impression on geneticists and biochemists, in part because the chemical nature of enzymes had meanwhile been revealed. Between 1926 and 1930 James B. Sumner of Cornell University and John H. Northrop of the Rockefeller Institute had shown that enzymes are proteins. The biochemical function of the gene was now to be looked for in the more general function of protein synthesis. It was firmly demonstrated in the early 1950's, when other examples of the determination of protein structure by genes were found in many animals, fungi and bacteria.

But the decisive new experiments were again, as in Garrod's time, on man—on human hemoglobin. Beginning with the hemoglobin of sickle-cell anemia, studied by Linus Pauling and his colleagues at the California Institute of Technology, quite a number of hereditary hemoglobin anomalies were discovered in an enterprising world-wide search. In several cases it was found that a gene mutation produces a single amino acid substitution in one location of the peptide chains, containing about 150 amino acids, that make up the hemoglobin molecule. So much precise information has by now been acquired linking genes and the amino acid composition in hemoglobins that there is no longer any doubt that genes determine protein synthesis.

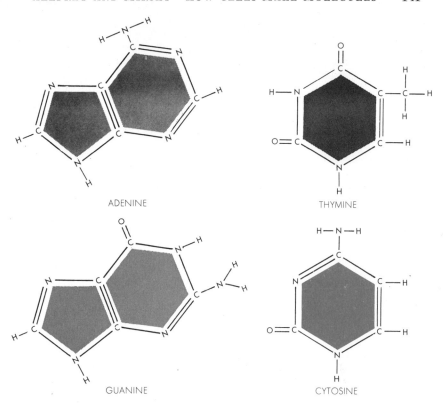

DNA, the carrier of genetic information that serves as the "blueprint" for protein synthesis, has as its key components four nitrogenous bases. Their formulas are diagramed above and their arrangement in the complex DNA molecule is illustrated on the following two pages.

One of the best examples of the control of protein synthesis by the gene concerns the synthesis of DNA, the substance of the gene itself. When a bacterial virus that contains DNA enters a bacterial cell, a large quantity of virus DNA is soon synthesized. At the same time there is a five-to-ten-fold increase in activity of the enzyme polymerase, the enzyme that polymerizes mononucleotides and so produces DNA. But the polymerase in this case is the virus-induced polymerase, and it is distinctly different from the polymerase present in the bacterium before entry of the virus. This is to be expected since the genetic information for the new enzyme comes from virus DNA and not from the bacterial genes.

The understanding of how DNA directs protein synthesis in the cell involves among other things a vast lore of knowledge about protein synthesis and structure gained without reference to the cell. But the primary action of DNA also raises a logical problem that should be dealt with at the outset. Only four different nucleotides make up the long chains of DNA. Some 20 amino acids make up the long polypeptide chains of proteins. The genetic information in DNA is therefore spelled out in a four-letter alphabet. But the information in this molecular script is communicated to another in which the message must be translated into a 20-letter alphabet, the letters of which are entirely different from those in the four-letter alphabet. How is the information encoded and conveyed? The most plausible answer is that groups of three or four nucleotides are arranged in different sequences, each corresponding to a particular amino acid. By such an arrangement the four-letter DNA alphabet is able to determine the spelling out of protein structure in the 20-letter alphabet.

All cells synthesize protein, some continuously, others for only a part of their life cycle. The proteins they make are enormously varied, differing in size, shape, over-all chemical composition and physical properties. But whatever their function, and regardless of their size, shape, solubility or enzyme activity, all proteins have an underlying similarity in constitution: they are all made up of the relatively simple molecular units of amino acid. The synthesis of a protein from these smaller units is conceptually a simple process, involving the joining of the individual amino acids to form long chains. The length of the chain and the sequence of the amino acids vary, of course, from one protein to another. But the essential unit of structure, the link that prolongs the chain, is ubiquitous; it is the peptide bond, the chemical union between the carboxyl group (COOH) of one amino acid and the amino group (NH_2) of the next amino acid in the chain.

These bonds, connecting amino acids in various sequences, are the key linkages to be created in carrying out the synthesis of proteins or smaller polypeptides, either in the cell or in the laboratory. Since peptide bonds do not form spontaneously when amino acids are mixed, other chemical means have to be used to drive the formation of the bonds. The synthetic system of the cell begins with a reaction that "activates" the carboxyl group of the amino acids.

This reaction, originally discovered by Mahlon B. Hoagland of the Harvard Medical School in 1955, derives the energy necessary for the activation from adenosine triphosphate (ATP), the main energy currency of the cell [see "How Cells Transform Energy," on page 85]. The ATP is cleaved to release two of its three phosphate groups; the remaining fragment, adenosine monophosphate (AMP), joins up with the acid group of the amino acid. In this way the amino acid is potentiated for the formation of the peptide bond. The enzymes that carry out this activation have great specificity: in general they react with only one type of amino acid. It is probable that most cells have activating enzymes for at least 20 amino acids. In animal cells these enzymes have been found in the nucleus as well as in the cytoplasm,

and there is good evidence that they play a role in the synthesis of the nuclear proteins, including the proteins of the chromosomes.

The activating enzymes mediate only the first step in a very complex and precisely ordered chain of reactions. The order is supplied by the information encoded in the DNA molecule. To convey the information from the DNA in the nucleus to the site of most protein synthesis in the cytoplasm, the cell employs another polynucleotide—ribonucleic acid (RNA)—as an intermediary. In RNA the sugar is ribose instead of deoxyribose; the main RNA bases are adenine, guanine, cytosine and, instead of thymine, uracil. As in DNA, the different nucleotides are linked together through their phosphate groups to form long chains.

The need for ribonucleic acid in protein synthesis was suggested 20 years ago, when Torbjörn O. Caspersson in Stockholm and Jean Brachet in Brussels showed that tissues that synthesize large amounts of protein, whether for growth or multiplication, are always rich in RNA. One of the highest RNA concentrations is found in the cells of the spinning gland of silkworms, which produce the proteins fibroin and sericin of the silk thread. In mammals, high RNA concentrations occur in such specialized cells as those of the pancreas and liver, in which many proteins are synthesized for transport to other parts of the organism.

When cells synthesize protein on this scale, it is usually observed that the synthetic machinery is highly organized into a lamellar network of membranes and RNA-rich particles. The biochem-

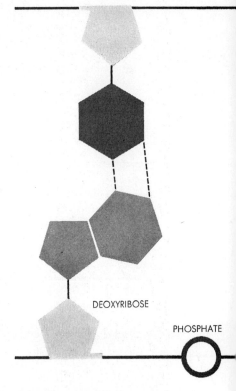

DNA MOLECULE is in the form of a long double chain of nucleotides—phosphate-linked deoxyribose sugar groups to each of

ical knowledge of these structures comes largely from studies of cell homogenates, from which different intracellular structures can be isolated by centrifuging at different speeds; that knowledge is now significantly supplemented by electron micrographs of the cytoplasmic membrane system [*see illustration on page 140*]. Particular interest attaches to the particles, which, in a liver cell, are

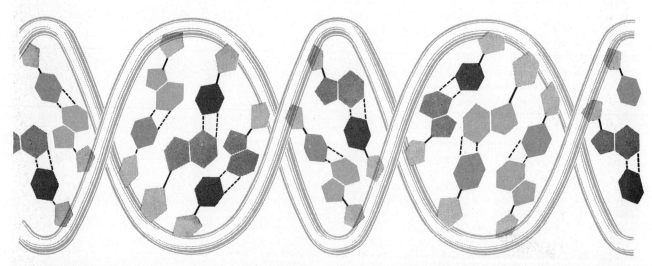

DNA MOLECULE, shown at the top of these pages as a straight ladder, is actually twisted into a double helix, according to the generally accepted Watson-Crick model. In this drawing the phosphate groups are not shown. The sugar and base molecules are shown diagrammatically; in the actual three-dimensional model the base pairs that make up the crosslinks all lie in parallel planes.

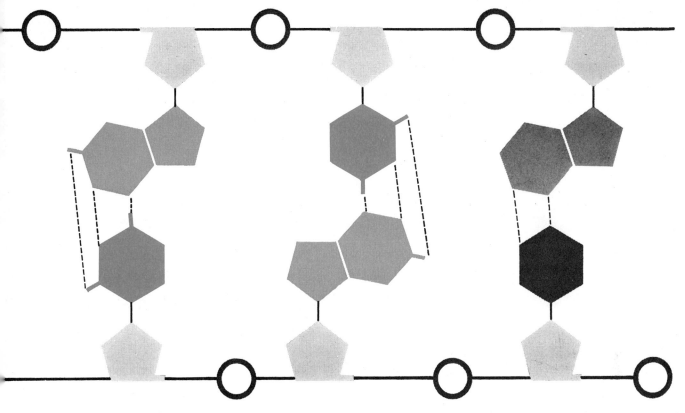

which one of the four bases is attached as a side group. Hydrogen bonds (*broken lines*) link pairs of bases to form the double chain. The bases are always paired as shown: adenine with thymine and guanine with cytosine. The sequence of pairs, however, can be varied infinitely. The sequence encodes the information that determines what kinds of protein shall be synthesized by the cell.

roughly spherical and about .000015 millimeter in diameter. In fully differentiated cells of the pancreas or liver, most of these particles are attached to membrane surfaces, but in embryonic tissues the particles appear free in the cytoplasm. George E. Palade of the Rockefeller Institute has suggested that the particles, not the membranes, are the primary sites of protein synthesis in these cytoplasmic systems. This view has been proved correct by biochemical studies of the particles themselves, isolated not only from pancreas and liver but also from tumor cells, plants, yeasts and bacteria. Most workers in the field now refer to the particles as ribosomes.

Whatever cell they come from, ribosomes are extraordinarily rich in RNA. Bacterial ribosomes, for example, have a molecular weight of nearly three million, of which at least 60 per cent is RNA. Ribosomes prepared from liver or yeast have a molecular weight of about four million, of which more than 40 per cent is RNA. Few purified ribosome preparations contain less RNA than this. Protein, often rich in basic amino acids, constitutes most of the remaining mass of the isolated ribosomes.

The first chemical indication that ribonucleoproteins play such a direct role in the synthesis of other proteins came from "tracer" experiments in which living animals were given isotopically labeled amino acids (that is, amino acids containing atoms of nitrogen 15 or carbon 14 instead of the nitrogen 14 or carbon 12 atoms usually present). Henry Borsook of the California Institute of Technology and Tore J. M. Hultin of the Wenner-Gren Institute in Stockholm in 1950 independently showed that when the cells of a labeled tissue were broken and fractionated, the highest concentration of labeled amino acids showed up in the microsome fraction. This fraction was subsequently shown by Philip Siekevitz and Palade to contain the ribosomal particles attached to membrane fragments. In 1953 Marie Maynard Daly and the authors made careful kinetic studies of the rate of nitrogen-15 amino acid uptake into different protein fractions of the pancreas cell. The results made it very likely that some of the protein attached to ribonucleic acid was a direct precursor of the enzyme proteins that are found free in the cell. Moreover, we found that an attack on the RNA by a specific enzyme, ribonuclease, stopped protein synthesis in isolated subcellular fractions. The experiments made on whole animals and in isolated cells were soon supplemented by studies of cell-free systems containing RNA, which would incorporate amino acids into protein in the test tube. The development of these systems was largely due to the experiments of Paul C. Zamecnik and his colleagues at the Massachusetts General Hospital.

Although many experiments suggested a direct role for RNA in protein synthesis, it has only recently been shown that the function of RNA is to supply the information necessary to organize the sequence of the amino acids in peptide chains. The argument is clinched by experiments in which a modification of the RNA has brought a change in the protein product. Working with a plant virus, Gerhard Schramm and his co-workers at the University of Tübingen succeeded in substituting a hydroxyl for an amino group in the viral RNA; the substitution led to the formation of a different viral protein. By growing bacteria in the presence of 5-fluoro-uracil, François Gros of the Pasteur Institute in Paris has caused this substance to replace the normal base uracil in the bacterial RNA; the bacteria thereupon synthesized an abnormal protein instead of the enzyme *beta*-galactosidase.

From all that is known, it is now supposed that RNA's in the ribosomes act as templates that determine the se-

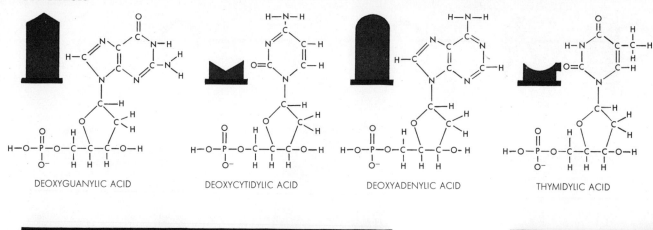

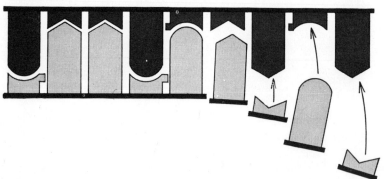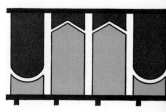

"BASE PAIRING" is the process by which DNA is replicated in cell division and by which it makes RNA, the closely related nucleic acid that in turn synthesizes proteins. The four nucleotides of DNA are symbolized (*upper left*) by four building blocks shaped as complementary pairs. RNA (*upper right*) differs only slightly: its sugar has an extra oxygen atom, and uridylic acid replaces DNA's thymidylic acid. The lower diagrams show how DNA can either replicate or form RNA carrying the same genetic information.

quence in which amino acids are linked together in protein chains. They reproduce in their nucleotide sequences the information encoded in the master templates of the DNA molecules in the cell nucleus.

But how do amino acids get to the RNA templates from their activating enzymes? Here too ribonucleic acids play a role. In 1957 several laboratories announced the discovery of low-molecular-weight RNA's that transfer activated amino acids to ribosomes. The function of these "transfer" RNA's is to get amino acids properly lined up on the RNA template of the ribosome. This view, put forward by Crick and others, assumes that there are 20 or so transfer RNA's, each specific for a particular amino acid and also capable of recognizing certain specific sites on the template. It is, of course, the nucleotide sequence in the shorter chains of these low-molecular-weight RNA's that capacitates them for

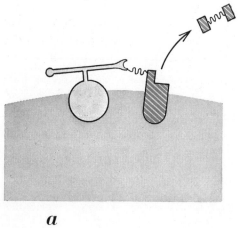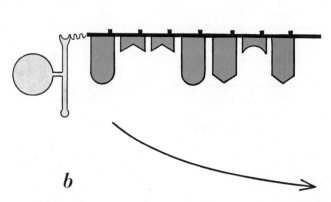

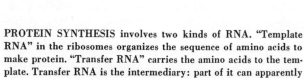

PROTEIN SYNTHESIS involves two kinds of RNA. "Template RNA" in the ribosomes organizes the sequence of amino acids to make protein. "Transfer RNA" carries the amino acids to the template. Transfer RNA is the intermediary: part of it can apparently recognize a specific amino acid and part is coded to seek the proper site on the template. The process begins (*a*) with activation of an amino acid by adenosine triphosphate (ATP), the cellular energy carrier (*hatched*). Two phosphate groups drop from ATP and, with

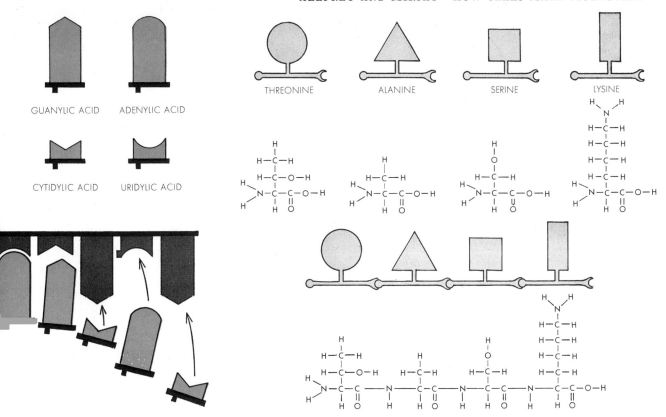

A single strand of DNA (*dark gray*) assembles new DNA nucleotides (*light gray*) or RNA nucleotides (*color*) that match its complementary strand.

AMINO ACIDS are the constituents of proteins. Here four amino acids (of 20-odd that are known) are represented by building blocks. In the top row the symbols and formulas of the four are shown separately; in the bottom row they have been linked by peptide bonds, in which H_2O is dropped from adjacent COOH and NH_2 groups to form a fragment of protein.

both of these highly specific reactions. According to this scheme the amino acids are transferred upon activation to the appropriate transfer-RNA chain. The transfer-RNA molecule then combines with its complementary sequence of nucleotides in the template RNA. This transfer reaction is known to be mediated by an enzyme requiring guanosine triphosphate. Other transfer RNA's, carrying other amino acids, take their appropriate places on the template, and the amino acids are now aligned and held in proper sequence to form the specific polypeptide chain of the protein.

Though this picture of transfer RNA as an adapter molecule is still tentative, something is known about the coupling of amino acids in the ribosome. Recent work on hemoglobin synthesis, by Richard Schweet of the University of Kentucky Medical School and by Howard M. Dintzis at the Massachusetts Institute of Technology, indicates that

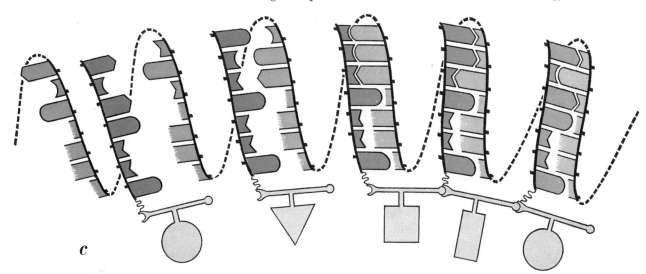

the help of an enzyme (*gray*), the amino acid is attached to the remaining adenylic acid by a high-energy bond (*wavy line*). Then a transfer RNA molecule moves in (*dark-colored symbol at "b"*). Each transfer RNA has an adenylic acid at one end; that end takes over the bond to the amino acid. Finally, the transfer RNA carries the selected amino acid to a ribosome (*c*). There the coded section of the transfer RNA finds its place on the template (*light color*) and positions the amino acid to join with others in a protein chain.

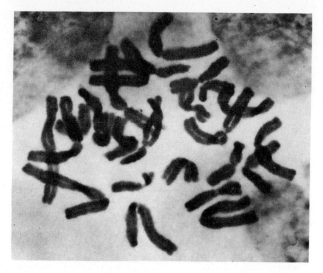
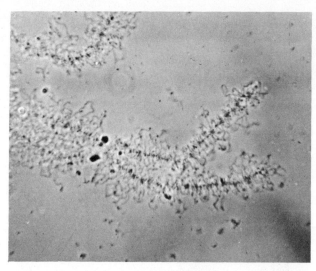

DNA IN CHROMOSOMES is an inert repository of genetic information when the chromosomes are tightly coiled rods during mitosis (*left*). Between cell divisions, when DNA is doing its genetic work, the chromosomes are greatly extended, as in the "lampbrush" configuration (*right*). These photomicrographs of newt chromosomes were taken by H. G. Callan of St. Andrews University.

peptide-bond formation proceeds much like a zipper, starting with the amino acid valine at one end of the chain and closing bond after bond until the protein molecule is finished. New amino acids are added to the growing chain at the rate of about two per second, finishing the protein molecule (of 150 amino acids) in 1.5 minutes. This impressive feat, the synthesis of a finished protein molecule in less than two minutes, is testimony to the efficiency of the protein-synthetic mechanism of the cell.

One question now remains: How is the information encoded in the master template, DNA, transmitted to ribonucleic acids? The most suggestive clues come from experiments on enzyme systems that synthesize RNA. Samuel Weiss at the University of Chicago and Jerard Hurwitz at New York University have described enzyme systems isolated from animal cells and bacteria that utilize all four nucleotides (as triphosphates) for RNA synthesis. Although the nucleic acid being synthesized is RNA, DNA must be present for synthesis to occur. What is more remarkable, the nucleotide composition of the DNA determines what kind of RNA will be formed. The base-pairing rules of the Watson-Crick DNA model appear to apply with equal rigor to the formation of these DNA-RNA hybrids. DNA templates induce the synthesis of complementary RNA molecules only.

This control over RNA synthesis by DNA is also seen in living cells. Elliot Volkin of the Oak Ridge National Laboratory found in 1958 that when bacterial viruses infect bacteria, an RNA is formed that resembles the virus DNA and not that of the host in its base composition. Benjamin D. Hall and Sol Spiegelman of the University of Illinois then showed that the sequence of the nucleotides in the new RNA molecule is complementary to that of the RNA of the virus.

So the story comes full circle. Specific genetic information resides in the nucleotide sequences in DNA. By means of base-pairing mechanisms these sequences are copied to produce either new DNA molecules for new cells or the RNA templates needed for protein synthesis. Specific nucleotide sequences in the ribosome templates encode the amino acid sequence for particular proteins. The transfer RNA's recognize these sequences and bring amino acids into the proper alignment. Peptide bonds then form with great specificity and rapidity, putting together the protein molecules characteristic of the species. These proteins, many of them enzymes, are the tools with which the cell synthesizes the host of other molecules (purines, pyrimidines, amino acids, carbohydrates, fats, sterols, pigments and so on) necessary to its structure and function.

What has been said so far makes it clear that there is a transmission of genetic information from the DNA in the chromosomes to the sites of protein synthesis. Textbooks often show this flow as an arrow leading from the nucleus to the cytoplasm. There is no corresponding arrow from the cytoplasm back to the nucleus. Absence of the return arrow might suggest to a biologist that there is something essential missing in the scheme, for in all biological systems that have been carefully studied there is a feedback. In the cell there is indeed evidence for a feedback control directed from the cytoplasm to the chromosomes. In some cases the feedback comes quickly and lasts for only a short time. In the pancreas, for example, when the cells are stimulated so that their cytoplasm synthesizes digestive enzymes, tracer experiments show that within a few minutes there is a rise in the uptake of amino acids into proteins of the chromosomes. There are also less immediate and more enduring cytoplasmic influences on chromosomes. Among these are the profound changes associated with cell differentiation [see "How Cells Specialize," page 222].

But although protein synthesis in chromosomes has been shown to be subject to feedback control, there is at present no evidence that the sequence of bases in DNA can be altered by feedback. The chromosomes of germ cells have changed in the course of evolution so that they carry genetic information that is effective in adapting an organism to its environment. The DNA of a germ cell has been shaped by evolution so that it can determine the synthesis of enzymes and other proteins that make for a viable organism. The important point here is that the changes that have taken place during the course of evolution in the DNA molecules of the germ cells of an organism are not the direct result of a feedback from the cytoplasm to the chromosomes in the nucleus. Changes in DNA itself, according to the generally held views of biologists today, are due to mutation and selection.

13

THE GENETIC CODE: II

by MARSHALL W. NIRENBERG   March 1963

Just 10 years ago James D. Watson and Francis H. C. Crick proposed the now familiar model for the structure of DNA (deoxyribonucleic acid), for which they, together with Maurice H. F. Wilkins, received a Nobel prize last year. DNA is the giant helical molecule that embodies the genetic code of all living organisms. In the October 1962 issue of *Scientific American* Crick described the general nature of this code. By ingenious experiments with bacterial viruses he and his colleagues established that the "letters" in the code are read off in simple sequence and that "words" in the code most probably consist of groups of three letters. The code letters in the DNA molecule are the four bases, or chemical subunits, adenine, guanine, cytosine and thymine, respectively denoted A, G, C and T.

This article describes how various combinations of these bases, or code letters, provide the specific biochemical information used by the cell in the construction of proteins: giant molecules assembled from 20 common kinds of amino acids. Each amino acid subunit is directed to its proper site in the protein chain by a sequence of code letters in the DNA molecule (or molecules) that each organism inherits from its ancestors. It is this DNA that is shaped by evolution. Organisms compete with each other for survival; occasional random changes in their information content, carried by DNA, are sometimes advantageous in this competition. In this way organisms slowly become enriched with instructions facilitating their survival.

The exact number of proteins required for the functioning of a typical living cell is not known, but it runs to many hundreds. The great majority, if not all, of the proteins act as enzymes, or biological catalysts, which direct the hundreds of different chemical reactions that go on simultaneously within each cell. A typical protein is a molecular chain containing about 200 amino acid subunits linked together in a specific sequence. Each protein usually contains all or most of the 20 different kinds of amino acids. The code for each protein is carried by a single gene, which in turn is a particular region on the linear DNA molecule. To describe a protein containing 200 amino acid subunits a gene must contain at least 200 code words, represented by a sequence of perhaps 600 bases. No one yet knows the complete base sequence for a single gene. Viruses, the smallest structures containing the blueprints for their own replication, may contain from a few to several hundred genes. Bacteria may contain 1,000 genes; a human cell may contain a million. The human genes are not strung together in

EXPERIMENT BEGINS when cells of the colon bacillus are ground in a mortar with finely divided aluminum oxide. "Sap" released from ruptured cells still synthesizes protein.

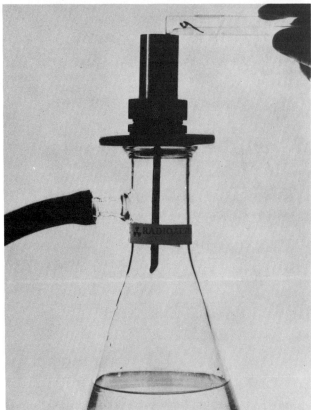
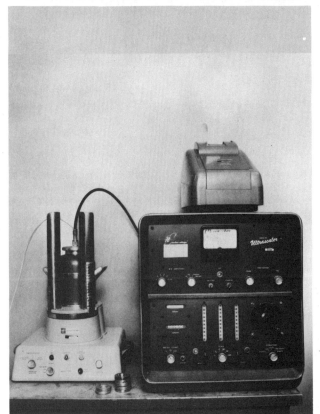

STEPS IN CODE BREAKING are shown in this sequence of photographs taken in the author's laboratory at the National Institutes of Health in Bethesda, Md. The open test tubes at upper left contain samples of the cell-free bacterial system capable of synthesizing protein when properly stimulated. The photograph shows stimulants being added. They include synthetic "messenger RNA" (ribonucleic acid) and amino acids, one of which is radioactive. The protein is produced when the samples are incubated 10 to 90 minutes. At upper right the protein is precipitated by the addition of trichloroacetic acid (TCA). At lower left the precipitate is transferred to filter-paper disks, which will be placed in carriers called planchettes. At lower right the planchettes are stacked in a radiation counting unit. Radiation measurement indicates how well a given sample of messenger RNA has directed amino acids into protein.

150 SYNTHESIS

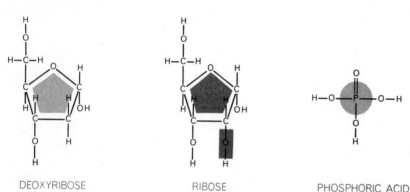

COMPONENTS OF DNA (deoxyribonucleic acid) are four bases adenine, guanine, thymine and cytosine (symbolized A, G, T, C), which act as code letters. Other components, deoxyribose and phosphoric acid, form chains to which bases attach (*see below*). In closely related RNA, uracil (U) replaces thymine and ribose replaces deoxyribose.

DNA STRUCTURE

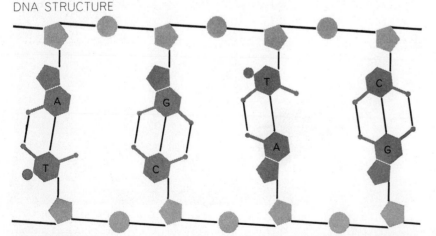

DNA MOLECULE resembles a chain ladder (actually twisted into a helix) in which pairs of bases join two linear chains constructed from deoxyribose and phosphate subunits. The bases invariably pair so that A links to T and G to C. The genetic code is the sequence of bases as read down one side of the ladder. The deoxyribose-phosphate linkages in the two linear chains run in opposite directions. DNA molecules contain thousands of base pairs.

one long chain but must be divided among at least 46 DNA molecules. The minimum number is set by the number of human chromosomes (46), which collectively carry the hereditary material. In fact, each chromosome apparently carries not one or two but several copies of the same genetic message. If it were possible to assemble the DNA in a single human cell into one continuous thread, it would be about a yard long. This three-foot set of instructions for each individual is produced by the fusion of egg and sperm at conception and must be precisely replicated billions of times as the embryo develops.

The bottom illustration at left shows how the bases in DNA form the cross links connecting two helical strands composed of alternating units of deoxyribose (a simple sugar) and phosphate. The bases are attached to the sugar units and always occur in complementary pairs: A joined to T, and G joined to C. As a result one strand of the DNA molecule, with its associated bases, can serve as the template for creating a second strand that has a complementary set of bases. The faithful replication of genes during cell division evidently depends on such a copying mechanism.

The coding problem centers around the question: How can a four-letter alphabet (the bases A, G, C and T) specify a 20-word dictionary corresponding to the 20 amino acids? In 1954 the theoretical physicist George Gamow, now at the University of Colorado, pointed out that the code words in such a dictionary would have to contain at least three bases. It is obvious that only four code words can be formed if the words are only one letter in length. With two letters 4×4, or 16, code words can be formed. And with three letters $4 \times 4 \times 4$, or 64, code words become available—more than enough to handle the 20-word amino acid dictionary [*see top illustration on page 156*]. Subsequently many suggestions were made as to the nature of the genetic code, but extensive experimental knowledge of the code has been obtained only within the past 18 months.

The Genetic Messenger

It was recognized soon after the formulation of the Watson-Crick model of DNA that DNA itself might not be directly involved in the synthesis of protein, and that a template of RNA (ribonucleic acid) might be an intermediate in the process. Protein synthesis is conducted by cellular particles called ribosomes, which are about half protein and half RNA (ribosomal RNA). Several years ago Jacques Monod and François Jacob of the Pasteur Institute in Paris coined the term "messenger RNA" to describe the template RNA that carried genetic messages from DNA to the ribosomes.

A few years ago evidence for the enzymatic synthesis of RNA complementary to DNA was found by Jerard Hurwitz of the New York University School of Medicine, by Samuel Weiss of the University of Chicago, by Audrey Stevens of St. Louis University and their respective collaborators [see "Messenger RNA," by Jerard Hurwitz and J. J. Furth, Offprint #119, for more information]. These groups, and others, showed that an enzyme, RNA polymerase, catalyzes the synthesis of strands of RNA on the pattern of strands of DNA.

RNA is similar to DNA except that RNA contains the sugar ribose instead of deoxyribose and the base uracil instead of thymine. When RNA is being formed on a DNA template, uracil appears in the RNA chain wherever adenine appears at the complementary site on the DNA chain. One fraction of the RNA formed by this process is messenger RNA; it directs the synthesis of protein. Messenger RNA leaves the nucleus of the cell and attaches to the ribosomes. The sequence of bases in the messenger RNA specifies the amino acid sequence in the protein to be synthesized.

The amino acids are transported to the proper sites on the messenger RNA by still another form of RNA called transfer RNA. Each cell contains a specific activating enzyme that attaches a specific amino acid to its particular transfer RNA. Moreover, cells evidently contain more than one kind of transfer RNA capable of recognizing a given amino acid. The significance of this fact will become apparent later. Although direct recognition of messenger RNA code words by transfer RNA molecules has not been demonstrated, it is clear that these molecules perform at least part of the job of placing amino acids in the proper position in the protein chain. When the amino acids arrive at the proper site in the chain, they are linked to each other by enzymic processes that are only partly understood. The linking is accomplished by the formation of a peptide bond: a chemical bond created when a molecule of water is removed from two adjacent molecules of amino acid. The process requires a transfer enzyme, at least one other enzyme and a cofactor: guanosine triphosphate. It appears that amino acid subunits are bonded into the growing protein chain one at a time, starting at the end of the chain carrying an amino group (NH_2) and proceeding toward the end that terminates with a carboxyl group (COOH).

The process of protein synthesis can be studied conveniently in cell-free extracts of the colon bacillus (*Escherichia coli*). The bacteria grow rapidly in suitable nutrients and are harvested by sedimenting them out of suspension with a centrifuge. The cells are gently broken open by grinding them with finely powdered alumina [*see illustration on page 148*]; this releases the cell sap, containing DNA, messenger RNA, ribosomes, enzymes and other components. Such extracts are called cell-free systems, and when they are fortified with energy-rich substances (chiefly adenosine triphosphate), they readily incorporate amino acids into protein. The incorporation process can be followed by using amino acids containing carbon 14, a radioactive isotope of carbon.

Optimal conditions for protein synthesis in bacterial cell-free systems were determined by workers in many laboratories, notably Alfred Tissières of Harvard University, Marvin Lamborg and Paul C. Zamecnik of the Massachusetts General Hospital, G. David Novelli of the Oak Ridge National Laboratory and Sol Spiegelman of the University of Illinois. When we began our work at the National Institutes of Health, our

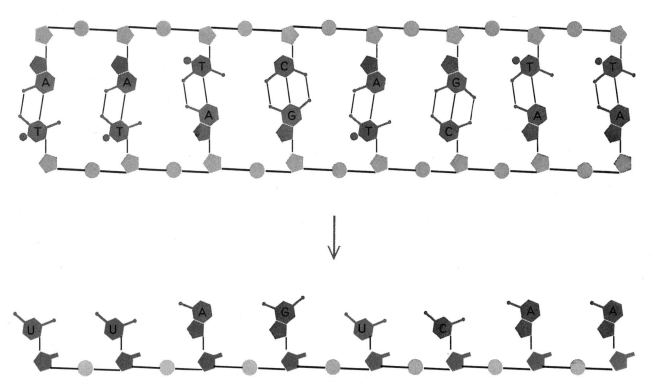

MESSENGER RNA is the molecular agent that transcribes the genetic code from DNA and carries it to the sites in the cell (the ribosomes) where protein synthesis takes place. The letters in messenger RNA are complementary to those in one strand of the DNA molecule. In this example UUAGUCAA is complementary to AATCAGTT. The exact mechanism of transcription is not known.

progress was slow because we had to prepare fresh enzyme extracts for each experiment. Later my colleague J. Heinrich Matthaei and I found a way to stabilize the extracts so that they could be stored for many weeks without appreciable loss of activity.

Normally the proteins produced in such extracts are those specified by the cell's own DNA. If one could establish the base sequence in one of the cell's genes—or part of a gene—and correlate it with the amino acid sequence in the protein coded by that gene, one would be able to translate the genetic code. Although the amino acid sequence is known for a number of proteins, no one has yet determined the base sequence of a gene, hence the correlation cannot be performed.

The study of cell-free protein syn-

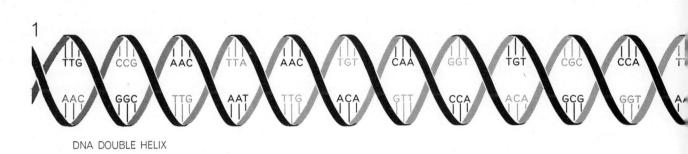

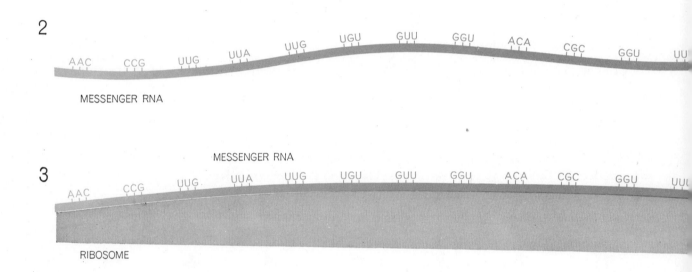

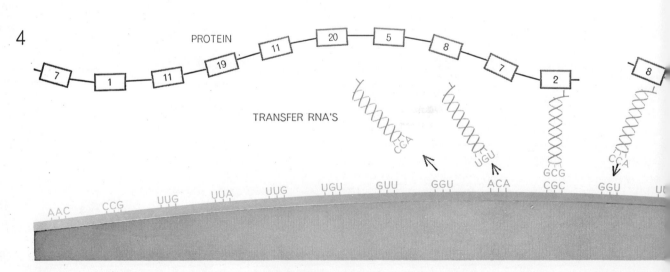

SYNTHESIS OF PROTEIN begins with the genetic code embodied in DNA (*1*). The code is transcribed into messenger RNA (*2*). In the diagram it is assumed that the message has been derived from the DNA strand bearing dark letters. The messenger RNA finds its way to a ribosome (*3*), the site of protein synthesis. Amino acids, indicated by numbered rectangles, are carried to proper sites on the messenger RNA by molecules of transfer RNA (*see illustration on opposite page*). Bases are actually equidistant, not

thesis provided an indirect approach to the coding problem. Tissières, Novelli and Bention Nisman, then at the Pasteur Institute, had reported that protein synthesis could be halted in cell-free extracts by adding deoxyribonuclease, or DNAase, an enzyme that specifically destroys DNA. Matthaei and I also observed this effect and studied its characteristics. It seemed probable that protein synthesis stopped after the messenger RNA had been depleted. When we added crude fractions of messenger RNA to such extracts, we found that they stimulated protein synthesis. The development of this cell-free assay for messenger RNA provided the rationale for all our subsequent work.

We obtained RNA fractions from various natural sources, including viruses, and found that many of them were highly active in directing protein synthesis in the cell-free system of the colon bacillus. The ribosomes of the colon bacillus were found to accept RNA "blueprints" obtained from foreign organisms, including viruses. It should be emphasized that only minute amounts of protein were synthesized in these experiments.

It occurred to us that synthetic RNA containing only one or two bases might direct the synthesis of simple proteins containing only a few amino acids. Synthetic RNA molecules can be prepared with the aid of an enzyme, polynucleotide phosphorylase, found in 1955 by Marianne Grunberg-Manago and Severo Ochoa of the New York University School of Medicine. Unlike RNA polymerase, this enzyme does not follow the pattern of DNA. Instead it forms RNA polymers by linking bases together in random order.

A synthetic RNA polymer containing only uracil (called polyuridylic acid, or poly-U) was prepared and added to the active cell-free system together with mixtures of the 20 amino acids. In each mixture one of the amino acids contained radioactive carbon 14; the other 19 amino acids were nonradioactive. In this way one could determine the particular amino acid directed into protein by poly-U.

It proved to be the amino acid phenylalanine. This provided evidence that the RNA code word for phenylalanine was a sequence of U's contained in poly-U. The code word for another amino acid, proline, was found to be a sequence of C's in polycytidylic acid, or poly-C. Thus a cell-free system capable of synthesizing protein under the direction of chemically defined preparations of RNA provided a simple means for translating the genetic code.

The Code-Word Dictionary

Ochoa and his collaborators and our group at the National Institutes of

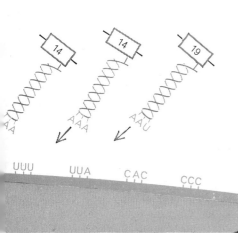

grouped in triplets, and mechanism of recognition between transfer RNA and messenger RNA is hypothetical. Linkage of amino acid subunits creates a protein molecule.

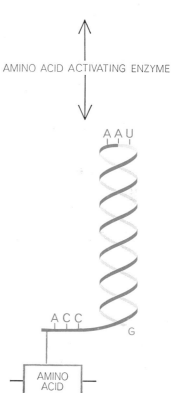

TRANSFER RNA is a special helical form of RNA that transports amino acids to their proper site in the protein chain. There is at least one transfer RNA for each of the 20 common amino acids. All, however, seem to carry the bases ACC where the amino acids attach and G at the opposite end. The attachment requires a specific enzyme and energy supplied by adenosine triphosphate. Unpaired bases in transfer RNA (AAU in the example) may provide the means by which the transfer RNA "recognizes" the place to deposit its amino acid package.

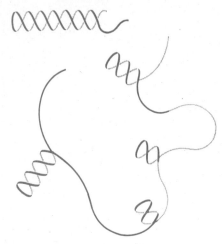

RNA STRUCTURE can take various forms. Transfer RNA (*top*) seems to be a fairly short double helix (probably less perfect than shown) that is closed at one end. Some RNA molecules contain a mixture of coiled and uncoiled regions (*bottom*).

Health, working independently, have now synthesized and tested polymers containing all possible combinations of the four RNA bases A, G, C and U. In the initial experiments only RNA polymers containing U were assayed, but recently many non-U polymers with high template activity have been found by M. Bretscher and Grunberg-Managó of the University of Cambridge, and also by Oliver W. Jones and me. All the results so far are summarized in the table at the bottom of pages 156 and 157. It lists the RNA polymers containing the minimum number of bases capable of stimulating protein formation. The inclusion of another base in a polymer usually enables it to code for additional amino acids.

With only two kinds of base it is possible to make six varieties of RNA polymer: poly-AC, poly-AG, poly-AU, poly-CG, poly-CU and poly-GU. If the ratio of the bases is adjusted with care, each variety can be shown to code with great specificity for different sets of amino acids. The relative amount of one amino acid directed into protein compared with another depends on the ratio of bases in the RNA. Assuming a random sequence of bases in the RNA, the theoretical probabilities of finding particular sequences of two, three or more bases can be calculated easily if the base ratio is known. For example, if poly-UC contains 70 per cent U and 30 per cent C, the probability of the occurrence of the triplet sequence UUU is $.7 \times .7 \times .7$, or .34. That is, 34 per cent of the triplets in the polymer are expected to be UUU. The probability of obtaining the sequence UUC is $.7 \times .7 \times .3$, or .147. Thus 14.7 per cent of the triplets in such a polymer are probably UUC. This type of calculation, however, assumes randomness, and it is not certain that all the actual polymers are truly random.

It had been predicted by Gamow, Crick and others that for each amino acid there might be more than one code word, since there are 64 possible triplets and only 20 amino acids. A code with multiple words for each object coded is termed degenerate. Our experiments show that the genetic code is indeed degenerate. Leucine, for example, is coded by RNA polymers containing U alone, or U and A, or U and C, or U and G.

It must be emphasized that degeneracy of this sort does not imply lack of specificity in the construction of proteins. It means, rather, that a specific amino acid can be directed to the proper site in a protein chain by more than one code word. Presumably this flexibility of coding is advantageous to the cell in ways not yet fully understood.

A molecular explanation of degeneracy has been provided recently in a striking manner. It has been known that some organisms contain more than one species of transfer RNA capable of recognizing a given amino acid. The colon bacillus, for example, contains two readily distinguishable species that transfer leucine. Bernard Weisblum and Seymour Benzer of Purdue University and Robert W. Holley of Cornell University separated the two leucine-transfer species and tested them in cell-free systems. They found that one of the species recognizes poly-UC but not poly-UG. The other species recognizes poly-UG but not poly-UC [*see top illustration on page 155*]. Although the number of transfer RNA species per cell is unknown, it is possible that each species corresponds to a different code word.

There is, however, the possibility of

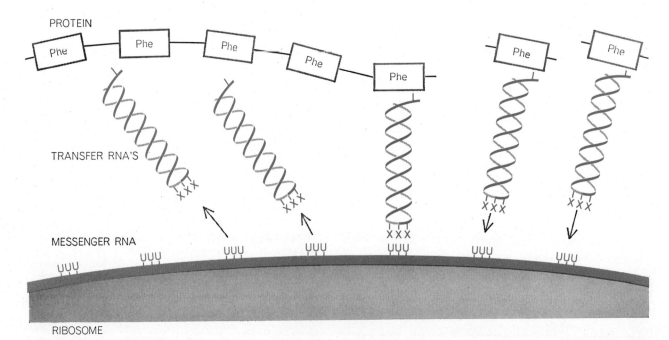

FIRST BREAK IN GENETIC CODE was the discovery that a synthetic messenger RNA containing only uracil (poly-U) directed the manufacture of a synthetic protein containing only one amino acid, phenylalanine (*Phe*). The finding was made by the author and J. Heinrich Matthaei. The X's in transfer RNA signify that the bases that respond to code words in messenger RNA are not known.

real ambiguity in protein synthesis. This would occur if one code word were to direct two or more kinds of amino acid into protein. So far only one such ambiguity has been found. Poly-U directs small amounts of leucine as well as phenylalanine into protein. The ratio of the two amino acids incorporated is about 20 or 30 molecules of phenylalanine to one of leucine. In the absence of phenylalanine, poly-U codes for leucine about half as well as it does for phenylalanine. The molecular basis of this ambiguity is not known. Nor is it known if the dual coding occurs in living systems as well as in cell-free systems.

Base sequences that do not encode for any amino acid are termed "nonsense words." This term may be misleading, for such sequences, if they exist, might have meaning to the cell. For example, they might indicate the beginning or end of a portion of the genetic message. An indirect estimate of the frequency of nonsense words can be obtained by comparing the efficiency of random RNA preparations with that of natural messenger RNA. We have found that many of the synthetic polymers containing four, three or two kinds of base are as efficient in stimulating protein synthesis as natural polymers are. This high efficiency, together with high coding specificity, suggests that relatively few base sequences are nonsense words.

In his recent article in *Scientific American* Crick presented arguments for believing that the coding ratio is either three or a multiple of three. Recently we have determined the relative amounts of different amino acids directed into protein by synthetic RNA preparations of known base ratios, and the evidence suggests that some code words almost surely contain three bases. Yet, as the table at the bottom of the next two pages shows, 18 of the 20 amino acids can be coded by words containing only two different bases. The exceptions are aspartic acid and methionine, which seem to require some combination of U, G and A. (Some uncertainty still exists about the code words for these amino acids, because even poly-UGA directs very little aspartic acid or methionine into protein.) If the entire code indeed consists of triplets, it is possible that correct coding is achieved, in some instances, when only two out of the three bases read are recognized. Such imperfect recognition might occur more often with synthetic RNA polymers containing only one or two bases than it does with natural messenger RNA, which always contains a mixture of all four. The results obtained with synthetic RNA may dem-

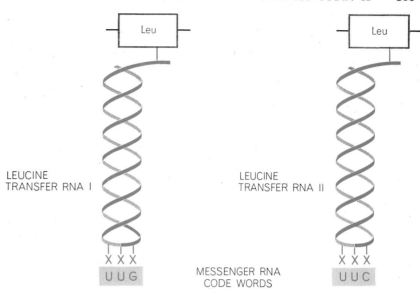

TWO KINDS OF TRANSFER RNA have been found, each capable of transporting leucine (*Leu*). One kind (*left*) recognizes the code word UUG; the other (*right*) recognizes UUC.

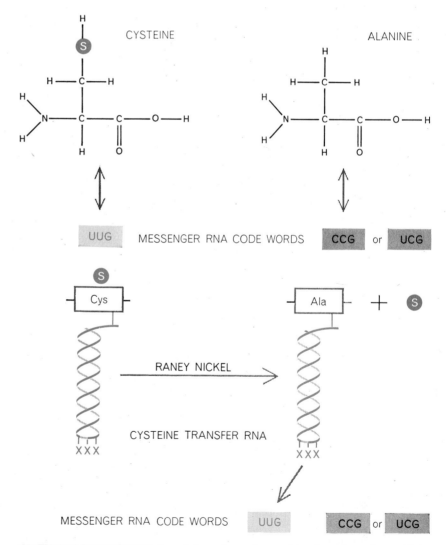

INGENIOUS EXPERIMENT showed that code-word recognition depends on the specificity of transfer RNA, not on the structure of the amino acid being transported. Cysteine is coded by UUG, alanine by CCG or UCG. Cysteine was hooked to its specific transfer RNA and sulfur was removed by a catalyst (Raney nickel). With sulfur removed from the molecule, cysteine became alanine. It was still directed into protein, however, as if it were cysteine.

SINGLET CODE (4 WORDS)	DOUBLET CODE (16 WORDS)				TRIPLET CODE (64 WORDS)			
					AAA	AAG	AAC	AAU
					AGA	AGG	AGC	AGU
					ACA	ACG	ACC	ACU
					AUA	AUG	AUC	AUU
					GAA	GAG	GAC	GAU
					GGA	GGG	GGC	GGU
A	AA	AG	AC	AU	GCA	GCG	GCC	GCU
G	GA	GG	GC	GU	GUA	GUG	GUC	GUU
C	CA	CG	CC	CU	CAA	CAG	CAC	CAU
U	UA	UG	UC	UU	CGA	CGG	CGC	CGU
					CCA	CCG	CCC	CCU
					CUA	CUG	CUC	CUU
					UAA	UAG	UAC	UAU
					UGA	UGG	UGC	UGU
					UCA	UCG	UCC	UCU
					UUA	UUG	UUC	UUU

CODE-LETTER COMBINATIONS increase sharply with the length of the code word. Since at least 20 code words are needed to identify the 20 common amino acids, the minimum code length is a sequence of three letters, assuming that all words are the same length.

onstrate the coding potential of the cell; that is, it may reveal code words that function routinely in the living cell and potential words that would be recognized if appropriate mutations were to occur in the cellular DNA. The table on page 12 summarizes the code-word dictionary on the assumption that all code words are triplets.

The Universality of the Code

Does each plant or animal species have its own genetic code, or is the same genetic language used by all species on this planet? Preliminary evidence suggests that the code is essentially universal and that even species at opposite ends of the evolutionary scale use much the same code. For instance, a number of laboratories in the U.S. and England have recently reported that synthetic RNA polymers code the same way in mammalian cell-free systems as they do in the bacterial system. The base compositions of mammalian code words corresponding to about six amino acids have been determined so far. It nevertheless seems probable that some differences may be found in the future. Since certain amino acids are coded by multiple words, it is not unlikely that one species may use one word and another species a different one.

An indirect check on the validity of code words obtained in cell-free systems can be made by studying natural proteins that differ in amino acid composition at only one point in the protein chain. For example, the hemoglobin of an individual suffering from "sickle cell" anemia differs from normal hemoglobin in that it has valine at one point in the chain instead of glutamic acid. Another abnormal hemoglobin has lysine at the same point. One might be able to show, by examining the code-word dictionary, that these three amino acids—glutamic acid, valine and lysine—have similar code words. One could then infer that the two abnormal hemoglobins came into being as a result of a mutation that substituted a single base for another in the gene that controls the production of hemoglobin. As a matter of fact, the code-word dictionary shows that the code words are similar enough for this to have happened. One of the code groups for glutamic acid is AGU. Substitution of a U for A produces UGU, the code group for valine. Substitution of an A for a U yields AGA, one of the code groups for lysine. Similar analyses have been made for other proteins in which amino acid substitutions are known, and in most cases the substitutions can be explained by alteration of a single base in code-word triplets. Presumably more code words will be found in the future and the correlation between genetic base sequences and amino acid sequences can be made with greater assurance.

The Nature of Messenger RNA

Does each molecule of messenger RNA function only once or many times in directing the synthesis of protein? The question has proved difficult because most of the poly-U in the experimental system is degraded before it is able to function as a messenger. We have found, nevertheless, that only about 1.5 U's in poly-U are required to direct the incorporation of one molecule of phenylalanine into protein. And George Spyrides and Fritz A. Lipmann of the Rockefeller Institute have reported that only about .75 U's are required per molecule of amino acid in their studies. If the coding is done by triplets, three U's would be required if the messenger functioned only once. Evidently each poly-U molecule directs the synthesis of more than one long-chain molecule of polyphenylalanine. Similar results have been obtained in intact cells. Cyrus Levinthal and his associates at the Massachusetts Institute

AMINO ACIDS CODED	U	A	C	G
	PHENYLALANINE	LYSINE	PROLINE●	
	LEUCINE■			

■ POLY U CODES PREFERENTIALLY FOR PHENYLALANINE
● REPORTED BY ONLY ONE LABORATORY; STILL TO BE CONFIRMED
▲ REQUIRES ONLY FIRST OF TWO BASES LISTED
△ REQUIRES ONLY SECOND OF TWO BASES LISTED

SPECIFICITY OF CODING is shown in this table, which lists 18 amino acids that can be coded by synthetic RNA polymers containing no more than one or two kinds of base. The only amino acids that seem to require more than two bases for coding are aspartic acid and methionine, which need U, A and G. The relative amounts of amino acids directed into proof Technology inhibited messenger RNA synthesis in living bacteria with the antibiotic actinomycin and found that each messenger RNA molecule present at the time messenger synthesis was turned off directed the synthesis of 10 to 20 molecules of protein.

We have observed that two factors in addition to base sequence have a profound effect on the activity of messenger RNA: the length of the RNA chain and its over-all structure. Poly-U molecules that contain more than 100 U's are much more active than molecules with fewer than 50. Robert G. Martin and Bruce Ames of the National Institutes of Health have found that chains of poly-U containing 450 to 700 U's are optimal for directing protein synthesis.

There is still much to be learned about the effect of structure on RNA function. Unlike DNA, RNA molecules are usually single-stranded. Frequently, however, one part of the RNA molecule loops back and forms hydrogen bonds with another portion of the same molecule. The extent of such internal pairing is influenced by the base sequence in the molecule. When poly-U is in solution, it usually has little secondary structure; that is, it consists of a simple chain with few, if any, loops or knots. Other types of RNA molecules display a considerable amount of secondary structure [see top illustration on page 154].

We have found that such a secondary structure interferes with the activity of messenger RNA. When solutions of poly-U and poly-A are mixed, they form double-strand (U-A) and triple-strand (U-A-U) helices, which are completely inactive in directing the synthesis of polyphenylalanine. In collaboration with Maxine F. Singer of the National Institutes of Health we have shown that poly-UG containing a high degree of ordered secondary structure (possibly due to G-G hydrogen-bonding) is unable to code for amino acids.

It is conceivable that natural messenger RNA contains at intervals short regions of secondary structure resembling knots in a rope. These regions might signify the beginning or the end of a protein. Alternative hypotheses suggest that the beginning and end are indicated by particular base sequences in the genetic message. In any case it seems probable that the secondary structure assumed by different types of RNA will be found to have great influence on their biological function.

The Reading Mechanism

Still not completely understood is the manner in which a given amino acid finds its way to the proper site in a protein chain. Although transfer RNA was found to be required for the synthesis of polyphenylalanine, the possibility remained that the amino acid rather than the transfer RNA recognized the code word embodied in the poly-U messenger RNA.

To distinguish between these alternative possibilities, a brilliant experiment was performed jointly by François Chapeville and Lipmann of the Rockefeller Institute, Günter von Ehrenstein of Johns Hopkins University and three Purdue workers: Benzer, Weisblum and William J. Ray, Jr. One amino acid, cysteine, is directed into protein by poly-UG. Alanine, which is identical with cysteine except that it lacks a sulfur atom, is directed into protein by poly-CG or poly-UCG. Cysteine is transported by one species of transfer RNA and alanine by another. Chapeville and his associates enzymatically attached cysteine, labeled with carbon 14, to its particular type of transfer RNA. They then exposed the molecular complex to a nickel catalyst, called Raney nickel, that removed the sulfur from cysteine and converted it to alanine—without detaching it from cysteine-transfer RNA. Now they could ask: Will the labeled alanine be coded as if it were alanine or cysteine? They found it was coded by poly-UG, just as if it were cysteine [see bottom illustration on page 155]. This experiment shows that an amino acid loses its identity after combining with transfer RNA and is carried willy-nilly to the code word recognized by the transfer RNA.

The secondary structure of transfer RNA itself has been clarified further this past year by workers at King's College of the University of London. From X-ray evidence they have deduced that transfer RNA consists of a double helix very much like the secondary structure found in DNA. One difference is that the transfer RNA molecule is folded back on itself, like a hairpin that has been twisted around its long axis. The molecule seems to contain a number of unpaired bases; it is possible that these provide the means for recognizing specific code words in messenger RNA [see illustration at right on page 153].

There is still considerable mystery about the way messenger RNA attaches to ribosomes and the part that ribosomes play in protein synthesis. It has been known for some time that colon bacillus ribosomes are composed of at least two types of subunit and that under certain conditions they form aggregates consisting of two subunits (dimers) and four subunits (tetramers). In collaboration with Samuel Barondes, we found that the addition of poly-U to reaction mixtures initiated further ribosome aggregation. In early experiments only tetramers or still larger aggregates supported the synthesis of polyphenylalanine. Spyrides and Lipmann have shown that poly-U makes only certain "active" ribosomes aggregate and that the remaining monomers and dimers do not support polyphenylalanine syn-

BASES PRESENT IN SYNTHETIC RNA

UA	UC	UG	AC	AG	CG
PHENYLALANINE ▲	PHENYLALANINE ▲	PHENYLALANINE ▲	LYSINE ▲	LYSINE ▲	PROLINE ▲
LYSINE △	PROLINE △	LEUCINE	PROLINE △	GLUTAMIC ACID	ARGININE ●
TYROSINE	LEUCINE	VALINE	HISTIDINE	ARGININE ●	ALANINE ●
LEUCINE	SERINE	CYSTEINE	ASPARAGINE	GLUTAMINE ●	
ISOLEUCINE		TRYPTOPHAN	GLUTAMINE	GLYCINE ●	
ASPARAGINE ●		GLYCINE	THREONINE		

tein by RNA polymers containing two bases depend on the base ratios. When the polymers contain a third and fourth base, additional kinds of amino acids are incorporated into protein. Thus the activity of poly-UCG (an RNA polymer containing U, C and G) resembles that of poly-UC plus poly-UG. Poly-G has not been found to code for any amino acid. Future work will undoubtedly yield data that will necessitate revisions in this table. An RNA-code-word dictionary derived from the table appears on page 158.

AMINO ACID	RNA CODE WORDS			
ALANINE	CCG	UCG ■		
ARGININE	CGC	AGA	UCG ■	
ASPARAGINE	ACA	AUA		
ASPARTIC ACID	GUA			
CYSTEINE	UUG △			
GLUTAMIC ACID	GAA	AGU ■		
GLUTAMINE	ACA	AGA	AGU ■	
GLYCINE	UGG	AGG		
HISTIDINE	ACC			
ISOLEUCINE	UAU	UAA		
LEUCINE	UUG	UUC	UUA	UUU □
LYSINE	AAA	AAG ●	AAU ●	
METHIONINE	UGA ■			
PHENYLALANINE	UUU			
PROLINE	CCC	CCU ▲	CCA ▲	CCG ▲
SERINE	UCU	UCC	UCG	
THREONINE	CAC	CAA		
TRYPTOPHAN	GGU			
TYROSINE	AUU			
VALINE	UGU			

△ UNCERTAIN WHETHER CODE IS UUG OR GGU

■ NEED FOR U UNCERTAIN

□ CODES PREFERENTIALLY FOR PHENYLALANINE

● NEED FOR G AND U UNCERTAIN

▲ NEED FOR U A,G UNCERTAIN

GENETIC-CODE DICTIONARY lists the code words that correspond to each of the 20 common amino acids, assuming that all the words are triplets. The sequences of the letters in the code words have not been established, hence the order shown is arbitrary. Although half of the amino acids have more than one code word, it is believed that each triplet codes uniquely for a particular amino acid. Thus various combinations of AAC presumably code for asparagine, glutamine and threonine. Only one exception has been found to this presumed rule. The triplet UUU codes for phenylalanine and, less effectively, for leucine.

thesis.

A possibly related phenomenon has been observed in living cells by Alexander Rich and his associates at the Massachusetts Institute of Technology. They find that in reticulocytes obtained from rabbit blood, protein synthesis seems to be carried out predominantly by aggregates of five ribosomes, which may be held together by a single thread of messenger RNA. They have named the aggregate a polysome.

Many compelling problems still lie ahead. One is to establish the actual sequence of bases in code words. At present the code resembles an anagram. We know the letters but not the order of most words.

Another intriguing question is whether in living cells the double strand of DNA serves as a template for the production of a single strand of messenger RNA, or whether each strand of DNA serves as a template for the production of two different, complementary strands of RNA. If the latter occurs—and available evidence suggests that it does—the function of each strand must be elucidated.

Ultimately one hopes that cell-free systems will shed light on genetic control mechanisms. Such mechanisms, still undiscovered, permit the selective retrieval of genetic information. Two cells may contain identical sets of genes, but certain genes may be turned on in one cell and off in another in highly specific fashion. With cell-free systems the powerful tools of enzymology can be brought to bear on these and other problems, with the promise that the molecular understanding of genetics will continue to advance rapidly in the near future.

14

POLYRIBOSOMES

by ALEXANDER RICH   December 1963

A typical mammalian cell contains instructions for making many thousands of different proteins and has the capacity to turn out thousands of protein molecules every minute. To a very large extent the living cell is an expression of the particular kinds of proteins it manufactures. It has been known for several years that the site of protein synthesis within the cell is the particle called the ribosome. Visible only in the electron microscope, ribosomes are approximately spherical and can be seen throughout the substance of all living cells. Although the internal structure of these particles is obscure, it has been established that they are composed of protein and ribonucleic acid (RNA) in about equal amounts.

Within the past 18 months experiments in our laboratory at the Massachusetts Institute of Technology and elsewhere have led to the hypothesis that the protein "factories" of the cell are not single ribosomes working in isolation but collections of ribosomes working together in orderly fashion as if they were machines on an assembly line. We have called such collections polyribosomes, or simply polysomes. As we shall see, the polyribosome is not the usual kind of assembly line. In such an assembly line the product moves down the line and component parts are added to it. In the polyribosome assembly line the ribosomes move down the line and each one makes a complete product. There is much evidence that the ribosomes are all alike, or at least interchangeable. They can move from one assembly line to another, making whatever protein a given line happens to call for.

How this specification of a protein takes place has been fully described in the pages of Scientific American, most recently in the following articles: "The Genetic Code," by F. H. C. Crick [Offprint #123] and "The Genetic Code: II," by Marshall W. Nirenberg [see page 148]. The genetic code of the cell, which constitutes the instructions for the synthesis of the cell's proteins, is embodied in a double-chain molecular helix of deoxyribonucleic acid (DNA). The code itself consists of sequences of four different kinds of subunit called bases. The DNA of a bacterium may contain some five million pairs of bases, which are needed to specify several thousand different proteins. The DNA of a mammalian cell may contain nearly 100 times as many base pairs, which specify many more proteins.

Proteins consist of linear chains of amino acid subunits. Short chains or chains that lack full protein activity are called polypeptides. Polypeptide chains can be folded into a specific three-dimensional configuration, and they often combine to form complex proteins. For example, the protein hemoglobin, which carries oxygen in the blood, is composed of four polypeptide chains, each of which contains about 150 amino acid subunits. Protein chains are built up from about 20 different kinds of amino acid. Each chain must have the right sequence of amino acid subunits to make sense, just as a sentence must consist of the right sequence of letters, spaces and punctuation. It is evident that an enormous number of different polypeptide chains can be constructed from 20 different amino acids, just as an enormous number of different sentences can be composed from the 26 letters of our alphabet.

The kernel of the genetic coding problem was to discover how a sequence of four different bases in DNA could specify a sequence of 20 different amino acids in a protein. It now appears that a triplet code is employed: a sequence of three bases is needed to specify each amino acid. It has also been shown that DNA does not take part directly in protein synthesis. Instead the genetic code in the long double-chain molecule of DNA is transcribed into shorter single chains of RNA, which carry away the information needed to construct one kind of polypeptide chain, or perhaps in some cases several chains. Because these molecules of RNA bear the genetic code to the site of protein synthesis they are called messenger RNA.

How do the amino acid molecules get to the site of synthesis and find their proper place in the polypeptide chain? As a first step they must be "activated," a task performed by the energy-rich substance adenosine triphosphate (ATP). So activated, they can be "recognized" by still smaller RNA molecules, containing about 70 base subunits, called transfer, or soluble, RNA. There is a different kind of transfer RNA for each amino acid. The transfer RNA and amino acid are joined by a specific enzyme, a protein with catalytic activity. The transfer RNA then acts as an adapter for depositing a given amino acid at a position in the polypeptide chain specified by messenger RNA. Presumably the ultimate selection of an amino acid is determined by weak chemical bonds between a sequence of bases in messenger RNA and a complementary sequence in transfer RNA. By this mechanism, through the agency of the ribosome, the information coded in messenger RNA is translated into a polypeptide chain.

About a year and a half ago my colleagues and I began puzzling about one geometrical aspect of this system. Consider for a moment the problem of synthesizing one of the polypeptide chains of hemoglobin, which contains about 150 amino acid subunits. If each subunit is specified by a triplet code, the messenger RNA must contain 450 bases merely to specify the sequence of subunits in one chain. In most RNA molecules the bases are stacked on top of one another more or less like a pile of pennies. Since the bases have a thickness of 3.4 angstrom units, the messenger RNA strand for the hemoglobin polypeptide chains should have a molecular length of at least 1,500 angstroms. In other possible arrangements of the RNA molecule the length might be almost twice as great. By comparison, the individual ribosome has a diameter of only about 230 angstroms, and so we wondered how the long messenger molecule interacted with such a small particle to manufacture a polypeptide chain. Some investigators thought that the RNA chain might be wrapped around the outside of the ribosome, but it was hard to visualize how intimate contact between the two could be maintained. The wrapping problem would be still more difficult for RNA chains 20,000 angstroms long, which are found in many viruses. Alternative suggestions that the messenger RNA might somehow be coiled inside the ribosome seemed to present even more formidable topological problems.

It occurred to us that proteins might actually be made on groups of ribosomes, linked together somehow by messenger RNA. There was already a little evidence pointing in this direction. Walter Gilbert of Harvard University, as well as other investigators, had found that when a synthetic RNA was added to a cell-free system of bacterial ribosomes, the ribosomes would tend to clump together. (In such experiments, initiated by Marshall Nirenberg at the National Institutes of Health, the ribosomes make synthetic polypeptides in accordance with instructions coded in the synthetic RNA. By comparing the base composition of the RNA with the amino acid composition of the polypeptide it is possible to compile a genetic code "dictionary.")

In some of the initial experiments in our laboratory Jonathan R. Warner, then a graduate student, tried to find in bacterial cells structures larger than single ribosomes. He was initially unsuccessful because, as we later realized, the vigorous grinding needed to break open the bacteria also destroys the delicate polyribosome structure.

At the same time Paul M. Knopf, a post-doctoral fellow in our group, was working with reticulocytes—the cells that make hemoglobin—from rabbits. Since the cells were readily available, we began looking for multiple ribosomal structure in them. The reticulocyte is a cell that has lost its nucleus but retains

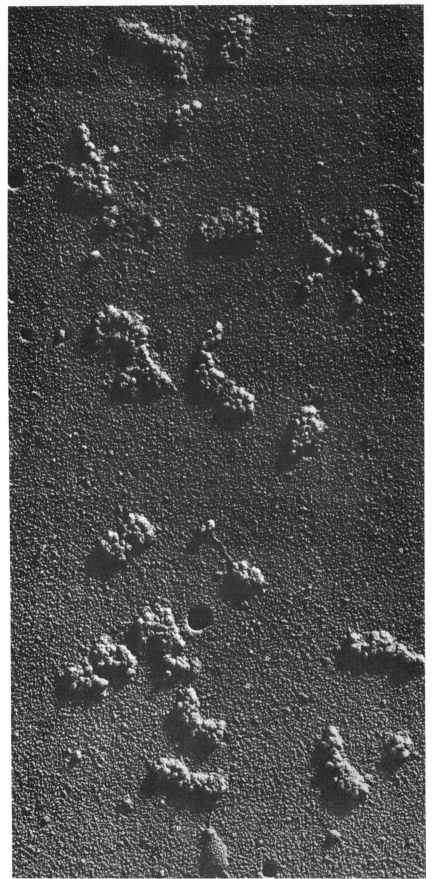

LARGE POLYRIBOSOMES obtained from a culture of human tumor cells are enlarged 100,000 diameters in this electron micrograph made by the author and his colleagues. Individual globular units in the clusters are ribosomes, believed to be held together by strands of messenger RNA (ribonucleic acid). Polyribosomes are the site of protein synthesis.

SYNTHESIS

CELL COMPONENT	STRUCTURE	FUNCTION
DNA (DEOXYRIBONUCLEIC ACID)	A polymer molecule in the form of a double-strand helix containing many thousands of subunits.	Contains genetic information coded in sequences subunits called bases.
MESSENGER RNA (A FORM OF RIBONUCLEIC ACID)	A single-strand polymer molecule containing hundreds of subunits.	Transcribes from DNA the information needed t make a protein molecule and carries it to site of pr tein synthesis.
TRANSFER, OR SOLUBLE, RNA (A FORM OF RIBONUCLEIC ACID)	A single-strand polymer molecule containing about 70 subunits. May be folded into a double helix in some regions.	Conveys specific amino acids to site of protein sy thesis. Each amino acid has its own type of transf RNA.
RIBOSOME	A globular structure consisting of 40 per cent protein and 60 per cent RNA.	Collaborates with messenger RNA to link togeth amino acids delivered by transfer RNA, thereby crea ing proteins.
POLYRIBOSOME OR POLYSOME	Strings of ribosomes temporarily held together by messenger RNA.	Provides actual mechanism of protein synthesis.

GLOSSARY OF CELL COMPONENTS required for protein synthesis describes their structure, function and size. The end result of the collaboration among these components is to produce protein molecules whose composition has been specified by the genetic code

the molecular apparatus for producing hemoglobin molecules. It is also a highly specialized cell: hemoglobin is virtually the only protein it manufactures. For this reason the reticulocyte offers many advantages for studying protein synthesis. Using this cell, for example, Howard M. Dintzis was able to show at M.I.T. that the polypeptides in hemoglobin are assembled by the sequential addition of amino acids, starting at one end of the polypeptide chain and proceeding to the other.

The choice of reticulocytes for our search proved fortunate because they can be broken open by gentle methods. The cells are suspended in a medium whose salt concentration is lower than that within the cells. Water flows into the cell, making it swell until it bursts. A series of experiments demonstrated that protein synthesis is carried out not on individual ribosomes but on ribosome clusters. At about the same time Alfred Gierer, working independently at the Max Planck Institute in Tübingen, made similar observations with rabbit reticulocytes. A short time later F. O. Wettstein, Theophil Staehelin and Hans Noll of the University of Pittsburgh found ribosome clusters in liver tissues.

The basic technique we used in our work was sucrose-gradient centrifuga-

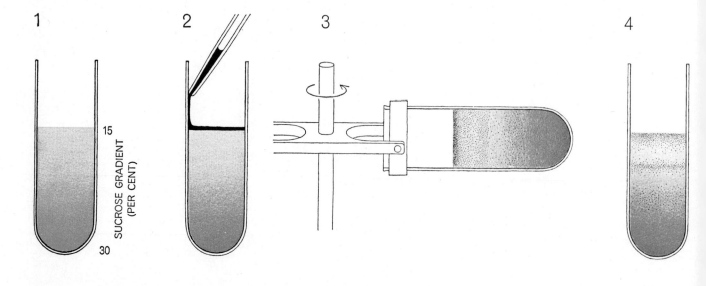

SUCROSE-GRADIENT TECHNIQUE provides a simple way to separate cell components that sediment at different rates when centrifuged. The gradient consists of ordinary sugar dissolved in a test tube (1). In a typical experiment rabbit reticulocytes (red-blood cells) are incubated 45 seconds with amino acids containing radioactive carbon 14. Ribosomes from the cells are layered on the sucrose gradient (2) and spun in a centrifuge (3). Separated fractions are removed in sequence (5) and analyzed. Ribosomes reveal their presence by strongly absorbing ultraviolet radiation at 2,600 angstrom units (6). A radiation counter determines the presence of newly synthesized polypeptide chains containing carbon 14 (7). These chains turn out to be in the faster sedimenting fractions.

SIZE

Diameter: 20 angstrom units
Length: several thousand angstroms up to several millimeters

Diameter: 10 to 15 angstroms
Length: 1,000 to several thousand angstroms

Length: 250 angstroms unfolded

Diameter: about 230 angstroms

Length: varies with length of messenger RNA holding ribosomes together

contained in DNA. Proteins are built up from about 20 varieties of amino acid.

tion, which enables one to separate materials that sediment at different speeds in a strong gravitational field. In this technique a plastic centrifuge tube is filled with a sugar solution that varies smoothly from a concentration of 30 per cent at the bottom of the tube to 15 per cent at the top. The gradient is obtained simply by slowly filling the tube from two reservoirs containing 15 and 30 per cent sucrose. The sample material, containing molecules of different sizes, is carefully deposited in a layer on top of the sugar solution; the tube is then placed in a centrifuge with a swinging-bucket rotor. The sucrose gradient is preserved during the centrifugation and is still maintained after the run by gravity. During the run molecules that sediment at different speeds travel different distances and remain separated when the run is ended. The plastic tube is removed from the centrifuge and its bottom is punctured to allow the collection of a sequence of fractions from bottom to top. These fractions can now be analyzed in various ways.

We designed the following simple experiment. A suspension of rabbit reticulocytes was incubated in a nutrient medium and then fed for 45 seconds with amino acids containing the radioactive isotope carbon 14. The time was kept short because we were interested in looking at the early stages of protein synthesis. After 45 seconds the cells were chilled to stop further metabolic activity, gently broken open and placed on a sucrose gradient. After centrifugation the fractions collected from the sucrose gradient were treated in two ways. The optical density, or amount of absorption, was read in the ultraviolet region at a wavelength of 2,600 angstroms, where nucleic acids strongly absorb radiation. Because ribosomes contain large amounts of ribonucleic acid, this is a sensitive method for determining their presence. In addition the radioactivity of the various fractions was measured. This measurement, by indicating the presence of amino acids containing carbon 14, told us which fractions contained polypeptide chains that were still growing.

The results are shown in the top illustration on the next page. It can be seen that two ultraviolet-absorbing peaks have migrated from the top of the tube. The first, or slow-moving, peak is typical of the peak for single ribosomes. Its speed of movement is represented by the sedimentation constant 74. The fast-moving peak has traveled about two and a half times farther and is much broader. Furthermore, the radioactivity in the growing hemoglobin chains was associated with the fast-moving peak and not with the peak containing single ribosomes. This clearly suggested that the fast-moving peak rather than the single-ribosome peak was the site of protein synthesis.

We then set about analyzing the fast-moving peak. It seemed plausible that it might contain clusters of ribosomes held together by one or more strands of messenger RNA. If this were so, it should be possible to free the ribosomes by subjecting the cell-free medium to ribonuclease, an enzyme that specifically

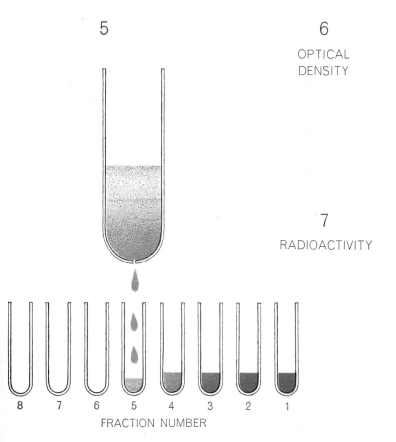

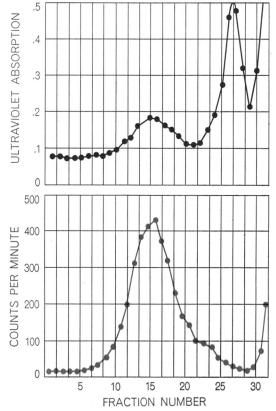

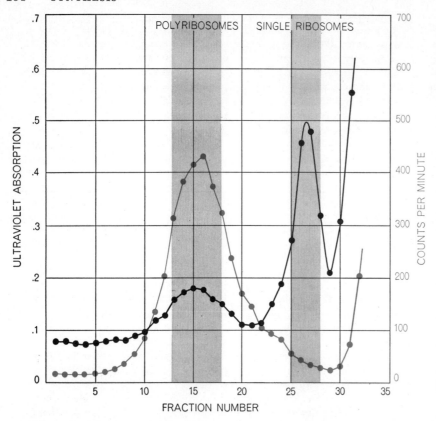

NORMAL RIBOSOME DISTRIBUTION in rabbit reticulocytes consists of a fast-sedimenting fraction of polyribosomes and a slow-sedimenting fraction of single ribosomes. High radioactivity (*color*) indicates that polyribosomes contain newly synthesized polypeptides.

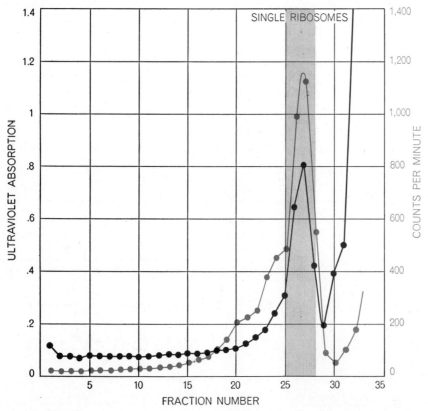

AFTER ADDITION OF RIBONUCLEASE, an enzyme that breaks RNA chains, the ribosomes from reticulocytes no longer exhibit a fast-sedimenting fraction. This implies that polyribosomes are held together by RNA, which, on breaking, releases single ribosomes.

breaks RNA chains. In fact, when a very small amount of ribonuclease was added to the medium before centrifugation, the fast-moving peak did not appear. Both its optical density and its radioactivity were transferred to the peak containing single ribosomes [*see bottom illustration on this page*]. This confirmed the hypothesis that the fast-moving peak represented ribosomes held together by RNA.

Further experiments told us more about this fast-moving component. We learned, for example, that it is fairly fragile. When we subjected the gently opened cells to a modest amount of shearing in a homogenizer, the sucrose-gradient pattern changed dramatically [*see illustration on opposite page*]. The broad peak containing the fast-moving component disappeared and was replaced by a series of peaks. The sedimentation pattern again told us the slow-moving first peak contained single ribosomes. We speculated that the second peak might contain pairs of ribosomes, the third peak clusters of three and so on. This tentative hypothesis was readily confirmed by an electron-microscope examination conducted in the laboratory of Cecil E. Hall at M.I.T. A sample taken from the first peak showed single ribosomes. A sample from the third peak showed mainly clusters of three ribosomes, and the fifth peak showed mainly clusters of five. These initial observations showed us that hemoglobin synthesis is actually carried out on a group of ribosomes, which we named the polyribosome, or polysome.

Further analysis showed that hemoglobin synthesis takes place primarily in a polysome containing five units, as shown clearly in electron micrographs. The micrographs also show, however, a fair number of four-unit and six-unit polysomes [*see top illustration on page 166*]. We were quite sure that these were not artifacts and that they must reflect the mechanism of protein synthesis.

The fragility of the polysome when subjected to mechanical forces, as well as its sensitivity to small amounts of ribonuclease, suggested that the ribosomes are held together by a single strand of RNA. This impression was strongly reinforced by more specialized electron micrographs made by Henry S. Slayter of M.I.T. The technique called negative staining shows that the ribosomes in a polysome are separated by gaps of 50 to 150 angstroms. Positive staining with uranyl acetate reveals that the ribosomes are connected by a thin thread 10 to 15 angstroms in diameter, which is about the thickness of a single strand of RNA. From the size of the gap between ribosomes one can compute that the over-all length of a five-unit polysome is near 1,500 angstroms. (The five ribosomes have a total diameter of 5×230, or about 1,150 angstroms, and there are five inter-ribosomal gaps of 50 to 150 angstroms each.) These measurements of total polysome length are thus near the length that we concluded must be needed to specify the information in a hemoglobin polypeptide chain of 150 amino acid subunits. In other words, the messenger RNA for a hemoglobin polypeptide chain is about the right length to hold together a five-unit polysome.

These various observations led us to the following picture of how the polysome functions. The fact that the ribosomes are separated by a considerable distance makes it seem unlikely that they cooperate in synthesizing a single polypeptide chain. Furthermore, if a ribosome is to have access to all the information coded in messenger RNA, it must "read" the strand from one end to the other. As it travels it must build up a polypeptide chain, adding one amino acid after another according to instructions. A similar conclusion was reached by Gilbert after he studied how transfer RNA is bound to the ribosome. The conclusion is also consistent with Dintzis' observation that hemoglobin synthesis proceeds in sequence.

Let us now imagine that the messenger RNA for hemoglobin contains not just one ribosome but five, all moving, say, from left to right [see bottom illustration on next two pages]. The ribosome at the extreme left has just attached itself to the strand and has started synthesizing a polypeptide chain. The other four ribosomes are proportionately further along in the synthesis process and the one at the extreme right has almost completed a polypeptide chain. Presumably the ribosomes are carried along by a ratchet-like mechanism that does not allow them to go backward. At each station along the way the appropriate amino acid, borne by transfer RNA, is selected from the cellular milieu and

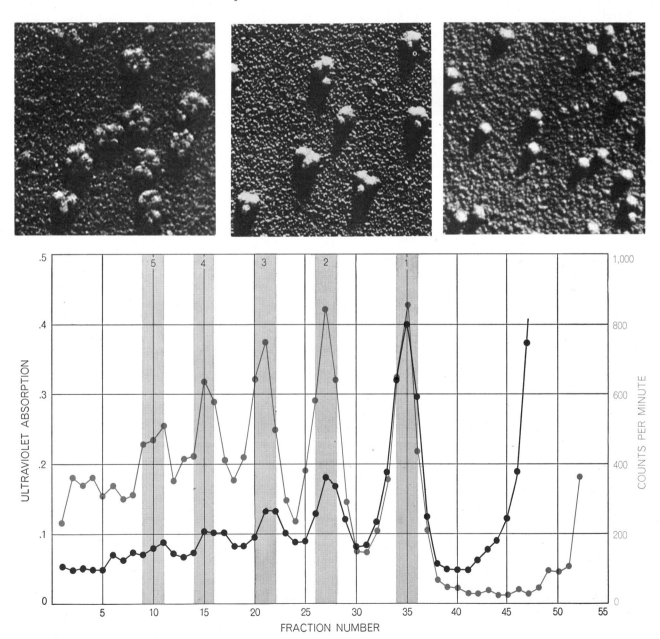

EFFECT OF GENTLE GRINDING is to produce a series of sedimentation peaks, indicating that the polyribosomes from reticulocytes have been broken up into smaller units. Electron micrographs (top) were made of samples obtained from the first, third and fifth peaks from the left. They contained respectively five-unit polyribosomes, three-unit polyribosomes and single ribosomes.

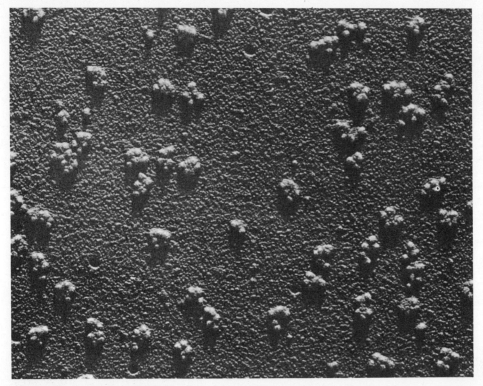
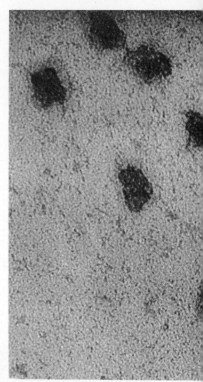

RETICULOCYTE POLYRIBOSOMES are shown at left shadowed with platinum and magnified 100,000 diameters in the electron microscope. Reticulyte polyribosomes at right have been positively stained with uranyl acetate and magnified 400,000 diameters by Henry

added to the growing polypeptide chain. When the synthesis is complete, the ribosome liberates the polypeptide chain and itself drops off the messenger strand. At about the same time another ribosome has found its way onto the messenger at the other end. The time needed for a single ribosome to traverse the messenger strand and produce a hemoglobin polypeptide chain has been estimated at one minute. In a bacterial cell the protein-synthesis time may be as little as 10 seconds.

In the reticulocyte the five-unit polysome is the most common species. The gaps between ribosomes vary somewhat, however, suggesting that the movement of ribosomes along the messenger strand has a statistical character. In some cases a ribosome will detach at one end before a new ribosome is attached at the other; this could account for the four-unit polysomes we see in some pictures. In other cases a ribosome may attach at one end before the fifth is released at the other end, thereby giving rise to a six-unit polysome. Such a statistical mode of operation would account for the distribution of polysome sizes observed in the reticulocyte.

Detroit might well envy the efficiency of the cell's protein factories. It is evident that protein synthesis is not really an

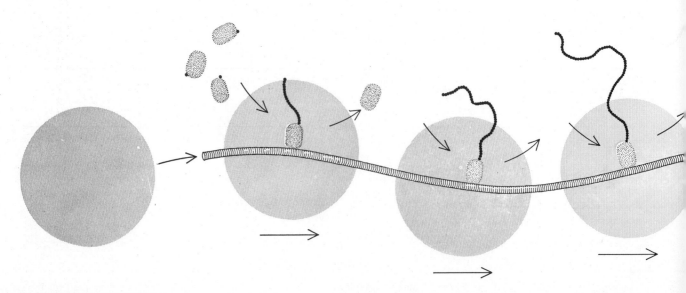

POLYRIBOSOME MECHANISM, as now visualized, consists of a long strand of messenger RNA to which single ribosomes attach themselves temporarily. As each ribosome travels along the strand it "reads" the information needed to synthesize a complete polypep-

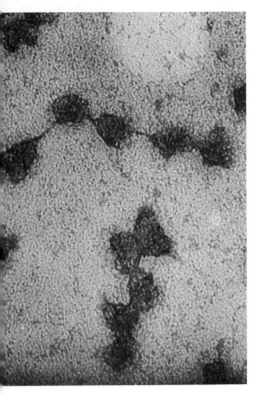

S. Slayter of the Massachusetts Institute of Technology. Note the connecting threads.

assembly line process as it is normally understood. It would be more appropriate to compare protein synthesis with the operation of a tape-controlled machine tool. The tool will turn out an object of any shape within its range of capabilities, in response to information coded on the input tape. In factories where such tools are used each tool is provided with its own tape, but if it served any purpose a single tape could easily be fed through a battery of identical tools. The living cell evidently makes one tape serve for many tools because this is an efficient way to do the job.

As the concept of the polysome became clearer we were naturally anxious to look for polysomes in other cells. It seemed likely that a variety of messenger lengths and polysome sizes would be found. This has turned out to be the case.

A human tumor cell known as the HeLa cell is widely grown in tissue culture and provides a convenient example of a mammalian cell that produces many kinds of protein. Polysomes from the HeLa cell were prepared at M.I.T. by Sheldon Penman, Yachiel Becker and James E. Darnell. When we subjected these polysomes to sucrose-gradient centrifugation, we obtained the curves plotted in the illustration at the top of the next page. The electron microscope shows that the most common polysome species is one containing five or six ribosomes, but the distribution is much broader than that in the reticulocyte. Some of the HeLa polysomes contain 30 or 40 ribosomes.

It is not surprising that the distribution of polysomes from another kind of mammalian cell is much broader than that found in the reticulocyte. The reticulocyte is highly specialized and predominantly makes a single protein. Other mammalian cells make a great variety of protein molecules to conduct a variety of metabolic activities. A broad distribution of polysome sizes implies that a cell contains messenger RNA of many different lengths. Presumably their length is proportional to the lengths of the polypeptide chains being synthesized, but this may not be the only interpretation. Some of the long messenger RNA strands associated with polysomes that contain 20 or more ribosomes may contain information for making more than one kind of polypeptide chain.

This is almost certainly true of polysomes consisting of 50 to 70 ribosomes, which are found in cells infected by the virus of poliomyelitis. The long chain of RNA that bears the genetic code of this virus evidently serves as a strand of messenger RNA when it enters a mammalian host cell. Experiments by Penman, Darnell, Becker and Klaus Scherrer have shown that the polysomes that occur normally in a mammalian cell in tissue culture decrease sharply when the cell has been infected by polio virus. The rate of disappearance of the polysomes can be hastened by feeding the cells actinomycin D, an antibiotic that prevents the manufacture of messenger RNA. Thus about three hours after poliovirus infection and treatment with actinomycin D few polysomes can be found in the cell. Half an hour later, however, a new class of polysomes appears. The proteins synthesized on these polysomes are characteristic of the polio virus rather than of the mammalian cell. These virus-induced polysomes are among the largest we have seen in the electron microscope. They undoubtedly manufacture more than one kind of protein molecule; hence some additional features may have to be added to the simple polysome model I have described.

I shall mention briefly a few of the experiments we have designed to test our polysome model. The model as-

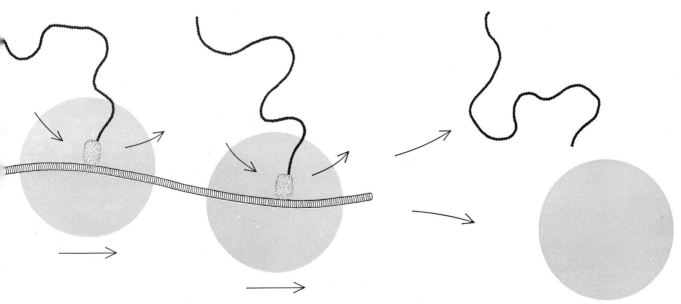

tide chain. The amino acids (*black dots*) for the chain are delivered by transfer RNA (*oblong shapes*). The polypeptide shown here contains 150 amino acid subunits, the number found in one chain of hemoglobin. The complete protein contains four chains.

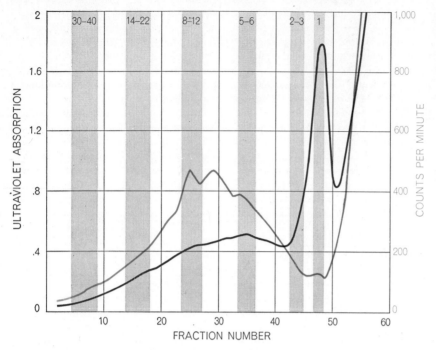

MAMMALIAN POLYRIBOSOMES appear as a very broad peak of fast-sedimenting material when analyzed by sucrose-gradient centrifugation. Electron micrographs show that one peak of the ultraviolet absorption coincides with polyribosomes containing five or six units.

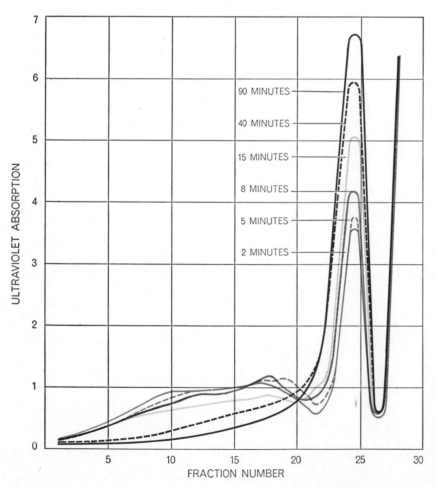

RELEASE OF SINGLE RIBOSOMES is demonstrated by incubating cell extracts with amino acids and an energy supply. Sucrose-gradient experiments show that the broad initial peak of polyribosomes gradually disappears and that the single-ribosome peak rises steadily.

sumes, for example, that an individual ribosome should be able to synthesize a polypeptide chain even though it normally works side by side with other ribosomes. This can be tested by saturating a reticulocyte extract with large amounts of external messenger RNA, such as the synthetic messenger RNA polyuridylic acid. This substance, which contains only one of the four bases normally found in messenger RNA, produces a synthetic polypeptide containing only one kind of amino acid: phenylalanine. By adding enough of the synthetic messenger to a reticulocyte extract one can obtain an extract in which there are as many messenger molecules as ribosomes. In this case most of the ribosomes should pair off with a messenger, leaving few ribosomes to form polysomes. The polysomes already in solution should be unaffected by the introduction of new messengers. Experiments of this type performed in our laboratory, as well as by Gierer in Tübingen, have shown that single ribosomes actively make the synthetic polypeptide polyphenylalanine when polyuridylic acid is added but that the polysomes themselves are inactive. Hence it is clear that individual ribosomes attached to single messenger strands can produce a polypeptide.

Our model also suggests that it should be possible to attach an additional ribosome to a polysome. We have postulated that a ribosome can attach itself to only one end of a messenger strand. If a ribosome could attach itself anywhere, chaos would result. In our simple model there should be only one attachment site on a polysome, whether it contains five ribosomes or 10. In two fractions containing equal numbers of ribosomes, however, there should be twice as many attachment sites in a fraction composed of five-unit polysomes as in a fraction composed of 10-unit polysomes.

Experiments to test this assumption were performed in our laboratory by Howard M. Goodman, a graduate student. Using a culture of HeLa cells, he produced single ribosomes labeled with the radioactive isotope hydrogen 3, or tritium. These single ribosomes were extracted and added to an unlabeled HeLa extract that contained a normal distribution of single ribosomes and polysomes. After a short period of incubation the extract was subjected to sucrose-gradient centrifugation. A test for radioactivity showed that some of the tritium-labeled single ribosomes had indeed become attached to polysomes. Moreover, in accordance with the prediction, twice as many single ribosomes were attached to five-unit polysomes as to 10-unit polysomes when the total number of ribosomes in each fraction was equal.

Our model also makes predictions about events at the terminal end of the messenger strand. It indicates that both ribosomes and polypeptide chains should be released from polysomes that are incubated under protein-synthesizing conditions. This can readily be tested in cell-free extracts because the extracts do not fully reproduce the functions of the intact cell. In particular, the messenger RNA is not replaced and other substances are destroyed, so that in the course of 90 minutes to two hours the cell extract gradually loses its ability to initiate the synthesis of protein. We are still, however, able to test for the release of ribosomes and polypeptide chains from polysomes.

To determine if single ribosomes are released from polysomes as protein synthesis proceeds we incubated cell extracts for varying periods before subjecting them to sucrose-gradient centrifugation. The results are plotted in the bottom illustration on the opposite page. At the beginning there is a large polysomal peak and a modest peak of single ribosomes, representing the normal distribution in the mammalian HeLa cell. As incubation proceeds there is a gradual decrease in the number of polysomes and a decrease in their size. At the same time there is an increase in the number of single ribosomes. At the end of 90 minutes of incubation most of the polysomes have disappeared, having been converted into single ribosomes. We have established that this release of single ribosomes takes place only if the energy necessary for protein synthesis is added to the reaction mixture. In other words, the system is not degraded simply by the passage of time.

To determine if polypeptides are released as incubation proceeds we devised the following experiment. A suspension of living mammalian cells was incubated for a minute and a half with carbon-14-labeled amino acids and then the cells were chilled to halt protein synthesis. This process loaded the polysomes with labeled amino acids that were linked into still unfinished polypeptide chains. Now the cells were broken open and the ribosomes and polysomes were isolated by centrifuging them into a pellet. The liquid on top of the pellet was poured off in order to get rid of the labeled amino acids floating around in the cell extract. The ribosomes and polysomes were then resuspended in a fresh cell extract identical with that removed except that it contained normal rather than radioactive amino acids.

This suspension was incubated under protein-synthesizing conditions, and radioactivity was measured as a function of time in the polysome fraction as well as in the soluble-protein fraction floating at the top of the sucrose gradient. As incubation proceeded, the radioactivity decreased in the former fraction and rose in the latter, showing that most of the labeled amino acids, originally held in the polysomes, were ultimately released as soluble protein. In sum, these three groups of experiments show that it is possible to attach ribosomes to polysomes, to detach ribosomes from polysomes and to liberate polypeptide chains under the conditions of protein synthesis.

I shall mention just one more experiment that supports our polysome model. This experiment, performed in our laboratory by Warner, established the average length of the incomplete polypeptide chains in the polysomes of the reticulocyte. In a complete polypeptide chain found in hemoglobin there are 17 subunits of the amino acid leucine. Warner incubated intact reticulocytes with carbon-14-labeled leucine and determined the number of leucine subunits in the polysome fraction. Knowing the number of ribosomes per polysome, he could easily calculate the number of leucine subunits per ribosome. He found the average number was 7.4. This implies that on the average there is almost half of a complete polypeptide chain on each ribosome in the polysome region. This is consistent with our proposed mechanism, which suggests that there is one growing polypeptide chain for each ribosome in the polysome.

Whether or not polysomes exist in all living cells is still to be determined. To date polysomes have been isolated from several species of bacteria, from the primitive plantlike organisms known as slime molds, from unicellular protozoa and from much more complex cells, including those of man. Therefore I believe that polyribosomes may be the general method used by nature for assembling amino acids into most proteins, and that protein synthesis does not usually occur on single ribosomes.

The discovery of polysomes represents the latest addition to the rapidly growing body of knowledge that describes at the molecular level how genetic information coded in DNA is eventually expressed in terms of active proteins that govern the metabolism and structure of the cell. One of the key problems still to be explained is how complicated globular proteins are put together to form a biologically active molecule. Some of these proteins have more than one polypeptide chain, and it may be that the polysomes play an active role in this next step of protein synthesis.

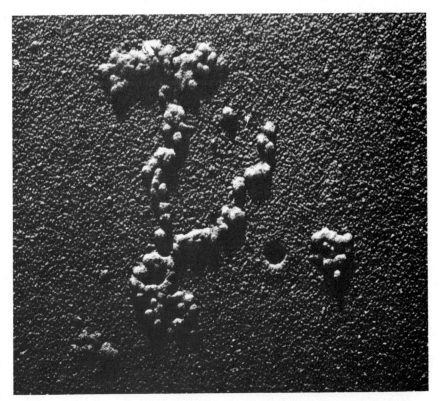

POLIO VIRUS POLYRIBOSOMES, the largest yet observed in the electron microscope, contain at least 50 individual ribosomes. These have been enlarged 115,000 diameters.

15

THE THREE-DIMENSIONAL STRUCTURE OF A PROTEIN MOLECULE

by JOHN C. KENDREW   December 1961

When the early explorers of America made their first landfall, they had the unforgettable experience of glimpsing a New World that no European had seen before them. Moments such as this—first visions of new worlds—are one of the main attractions of exploration. From time to time scientists are privileged to share excitements of the same kind. Such a moment arrived for my colleagues and me one Sunday morning in 1957, when we looked at something no one before us had seen: a three-dimensional picture of a protein molecule in all its complexity. This first picture was a crude one, and two years later we had an almost equally exciting experience, extending over many days that were spent feeding data to a fast computing machine, of building up by degrees a far sharper picture of this same molecule. The protein was myoglobin, and our new picture was sharp enough to enable us to deduce the actual arrangement in space of nearly all of its 2,600 atoms.

We had chosen myoglobin for our first attempt because, complex though it is, it is one of the smallest and presumably the simplest of protein molecules, some of which are 10 or even 100 times larger. The purpose of this article is to indicate some of the reasons why we thought it important to elucidate the three-dimensional architecture of a protein, to explain something of the methods we used and to describe our results.

In a real sense proteins are the "works" of living cells. Almost all chemical reactions that take place in cells are catalyzed by enzymes, and all known enzymes are proteins; an individual cell contains perhaps 1,000 different kinds of enzyme, each catalyzing a different and specific reaction. Proteins have many other important functions, being constituents of bone, muscle and tendon, of blood, of hair and skin and membranes. In addition to all this it is now evident that the hereditary information, transmitted from generation to generation in the nucleic acid of the chromosomes, finds its expression in the characteristic types of protein molecule synthesized by each cell. Clearly to understand the behavior of a living cell it is necessary first to find out how so wide a variety of functions can be assumed by molecules all made up for the most part of the same few basic units.

These units are amino acids, about 20 in number, joined together to form the long molecular chains known as polypeptides. Each link in a chain consists of the group —CO—CHR—NH—, where C, O, N and H represent atoms of carbon, oxygen, nitrogen and hydrogen respectively, and R represents any of the various groups of atoms in a side chain that differs for each of the 20 amino acids. All protein molecules contain polypeptide chains, and some of them contain no other constituents; in others there is an additional group of a different kind. For example, the hemoglobin in red blood corpuscles contains four polypeptide chains and four so-called heme groups: flat assemblages of atoms with an iron atom at the center. The function of the heme group is to combine reversibly with a molecule of oxygen, which is then carried by the blood from the lungs to the tissues. Myoglobin is, as it were, a junior relative of hemoglobin, being a quarter its size and consisting of a single polypeptide chain of about 150 amino acid units together with a single heme group. Myoglobin is contained within the cells of the tissues, and it acts as a temporary storehouse for the oxygen brought by the hemoglobin in the blood.

Following the classic researches on the insulin molecule by Frederick Sanger at the University of Cambridge, several groups of investigators have been able to discover the order in which the amino acids are arranged in the polypeptide chains of a number of proteins [see "The Chemical Structure of Proteins," by William H. Stein and Stanford Moore, Offprint #80, for more information]. This laborious task does not, however, provide the whole story. A polypeptide chain of perhaps hundreds of links could be arranged in space in an almost infinite number of ways. Chemical methods give only the order of the links; equally important is their arrangement in space, the way in which particular side chains form crosslinks to bind the whole structure together into a nearly spherical object (as most proteins are known to be). Also of equal importance is the way in which certain key amino acid units, perhaps lying far apart in the sequence, are brought together by the three-dimensional folding to form a particular constellation of precise configuration—the so-called active site of the molecule—that enables the protein to perform its special functions. How is it possible to discover the three-dimensional arrangement of a molecule as complicated as a protein?

The key to the problem is that many proteins can be persuaded to crystallize, and often their crystals are as regular and as nearly perfect in shape as the crystals of simpler compounds. The fact that pro-

CRYSTALS OF MYOGLOBIN prepared from sperm-whale muscle are enlarged some 50 diameters. In them the molecules of myoglobin are stacked in regular array. By directing a beam of X rays at a single crystal and analyzing the pattern of the reflected rays, the author and his colleagues were able to plot the density of electrons in the molecule and thereby to locate the atoms in it.

teins crystallize is interesting in itself, for crystallization implies a regular three-dimensional array of identical molecules. If all the molecules did not have the same detailed shape, they could not form the repeating arrays that are necessary if the aggregate is to possess the regular external shape of a crystal. Therefore it appears that all protein molecules of a given type are identical—that is, they are not simply "colloidal" aggregates of indefinite shape. The existence of protein crystals means, in fact, that proteins do have a definite three-dimensional structure to solve. And the most powerful techniques for studying the structures of crystals are those of X-ray crystallography.

The X-Ray Approach

In 1912 Walter Friedrich, C. M. Paul Knipping and Max von Laue discovered that if a crystal is turned in various directions while a beam of X rays is sent through it, some of the X rays do not travel in a straight line. When the transmitted rays fall on a photographic plate, they produce not only a dark central spot but also a pattern of fainter spots around it. The reason for this diffraction pattern is that X rays are scattered, or reflected, by the electrons that form the outer part of each atom in the crystal.

The atoms are arranged in an orderly array, something like the trees in a regularly planted orchard. As one drives past an orchard in an automobile and looks into it along different directions, one sees one set after another of lines of trees coming into view end on. Similarly, if one could look at the atoms of a crystal, one would see different planes of atoms in different directions. The X-ray beam is reflected by these sets of planes much as light is reflected from the surface of a mirror; that is to say, the angle of reflection is equal to the angle of incidence. But it can be shown that because the reflection is a set of parallel planes rather than a single surface, as in a mirror, the reflected beam will "flash up" only at a particular angle of incidence between incident beam and planes, this angle becoming greater the closer together the planes of the set are. Thus each spot in the X-ray diffraction pattern corresponds to a particular set of planes; and the spots farthest out in the pattern (those made by X rays diffracted through the biggest angles) correspond to the most closely spaced sets of planes. In an X-ray camera the crystal is rotated in a predetermined manner so that one after another of the sets of planes comes into the correct reflecting position. As each set does so, the corresponding reflected beam flashes up and makes its imprint on the photographic plate.

Each type of crystal has its own characteristic arrangement of atoms and so will produce its own specific X-ray pattern, the features of which can be unambiguously, if tediously, predicted by calculation if the structure of the crystal is known. X-ray analysis involves the reverse calculation: Given the X-ray pattern, what is the crystal structure that must have produced it?

In analyzing complex crystals the calculation is carried out by applying a method known as Fourier synthesis to the repeating, three-dimensional configuration. To understand what this involves, consider first a one-dimensional analogy: a musical note. Physically, a steady musical note is a repeating sequence of rarefactions and condensations in the air between the listener and the instrument producing the note. If the den-

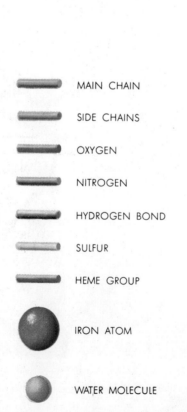

THREE-DIMENSIONAL MODEL of the myoglobin molecule is depicted in this painting by Irving Geis. The key to the model is at the left side of the painting. The molecule consists of some 150 amino acid units strung together in a single chain with a heme group attached to it. At the center of the heme group is a single atom of iron. Most of the amino acid units are arranged in helical sections such as the one running diagonally across the bottom of the painting. Each amino acid unit in the model is identified in the illustration on the following two pages. The model is the result of work by the author, R. E. Dickerson, B. E. Strandberg, R. G. Hart, D. R. Davies, D. C. Phillips, V. C. Shore and H. C. Watson.

— MAIN CHAIN
— SIDE CHAINS
— OXYGEN
— NITROGEN
— HYDROGEN BOND
— SULFUR
— HEME GROUP
● IRON ATOM
● WATER MOLECULE

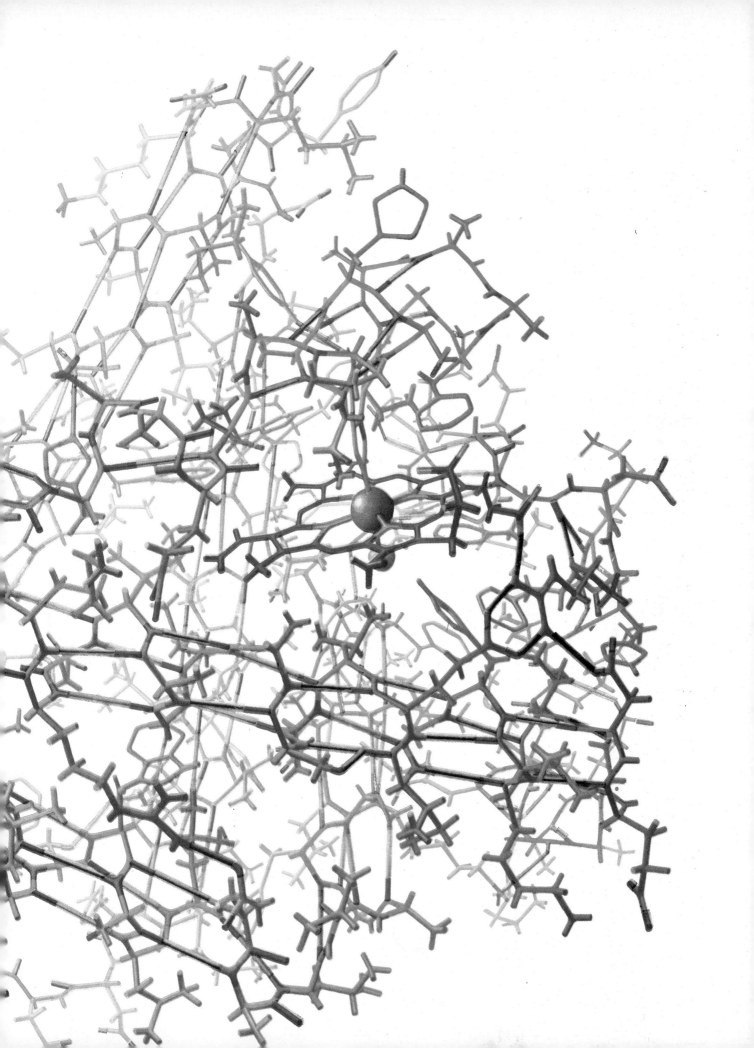

174 SYNTHESIS

sity of the air is plotted along the path, the graph is a complex but perfectly repetitive wave form. More than 150 years ago the French physicist Jean Baptiste Fourier discovered that any such wave form can be decomposed, or analyzed, into a set of harmonics that are pure sine waves of shorter and shorter wavelength and thus of higher and higher frequency [see "The Reproduction of Sound," by Edward E. David, SCIENTIFIC AMERICAN, August, 1961]. The reverse process — Fourier synthesis — consists in combining a series of pure sine waves of the proper relative amplitude in the proper relative phases (that is, in or out of step with one another to the correct extent) so as to reproduce the original wave form, or note. In practice it is not necessary to use all the components to obtain a reasonably faithful reproduction. The greater the number of higher harmonics that are included, however, the more nearly perfect is the rendering of the note.

To an X-ray beam a crystal is an extended electron cloud, the density of

ALANINE	ALA
ARGININE	ARG
ASPARTIC ACID OR ASPARAGINE	ASP
GLUTAMIC ACID OR GLUTAMINE	GLU
GLUTAMIC ACID	GLU.C
GLYCINE	GLY
HISTIDINE	HIS
ISOLEUCINE	ILEU
LEUCINE	LEU
LYSINE	LYS
METHIONINE	MET
PHENYLALANINE	PHE
PROLINE	PRO
SERINE	SER
THREONINE	THR
TYROSINE	TYR
VALINE	VAL

A- 1 VAL (AMINO END)
  2 ALA
  3 GLY
  4 GLU
  5 TYR
  6 SER
  7 GLU
  8 ILEU
  9 LEU
  10 LYS
  11 (NOT GLY)
  12 TYR
  13 (NOT GLY)
  14 LEU
  15 LEU
  16 GLU
AB- 1 (NOT GLY)
B- 1 LEU
  2 VAL OR THR
  3 ALA
  4 GLY
  5 HIS
  6 GLY
  7 LYS
  8 LEU
  9 THR
  10 LEU
  11 ILEU
  12 SER
  13 LEU
  14 PHE
  15 LYS
  16 SER

C- 1 HIS
  2 PRO
  3 GLU.C
  4 THR
  5 LEU
  6 GLU
  7 LYS
CD- 1 PHE
  2 ASP
  3 ARG
  4 PHE
  5 LYS
  6 HIS
  7 LEU
  8 LYS
D- 1 THR
  2 GLU.C
  3 ALA
  4 GLU.C
  5 MET
  6 LYS
  7 ALA
E- 1 SER
  2 GLU.C
  3 ASP
  4 LEU
  5 LYS
  6 VAL
  7 HIS
  8 GLY
  9 ILEU
  10 GLU
  11 VAL
  12 ASP
  13 (NOT ALA, GLY)
  14 ALA
  15 LEU
  16 GLY
  17 ALA
  18 ILEU
  19 ASP
  20 ARG
EF- 1 LYS
  2 LYS
  3 GLY
  4 LEU
  5 HIS
  6 (NOT GLY)
  7 (NOT GLY)
  8 GLU
F- 1 GLU
  2 ALA
  3 PRO
  4 THR
  5 ALA
  6 HIS
  7 SER
  8 HIS
  9 ALA

FG- 1 (NOT GLY)
  2 (NOT GLY)
  3 PHE
  4 (NOT ALA)
  5 ILEU
G- 1 PRO
  2 ILEU
  3 LYS
  4 TYR
  5 (NOT ALA, GLY)
  6 GLU
  7 HIS
  8 LEU
  9 SER
  10 (NOT GLY, ALA)
  11 ALA
  12 VAL OR THR
  13 ILEU
  14 HIS
  15 VAL
  16 ARG
  17 ALA
  18 THR
  19 LYS
GH- 1 HIS
  2 ASP
  3 ASP
  4 GLU
  5 PHE
  6 GLY
H- 1 ALA
  2 PRO
  3 ALA
  4 ASP
  5 GLY
  6 ALA
  7 MET
  8 GLY
  9 LYS
  10 ALA
  11 LEU
  12 GLU.C
  13 LEU
  14 PHE
  15 ARG
  16 LYS
  17 ASP.C
  18 ILEU
  19 ALA
  20 ALA
  21 LYS
  22 TYR
  23 LYS
  24 GLU.C
HC- 1 LEU
  2 GLY
  3 TYR
  4 GLY
  5 GLU.C (CARBOXYL END)

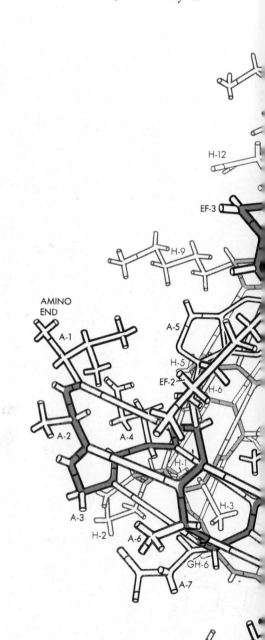

SEQUENCE OF AMINO ACID UNITS in the model of myoglobin is indicated by the letters and numbers in the illustration on these two pages. The amino acid unit represented by each symbol is given in the table above; the key to the abbreviations is at top left in the table. The brackets in the table indicate those amino acid units which form a helical section. The direction of the main chain is traced in color in the illustration; the chain begins at far left (*amino end*) and ends near the top (*carboxyl end*). Here the heme group is indicated in gray. Not all the amino acid units in the model have been positively identified. In some cases it has only been determined that they cannot be certain units. The over-all configuration of the molecule, however, is known with a considerable degree of confidence.

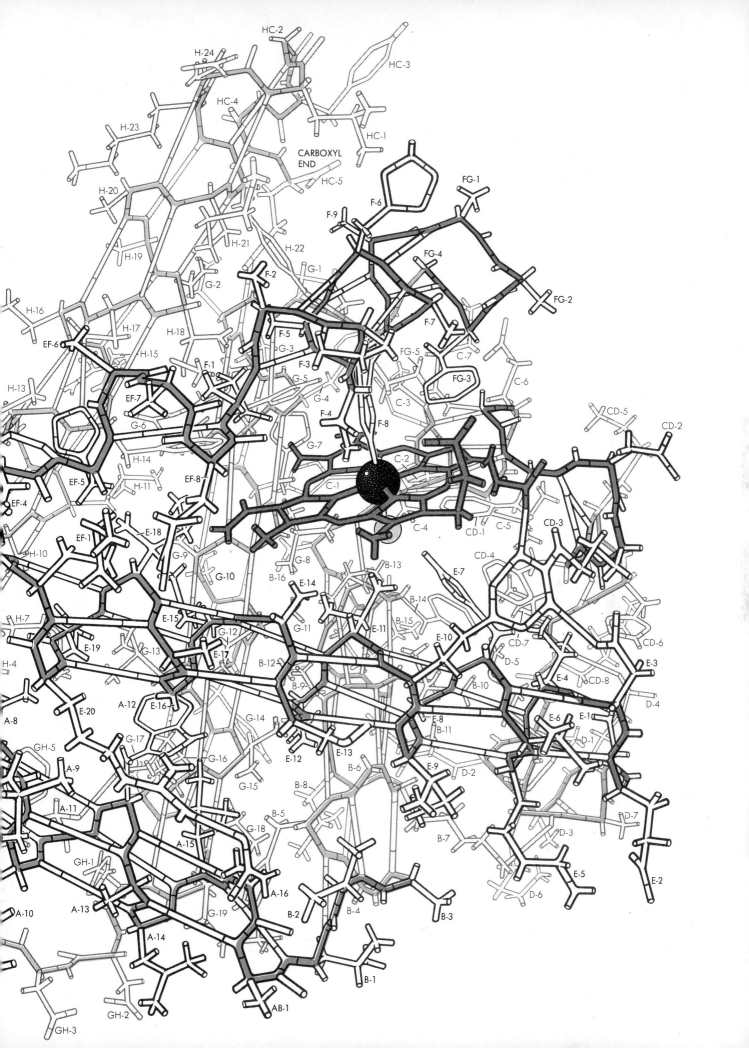

SPECIAL X-RAY CAMERA is used to make X-ray diffraction photographs of the myoglobin crystal. The crystal is contained in the thin glass tube exactly at the center of the photograph. The beam of X rays comes out of metal tube just to the right of glass one.

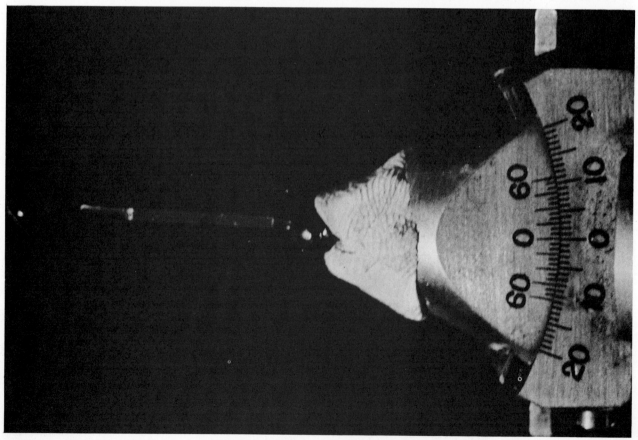

CRYSTAL OF MYOGLOBIN is the dark speck in the middle of the glass tube in this close-up. As the X-ray exposure is made the crystal is rotated so that the X rays reflected from planes of atoms in the crystal "flash up" to make spots on a photographic plate.

which varies from place to place in three dimensions but in a regular, repeating way. (The density at any point depends on the types of atom in the neighborhood and their spatial arrangement.) The crystal can therefore be thought of as a kind of three-dimensional sound wave consisting of rarefactions and condensations of electrons rather than of air particles. This wave too can be decomposed into harmonics—that is, simple, repetitive patterns of density variation; along any single direction the density of each harmonic varies sinusoidally.

It turns out that each harmonic corresponds to one particular spot in the X-ray diffraction pattern. The reason is that each set of possible atomic planes in the crystal constitutes one element, so to speak, of the over-all periodic structure. That is just what a harmonic is: a component of the over-all periodic structure. From the position and darkness of a spot, the "wavelength" (the spacing between high-density peaks) and "amplitude" (the value of the density at the peaks) of the corresponding harmonic can be computed. So the problem is reduced to calculating the harmonics from the spot pattern and then adding them together to arrive at the total structure.

There is, however, a serious catch: the diffraction pattern provides information on the wavelength and amplitude of the harmonics but not on their relative phases. In the case of sound, phase is not particularly important in synthesizing a wave; the ear is rather insensitive to phase difference and hears very nearly the same note so long as the relative amplitudes of the harmonics are correct. On the other hand, the shape of the wave as seen by the eye varies greatly when the relative phases of the components are shifted.

In deriving crystal structure the correct shape of the three-dimensional "wave" is precisely what one is looking for. But the X-ray picture contains only half of the information required; it contains the amplitudes but not the phases. In simpler structures crystallographers get around the difficulty by a method of trial and error; from a plausible model structure they calculate the phases and use these in conjunction with the measured amplitudes to calculate a Fourier synthesis, that is to say, an enlarged picture of the distribution of the electrons (and hence of the atoms) in the structure. The result should be a good deal closer to the real structure than the original model, and from it crystallographers can calculate a new and improved set of phases. If the original model was good enough to put them on

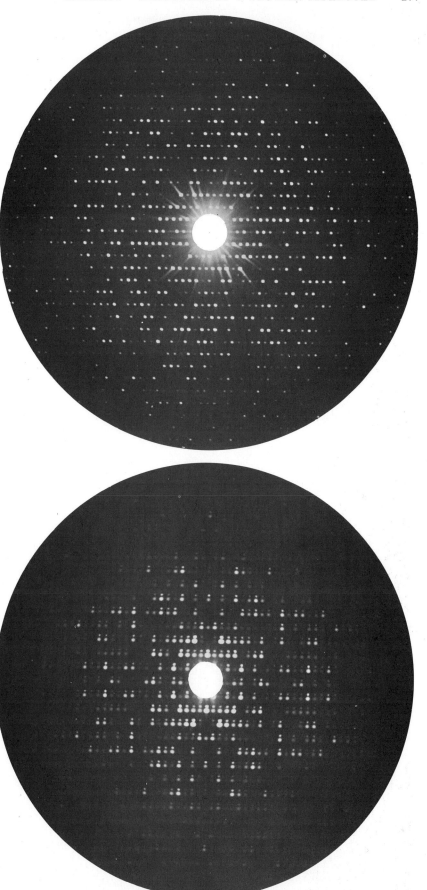

X-RAY PHOTOGRAPHS of myoglobin are patterns of spots. At top is a photograph of a normal crystal. At bottom is a photograph of a different type in which patterns of normal crystal and one labeled with heavy atoms are superimposed slightly out of register. Differences in density between two sets made it possible to determine phase of X-ray reflections.

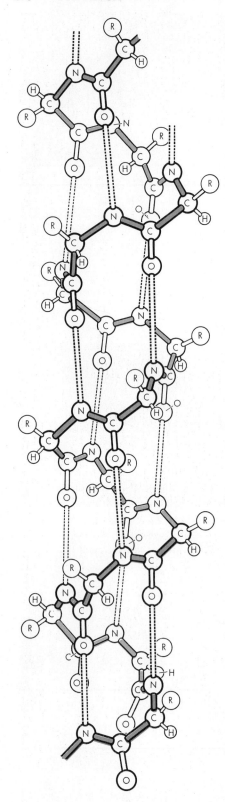

ALPHA HELIX of a protein molecule is a coiled chain of amino acid units. The backbones of the units form a repeating sequence of atoms of carbon (C), oxygen (O), hydrogen (H) and nitrogen (N). The R stands for the side chain that distinguishes one amino acid from another. The configuration of the helix is maintained by hydrogen bonds (*broken lines*). The hydrogen atom that participates in each of these bonds is not shown.

the right track, they gradually approach the true structure by a series of successive approximations, or refinements.

As in the case of the musical note, the greater the number of higher harmonics that are included, the sharper and more precise is the resulting picture. The higher harmonics of a musical note are, of course, the components of shortest wavelength (highest frequency); in a crystal structure the harmonics are correspondingly the reflections from the most closely spaced sets of planes. As has been mentioned, these reflections occur at the largest angles and show up as spots farthest out in the pattern. The resolution of the final picture—that is, the smallest scale of detail it can show—depends on the outer limit of the spots included in the analysis; the number of spots that have to be included goes up in proportion to the cube of the resolving power required.

The first X-ray photographs of protein crystals were made nearly 25 years ago, but for many years it was not possible even in principle to imagine how the structures of crystals so complex could be discovered. Their X-ray patterns contained many thousands of reflections, paralleling the complexity of the molecules themselves. There was no hope of proceeding by trial and error; the first model could never be good enough to provide a useful starting point. So although protein crystallographers discovered many interesting facts about protein crystals, they did not succeed in extracting much information bearing directly on the molecular structure.

In 1953 the whole prospect was transformed by a discovery of my University of Cambridge colleague Max F. Perutz, who had been studying hemoglobin crystals for many years. The hemoglobin molecule contains two free sulfhydryl groups (SH) of the amino acid cystine; by well-known reactions it is possible to attach atoms of mercury to these groups. Perutz found that if he made crystals of hemoglobin labeled with mercury, their X-ray pattern differed significantly from that of unlabeled crystals, even though the mass of a mercury atom is very small compared with that of a complete hemoglobin molecule. This made it possible to apply the so-called method of isomorphous replacement in the Fourier synthesis.

A full explanation of the method is beyond the scope of the present article. Suffice it to say that by comparing in detail the X-ray patterns of crystals with and without heavy atoms it is possible to deduce the phases of all the reflections, and this without any of the guesswork of the trial-and-error method. Thus Perutz' observation for the first time made it possible, in principle at least, to solve the complex X-ray pattern of a protein crystal and to produce a model of the structure of the molecule.

In our studies of myoglobin we could not follow Perutz' method for attaching mercury atoms, because myoglobin lacks free sulfhydryl groups. We were, however, successful in finding other ways to attach four or five kinds of heavy atom at different sites in the molecule, and we were then able to proceed to a study of the three-dimensional structure of the crystal. A complete solution would involve including all the reflections in the X-ray pattern in our calculation—some 25,000 in all. At the time this work began no computers in existence were fast enough or large enough to handle so great an amount of data; besides, we thought it better in the first instance to test the new method on a smaller scale.

The Six-Angstrom-Unit Picture

As has already been indicated, if we include only the central reflections of the pattern, we obtain a low-resolution, or crude, representation of the structure. The higher the resolution that is desired, the farther out in the pattern must the reflections be measured. We decided that in the first stage of the project we would aim for a resolution of six angstrom units (an angstrom unit is one hundred-millionth of a centimeter). This would be sufficient to reveal the general arrangement of the polypeptide chains in the molecule, but not the configuration of the atoms within the chains or that of the side chains surrounding them. To achieve a six-angstrom resolution we had to measure 400 reflections from the unlabeled protein and from each of five types of crystal containing heavy atoms. Our calculations, which were completed in the summer of 1957, gave us the density of electrons at a large number of points in the crystal, a high electron density being found where many atoms are concentrated. Crystallographers usually represent a three-dimensional density distribution, or Fourier synthesis, by cutting an imaginary series of parallel sections through the structure. The density distribution in each section is represented by a series of density contours drawn on a lucite sheet. When all the sheets are stacked together, they give a representation in space of the density throughout the molecule.

As soon as we had constructed our

ELECTRON-DENSITY MAP of the myoglobin crystal is made up of lucite sheets, on each of which are traced the contours of electron density at that depth. The dark band in the middle is the heme group. This map was made at a resolution of two angstrom units.

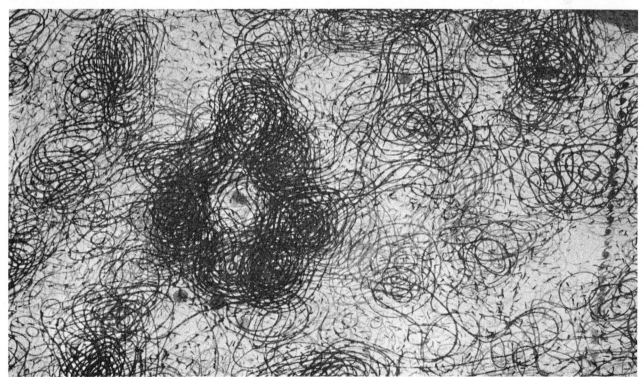

MAP SHOWS ALPHA HELIX when it is seen from the appropriate angle. Here the camera looks through the contours on a series of lucite sheets. Alpha helix is the dark ring at left center. Thus it is seen along its axis, as though it were a cylinder seen from the end.

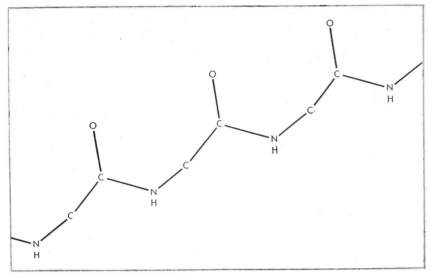

BACKBONE OF THE ALPHA HELIX is shown schematically in this diagram. The sequence of atoms in the helix is —CO—CHR—NH—. Here the HR attached to isolated C is omitted.

BACKBONE IS SUPERIMPOSED on contours made by plotting on a cylinder density along helix in crystal. Cylinder was then unrolled. Backbone thus appears to repeat.

first lucite density map of the myoglobin crystal, we could see at a glance that it contained the features we were looking for, namely a set of high-density rods of just the dimensions one would expect for a polypeptide chain. Closer examination showed that in fact it consisted of almost nothing but a complicated and intertwining set of these rods, sometimes going straight for a distance, then turning a corner and going off in a new direction. In addition to the rods we were able to see very dense peaks, which we took to be the heme groups themselves. The iron atom at the center of the heme group, being by far the heaviest atom in the molecule and therefore having the largest number of planetary electrons, would be expected to stand out as a prominent feature. It was not at all easy, however, to gain any impression of what the molecule was actually like, largely because the molecules are packed together in the crystal and it is hard to see where one begins and the next one ends.

Our next task was to dissect out a single molecule from the enlarged density map of the crystal so that we could look at it separately. Fortunately all protein crystals, including myoglobin, contain a good deal of liquid which fills up the interstices between neighboring molecules, and which at low resolution looks like a uniform sea of density, so that it can easily be distinguished from the irregular variations of density within the molecule itself. By looking for the liquid regions we were able to draw an outline surface around the molecule and so isolate it from its neighbors. Extracted in this way, the molecule stood forth as the complicated and asymmetrical object shown in the top illustration on page 183. The polypeptide chain winds irregularly around the structure, supporting the heme group in a kind of basket. For the most part the course of the polypeptide chain could be followed, but we could not be sure of its route everywhere, since its density became lower at the corners and it tended to fade into the background at those points. Our model did, however, give us a good general picture of the layout of the molecule, and it showed us that it was indeed much more complicated and irregular than most of the earlier theories of the structure of proteins had suggested.

The Two-Angstrom-Unit Picture

At a resolution of six angstroms we could not expect to see any details of the polypeptide chain or of the side chains attached to it. To see all the atoms of a structure as separate blobs of density it would be necessary to work at a resolution higher than 1.5 angstroms, for neighboring atoms attached to one another by chemical bonds lie only from one to 1.5 angstroms apart. A Fourier synthesis of myoglobin at 1.5 angstroms resolution would involve 25,000 reflections; we decided in the second stage of the work to limit our ambitions to a resolution of two angstroms. Even this required that we include in our calculations nearly 10,000 reflections for the unsubstituted crystal and the same number for each of the heavy-atom derivatives. It was necessary to measure a formidable number of X-ray photographs, a task that took a team of six people many months to complete. At the end of this time the mass of data that we had accumulated was so great that it could be handled only by a truly fast computer. We were fortunate that a machine of this class—EDSAC Mark II—had recently come into service at the University of Cambridge, and we were able to use it for deducing the phases of all the 10,000 reflections and for the ensuing calculations of the Fourier synthesis itself. Fast though it is, we taxed the powers of EDSAC II to the utmost, and it was clear that any further improvement in resolution would demand the use of still more powerful machines.

Once more our results were plotted in the form of a three-dimensional contour map [*see top illustration on page 179*]. Since we were looking for finer details, it was necessary to cut sections through the structure at closer intervals than before; in fact, this time we had about 50 sections compared to 16 in the six-angstrom map. To construct the new map it was necessary to calculate the electron density at 100,000 points in the molecule. Indeed, the amount of information contained in the final synthesis was so great that drawing and building the density map was in itself a lengthy task, amounting to some six man-months of work. The result was a complicated set of dense and less dense regions that at first sight seemed completely irregular. Our first step was to see what we could learn from it about the configuration of the polypeptide chains, which in our earlier synthesis had appeared merely as solid rods.

Here I shall digress briefly to consider some earlier work on the fibrous protein of hair. In fibrous proteins the polypeptide chains probably run parallel to the axis of the fiber for considerable distances. Such protein fibers were among the earliest biological macromole-

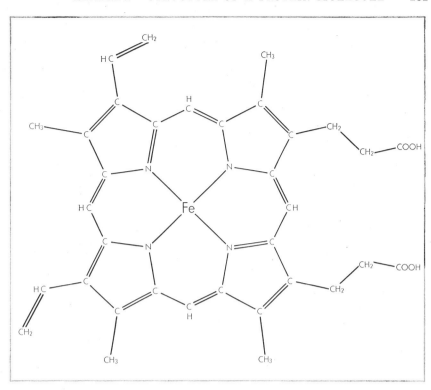

SKELETON OF THE HEME GROUP is outlined in this diagram. At the center of the group is the iron atom (*Fe*). There are four such groups in hemoglobin and one in myoglobin.

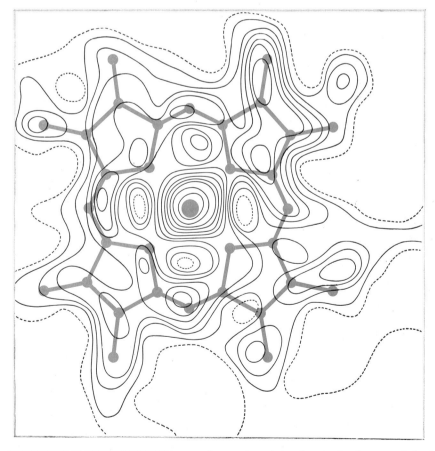

SKELETON IS SUPERIMPOSED on another section of the electron-density map of the myoglobin crystal. Here bonds to the iron atom are omitted to show contours around atom.

cules to be examined by X-ray methods. W. T. Astbury, in his classical work carried out at the University of Leeds in the early 1930's, showed that a human hair gave a characteristic X-ray pattern, which on stretching changed reversibly into quite a different pattern. He was able to show that the pattern of stretched hair—the so-called beta form—corresponded to the polypeptide chains being almost fully extended; it followed that in the unstretched, or alpha, form the chains must assume some kind of folded configuration. For many years different workers proposed a succession of more or less unsatisfactory models of the folded chain in unstretched hair, but finally in 1951 Linus Pauling and Robert B. Corey of the California Institute of Technology found the definitive solution, showing that the chain actually took up a helical or spiral shape, the now famous alpha helix [see illustration on page 178]. In this configuration the successive turns of the helix are held together by weak hydrogen bonds between NH groups on one turn and CO groups opposite them on the next turn. The alpha helix turned out to be present in several other fibrous proteins besides hair; and although there was no definite proof of the fact, indirect evidence indicated that the alpha helix, or something like it, could exist in the molecule of globular proteins too.

The first thing we wanted to do when we finished our Fourier synthesis at two angstroms resolution was to see whether or not there was anything to the idea of helical structures in a globular protein. Accordingly we looked through the stack of lucite sheets in a direction corresponding to the axis of one of the rods we had seen at low resolution. We were delighted to find that the dense rod now revealed itself as a hollow cylinder running dead straight through the structure [see bottom illustration on page 179]. Closer examination showed that the density followed a spiral course along the surface of the cylinder, indicating that the polypeptide chain indeed assumed a helical shape. Detailed measurement of the spiral density showed that it followed precisely the dimensions of the alpha helix deduced by Pauling and Corey 10 years earlier. In fact, it turned out that about three-quarters of the polypeptide chain in the molecule took the form of straight lengths of alpha helix, the helical segments being joined by irregular regions at the corners. In all there were eight such segments, varying in length from seven to 24 amino acid units. In each segment it was possible to fit the alpha helix exactly to the observed density in such a way that we could be reasonably sure of the placing of each atom, even though at this resolution we had not secured full separation between one atom and its neighbors.

The next object of interest was the heme group. Looking at the map from the appropriate angles, we now saw this group as a flat object with a region of high density at the iron atom in the center. A section through the map through the plane of the flat object shows a variation in density that closely follows the known chemical structure of the system of rings in the heme group [see illustrations on preceding page].

The Three-dimensional Model

When we came to study our structure in detail, we soon felt the need of a better way to represent the three-dimensional density distribution. We wanted some method that would enable us to fit actual atomic models to the features we could see. Our solution was to erect a forest of steel rods on which we placed colored clips to represent the distribution in space of points of high density, different colors representing different values of the density [see illustration on page 184]. The scale of this model was five centimeters per angstrom, so that the whole model would fit in a cube about six feet on a side. Each helical segment of polypeptide chain could be seen as a spiral of colored clips passing through the model, and we were then able to insert actual alpha helices made of skeleton-type models (similar to the familiar ball-and-spoke models but with the balls omitted) and to show that they precisely followed the dense trail of clips. In this way we were able to trace the polypeptide chain from beginning to end, right through the molecule, and to establish its configuration in each of the irregular corners joining neighboring helices. Once the course of the main chain had been delineated with atomic models, we were able to see the side chains emerging from it at appropriate intervals as dense branches of various sizes. At first we thought it unlikely that we would be able to identify many of the side chains, but after some practice we found that in fact we were able to do so surprisingly often. As mentioned earlier, side chains in proteins are of only 20 kinds (in myoglobin only 17), and they are of very different shapes and sizes, ranging from the one in glycine, which is only a single hydrogen atom (invisible to the crystallographer), to the chain in tryptophan, with a double-ring system of 10 carbon and nitrogen atoms. Our problem was reduced to deciding among 17 possible side chains in each case.

Some of our identifications were definite, others were tentative. Fortunately an independent check on our conclusions lay at hand. For several years A. B. Edmundson and C. H. W. Hirs, working in Stein and Moore's laboratory at the Rockefeller Institute, had been trying to establish the amino acid sequence of myoglobin by chemical methods. Their work is still incomplete, but they have broken down the molecule into a set of short pieces, or peptides, the compositions—and in some cases the internal sequences—of which they have determined. The order in which the peptides are arranged in the intact molecule has yet to be established chemically. We have been able, however, to place almost every one of the peptides with certainty in its correct position along the chain by comparing its composition with our X-ray identifications of the side chains. There are virtually no gaps left, nor are there peptides unplaced. Once assigned to their correct positions, the peptides often help to confirm doubtful X-ray identifications, and by putting the two types of evidence together we arrive at a nearly complete amino acid sequence for the whole molecule.

Simply to determine the amino acid sequence was not our main aim in undertaking the X-ray analysis of myoglobin. We were much more concerned with the three-dimensional arrangement of the side chains in the molecule and with the interactions between them that produce and maintain the molecule's characteristic configuration. To study these interactions we undertook to make a model of the whole molecule, with every side chain in place [see illustration on pages 172 and 173]. The result was an object still more complex than the low-resolution model, although of course all the features of the latter are still apparent in the former. We can now discern many of the types of interaction that protein chemists have postulated on the basis of physicochemical studies. For example, positively charged basic groups such as those of lysine and arginine are held by electrostatic attraction close to negatively charged acid groups such as aspartic or glutamic acid; several types of hydrogen-bond interaction can be seen, among them NH groups in the main chain bonded to the oxygen atom of serine or threonine; and everywhere we find a close interlocking of hydrocarbon groups such as CH_2 or CH_3, giving rise to the so-called van der Waals' attraction. The structure is not yet sufficiently complete in all details to allow a

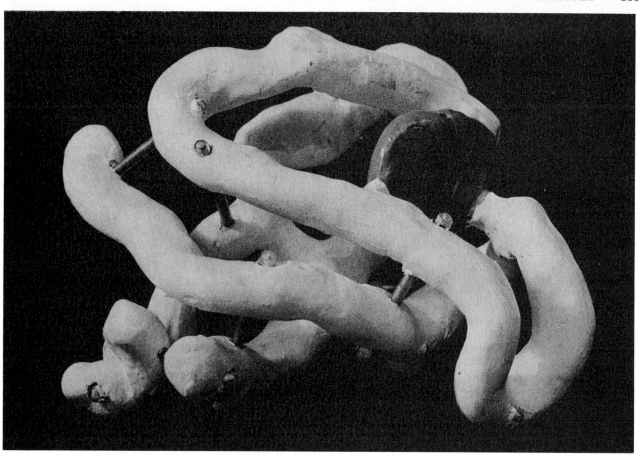

EARLY MODEL of the myoglobin molecule was made at a resolution of six angstrom units. This model has the same general configuration as that of the model depicted on pages 172 and 173, but it lacks detail. The heme group is the flat section at upper right.

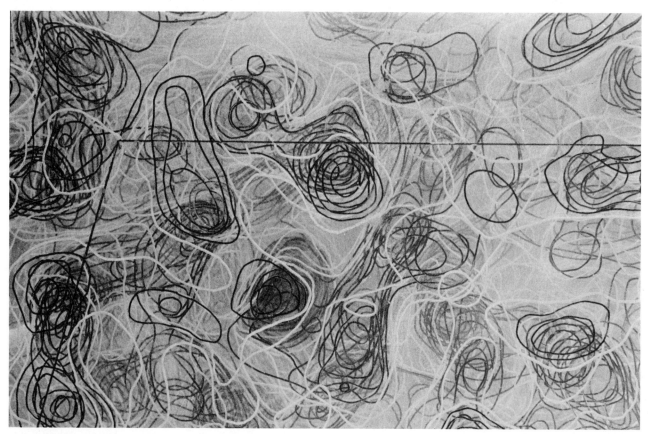

CLOSE-UP OF CONTOURS of map on which the six-angstrom model was based shows that contours are coarser than those in two-angstrom map. Early model was based on work of author, G. Bodo, H. M. Dintzis, R. G. Parrish, H. W. Wyckoff and D. C. Phillips.

full analysis of the interactions, but at least we can now see the general pattern of forces that maintains the integrity of the molecule.

We can also often see why helical segments end at a particular place; in many instances proline side chains are found at the ends of helices and, as was pointed out several years ago, proline is bound to interfere with helix formation because of its peculiar shape, unlike that of any other naturally occurring amino acid. Finally, we can examine the way in which the heme group itself is attached to the rest of the molecule; the iron atom is attached to a nitrogen in a histidine side chain (as had been suggested years ago on the basis of chemical evidence), and the flat ring system is stabilized by hydrocarbon side chains, especially ring side chains, lying parallel to it.

In similar studies of the larger hemoglobin molecule Perutz and his collaborators have shown that, at least to the resolution of 5.5 angstroms that they have so far achieved, there is an astonishing similarity between the three-dimensional structure of myoglobin and the structure of each of the four subunits formed by the individual polypeptide chains of hemoglobin. This is a most remarkable result considering that we are dealing with two distinct proteins, one found in muscle and the other in red blood cells, one derived from sperm whale and the other from horse. Furthermore, the amino acid compositions of the two proteins are known to differ substantially.

The amino acid sequences of the hemoglobin chains have been completely determined. We have found that when we lay the hemoglobin sequences alongside those of myoglobin, making appropriate allowances for slight differences in their length, there are many correspondences, often just at those points where a study of the myoglobin molecule indicates that a crucial stabilizing reaction takes place. We can even begin to find chemical explanations for some of the peculiarities of the congenitally abnormal hemoglobins present in individuals suffering from certain rare blood diseases.

Even in the present incomplete state of our studies on myoglobin we are beginning to think of a protein molecule in terms of its three-dimensional chemical structure and hence to find rational explanations for its chemical behavior and physiological function, to understand its affinities with related proteins and to glimpse the problems involved in explaining the synthesis of proteins in living organisms and the nature of the malfunctions resulting from errors in this process. It is evident that today students of the living organism do indeed stand on the threshold of a new world. Analyses of many other proteins, and at still higher resolutions (such as we hope soon to achieve with myoglobin), will be needed before this new world can be fully invaded, and the manifold interactions between the giant molecules of living cells must be comprehended in terms of well-understood concepts of chemistry. Nevertheless, the prospect of establishing a firm basis for an understanding of the enormous complexities of structure, of biogenesis and of function of living organisms in health and disease is now distinctly in view.

FOREST OF RODS was used to build up the two-angstrom model of the myoglobin molecule from electron-density map. Densities were indicated by clips on rods, and model was based on position of clips. Outline of heme group is visible at upper left center.

186 SYNTHESIS

COLLAGEN FIBRILS, carefully pulled away from human skin, show bands spaced about 700 angstrom units (A.) apart. It was first believed that this represented the length of the underlying collagen molecule; actually the length is about four times greater *(see illustration on page 189)*. This electron micrograph by the author is reproduced at a magnification of 42,000 diameters.

COLLAGEN

by JEROME GROSS   May 1961

Collagen is perhaps the most abundant protein in the animal kingdom. It is the major fibrous constituent of skin, tendon, ligament, cartilage and bone. Its properties are diverse and remarkable. In tendon it has a tensile strength equal to that of light steel wire; in the cornea it is as transparent as water. It accounts for the toughness of leather, the tenacity of glue and the viscousness of gelatin. It also underlies the development of crippling deformities associated with the rheumatic diseases and with a number of congenital defects of the skeleton, blood vessels and other connective tissue.

By piecing together information derived from X-ray diffraction, chemical analysis, electron microscopy and many other techniques, it is now possible to present a reasonable account of the way collagen fibers are built up from long-chain molecules. Particularly instructive was the discovery that tiny collagen fibers, or fibrils, can be dissociated into their constituent molecules and then reaggregated, outside the living organism, into their original form. This disassembly-reassembly technique, which was perhaps first clearly demonstrated with collagen, has since been successfully applied to other giant molecules and to even more complex biological systems [see "Tissues from Dissociated Cells," by A. A. Moscona; SCIENTIFIC AMERICAN, May, 1959]. The procedure of taking biological materials apart and putting them together again provides much more insight into dynamic mechanisms than does the traditional biochemical approach of breaking things to bits and analyzing the pieces.

Collagen appeared very early in evolution, at least as far back as the coelenterates (a phylum including jellyfishes and sea anemones) and sponges, and it seems to have changed very little in structure and composition since then. It usually appears as bundles of individual, nonbranching fibrils, varying greatly in diameter from tissue to tissue. In skin, under an ordinary light microscope, these bundles appear to be woven together at random, but a definite order emerges if larger areas of tissue are examined. In tendon, collagen fibers are arranged in long parallel bundles. In the cornea of the eye, transparency depends upon the orderly arrangement of collagen fibrils that probably have a refractive index identical to that of the substance in which they are embedded. (The same type of order appears in the translucent skin of the developing amphibian embryo, but after metamorphosis the skin acquires a somewhat random intermeshing structure.) In bone, the collagen fibrils are organized much like the struts and girders of a bridge; the mineralization of bone follows the detailed fine structure of the fibrils. In cartilage, which coats the inner surface of joints and which must have considerable elasticity and smoothness, the collagen fibrils are usually very thin, randomly oriented and embedded in a large volume of extracellular matrix.

Collagen is synthesized primarily by cells called fibroblasts. The basic collagen molecule is a group of three polypeptide chains each composed of about a thousand amino acid units linked together. In ordinary proteins the chains are assembled from the standard assortment of 22 amino acids, and once linked together end to end the acids are not further altered. In the synthesis of collagen, however, two unusual amino acids, hydroxyproline and hydroxylysine, seem to be formed *after* the molecular chain has been assembled; the new amino acids are created by addition of hydroxyl (OH) groups to some of the proline and lysine units in the chain. This alteration of the primary molecular structure has not been observed in the synthesis of other proteins.

Proline and hydroxyproline, which together make up as much as 25 per cent of the links in the collagen molecule, prevent easy rotation of the regions in which they are located, thus imparting rigidity and stability to the collagen molecule. The higher the content of proline and hydroxyproline, the higher the resistance of the molecule to heat or chemical denaturation. Preliminary studies suggest that collagens have another distinctive feature: in long stretches of the molecular chain every fourth position seems to be occupied by glycine, which is followed immediately by proline or hydroxyproline. In any case, all collagens studied so far, regardless of their source, contain about 30 per cent glycine, with a variation of less than 5 per cent.

Collagen owes its properties not only to its chemical composition but also to the physical arrangement of its individual molecules. The basic molecular chain is twisted into a left-handed helix, and three such helices are wrapped around each other to form a right-handed superhelix [see illustration on page 189]. The three chains appear to be held together by hydrogen bonds established between the oxygen atoms, located where amino acids are joined by peptide linkages in one chain, and the nitrogen atoms, located at peptide linkages in an adjacent chain. This picture of the structure was first proposed in 1954 by the Indian workers G. N. Ramachandran and G. Kartha, and later refined by two British groups: Alexander Rich and F. H. C. Crick in Cambridge and Pauline

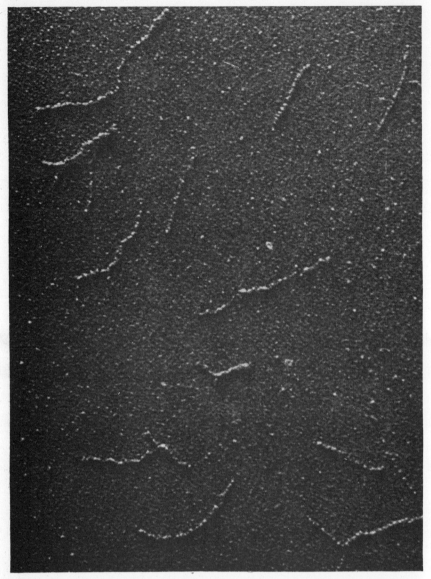

COLLAGEN MOLECULES, dissolved from the collagen of the fish swim bladder, are enlarged 140,000 diameters. They are 2,800 to 2,900 A. long and 14 to 15 A. wide. This electron micrograph was made by Cecil E. Hall of the Massachusetts Institute of Technology.

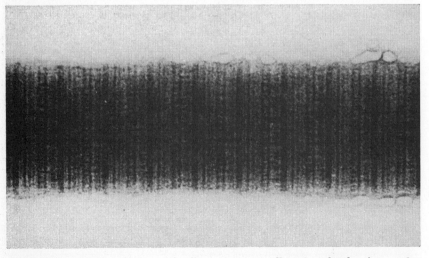

COLLAGEN FIBRIL is built up of collagen (or tropocollagen) molecules that overlap as shown on the opposite page. The intricate fine structure repeats about every 700 A.; the magnification is 250,000 diameters. Alan J. Hodge made electron micrograph at M.I.T.

M. Cowan and S. McGavin in London. The superhelix varies in cross section and electric charge along its length. There is convincing evidence from recent electron micrographs that these variations, or "bumps," are irregularly spaced. Alan J. Hodge and Francis O. Schmitt of the Massachusetts Institute of Technology have suggested that the molecule also has a short flexible "tail" at each end that participates in fibril formation.

For a long time it was believed that the collagen molecule was about 700 angstroms long (an angstrom is one ten-billionth of a meter). This length was inferred from early electron micrographs of collagen fibrils, which showed a series of regular bands with such a spacing. It was assumed that the bands marked the places where the molecules were joined end to end. As so often happens in science, things turned out to be more complex and more interesting than they had seemed at first.

The clue to the currently accepted value for length came from experimentally reconstituted collagen. It has been known since at least 1872 that if collagen fibers are dissolved in acid, reconstituted fibers will automatically appear when the acid is neutralized. The experiment remained little more than a curiosity until 1942, when it was repeated by Schmitt and his associates. It was this group's electron micrographs of natural collagen that had first shown the 700-angstrom periodicity. They were pleased but not greatly surprised to find that the same periodicity appeared in the reconstituted fibers [see top illustration on page 190]. The molecular length of 700 angstroms seemed to be confirmed.

Several years later V. N. Orekhovich and his associates in the U.S.S.R. found needle-like entities when they observed samples of reconstituted collagen under a light microscope. They believed that these entities, which they called procollagen, were newly formed collagen molecules capable of linking up to form the native fibril. Upon learning of this work Schmitt and his co-workers were naturally curious to examine the new material in the electron microscope. When Schmitt, John H. Highberger and I followed the Orekhovich method, we discovered a new type of reconstituted collagen fibril different in structure from any seen before. This new fibril showed bands spaced at about 2,800 angstroms, and a fine structure that surprised us by being symmetrical [see middle illustration on page 190]. We designed

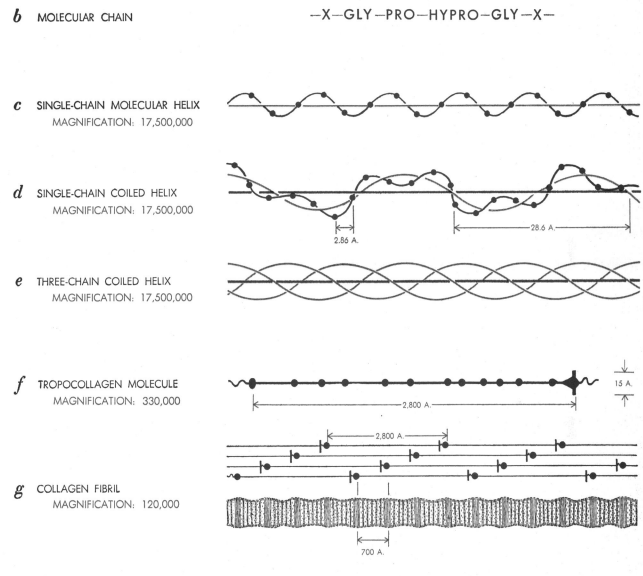

FORMATION OF COLLAGEN can be visualized in seven steps. The starting materials (*a*) are amino acids; the letter "R" in amino acid X represents any of some 20 different side chains. "Hypro" stands for hydroxyproline, created from proline after the molecular chain (*b*) has been formed. The chain twists itself into a left-handed helix; three chains then intertwine to form a right-handed superhelix, which is the tropocollagen molecule. Many molecules line up in staggered fashion (*g*), overlapping by one-quarter of their length, to form a fibril. Fibrils in tissue (*h*) are often stacked in layers with fibrils aligned at right angles.

RECONSTITUTED COLLAGEN fibrils form spontaneously when an acid solution of native collagen is neutralized. The reconstituted fibrils duplicate the native form. The three electron micrographs on this page are all at the same magnification: 70,000 diameters.

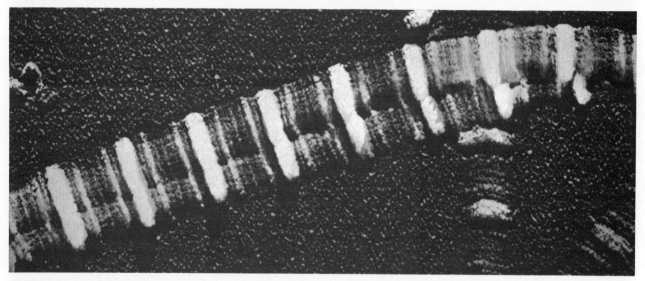

"FIBROUS LONG SPACING" form of collagen is produced by adding glycoprotein to an acid solution of native collagen. The chief feature of this form is its symmetrical intraperiod fine structure. The spacing of the period is about 2,800 angstroms.

"SEGMENT LONG SPACING" form of collagen is produced by adding adenosine triphosphoric acid to an acid solution of collagen. The fine structure is asymmetrical, reflecting the underlying asymmetry of the tropocollagen molecule. The molecular arrangements that produce these three forms of collagen are depicted in the illustration at the bottom of pages 192 and 193.

this new type of structure "fibrous long spacing," or FLS, simply to indicate that it was fibrous and had a long period. The FLS material can be produced in almost 100 per cent yield by adding a negatively charged large molecule (such as the alpha-one acid glycoprotein derived from blood serum) to a dilute acetic acid solution of purified collagen.

We soon discovered that collagen could be recrystallized into still a third type of structure, which lacked the characteristic fibrous or beltlike appearance of the two other forms. The new material looked superficially like isolated segments of FLS, but closer examination revealed that the numerous crossbands were asymmetrically spaced [see bottom illustration on opposite page]. This structure is composed of threadlike units running perpendicularly to the bands. We called the arrangement "segment long spacing," or SLS, to indicate a nonfibrous material with a long period. SLS can be obtained by adding adenosine triphosphate (ATP) to acid solutions of collagen.

Any of the three structurally different forms can be dissolved and converted into either of the other two forms. We concluded that all three forms were being created from threadlike units about 2,800 angstroms long and less than 50 angstroms wide. Since this was evidently the fundamental unit of collagen structure, we named it "tropocollagen" (from the Greek meaning "turning into collagen"). The tropocollagen molecule, as we now know, is composed of the three helical chains.

A more refined estimate of the dimensions of tropocollagen was subsequently made by Helga Boedtker and Paul Doty after they had studied collagen in solution. They estimate that the molecule is 2,900 angstroms long by 14 angstroms wide. Cecil E. Hall of M.I.T. has confirmed these dimensions by electron microscopy of individual molecules [see top illustration on page 190]. On the basis of these and other studies we can postulate how the three basic forms of collagen are assembled from the tropocollagen molecule.

In the native collagen fibril the molecules are lined up facing in the same direction and overlapping by about one-quarter of their length [see illustration at bottom of next two pages]. It is this overlapping that creates the periodicity of about 700 angstroms.

In the FLS form the molecules again lie side by side, but they are not all facing in the same direction and they do not overlap. Since there is no overlapping, the major periodicity measures

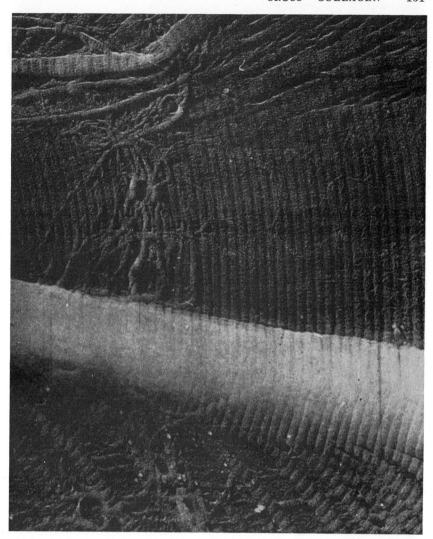

SHEET OF COLLAGEN precipitated from solution shows that tropocollagen molecules will lock themselves into an orderly two-dimensional array. Investigators are seeking conditions that will produce order in three-dimensions. Magnification is 40,000 diameters.

about 2,800 angstroms. The random positioning of "heads" and "tails" of the adjacent molecules accounts for the symmetrical fine structure, even though the discontinuities, or "bumps," along the length of the individual molecule are asymmetrically spaced. This can be understood by imagining two identical but nonsymmetrical molecules lying side by side and facing in opposite directions. If one were to take a blurred photograph, the two molecules would seem to blend into one and the two sets of irregularities would merge so that they would appear to be symmetrically spaced.

In the SLS form the molecules are also nonoverlapping but they all face the same way. As a result the discontinuities of the individual molecules are revealed in the electron microscope as an asymmetrical pattern of bands. Actually it is this pattern in electron micrographs that leads one to infer that the basic molecule contains irregularly spaced discontinuities.

Since these in vitro experiments employ acids in concentrations considered "unphysiological," they left us puzzled as to how fiber formation actually occurs in an animal. We began to gain insight into the natural process about eight years ago when we discovered that a certain fraction of collagen from young animals could be dissolved in cold neutral salt solution, and that simply by warming such solutions to body temperature we could make the dissolved collagen molecules polymerize spontaneously to form a typical crossstriated fibril with the native periodicity. The more rapidly the animal grows, the larger is the amount of collagen that can be extracted. If growth ceases as a result of starvation for a period as short as two days, this collagen fraction disappears from tissues. British investigators have demonstrated that the collagen extractable in cold salt solution is newly synthesized by the cells; it is still soluble because it is not yet tightly aggregated.

As the collagen becomes older, dilute acids are required to extract it; upon further aging it becomes insoluble even in acids.

A similar aging process can be demonstrated in vitro with collagen extracted in cold neutral salt solution. When the solution is warmed to body temperature, a gel composed of typical banded fibrils forms; if it is quickly cooled again, most of the fibrils redissolve. If, however, the gel is allowed to stand at body temperature for 24 hours, it no longer dissolves upon cooling. And if it is aged at body temperature for two weeks, it becomes completely insoluble in dilute acids as well. Since the collagen is highly purified, we do not believe that this time-dependent aging results from interaction with enzymes or other substances. The aging is probably a function of the highly specific structure of the molecule alone. It is likely that if two parallel collagen molecules overlap by a quarter of their length, the charge distribution and three-dimensional configuration of adjacent sections are complementary and therefore attract each other; the "bump" of one fits into the "groove" of the other. The decreasing solubility with time can be explained by the increasing perfection of fit between the molecules as they gradually pack together in a "lock and key" type of association along their length. The "glue" binding the molecules ever more tightly is fundamentally a secondary bond created by electric forces, which rise sharply in strength as surfaces are brought closer and closer together.

These studies suggest a reasonable picture for the first steps in the process by which the body produces collagenous or connective tissue. The fibroblast evidently synthesizes complete collagen molecules (the three-stranded tropocollagen) and extrudes them into the space outside the cells, where they polymerize into fibrils. Although polymerization may require nothing more than time and body heat, other substances in the extracellular environment may play a regulatory role.

We are having difficulty explaining what happens next. How do collagen fibrils become organized into the highly ordered patterns that can be seen in skin, bone and cornea? The cornea, in particular, has a plywood-like structure in which successive layers of fibrils are laid down at right angles to each other in near-crystalline array [see illustrations on page 194]. In studying sections of cornea, Marie A. Jakus of the Retina Foundation in Boston has observed that the fibrils of one layer lie along dark cross-striations in the fibrils below, which are oriented at right angles.

There is still no generally accepted mechanism to explain this intricate ordering. I am inclined to think that the tropocollagen molecules, after extrusion from the fibroblast, form a suspension of liquid crystals (loose semicrystalline aggregates). From such a suspension fibrils could be expected to condense in an orderly fashion. It has been shown that such suspensions of certain rod-shaped giant molecules will turn spontaneously into aggregates that display order in three dimensions. While we have not yet been able to duplicate this experiment with collagen, we have been able to precipitate collagen in broad sheets with a periodicity extending laterally over many square microns [see illustration on page 191]. A significant aspect of this pattern is the appearance of two degrees of order that are mutually perpendicular. Whereas the molecules themselves are laid down in parallel rows, other rows having like electric charge (visible as bands in the collagen fine structure) run at right angles to the molecular rows. One can imagine that with a suitable adjustment of conditions another layer of collagen molecules might precipitate on the first and follow not the molecular rows but the rows of electric charge.

There are a number of questions to answer: If collagen molecules are being secreted at a constant rate from the cells, why should they not form fibrils all

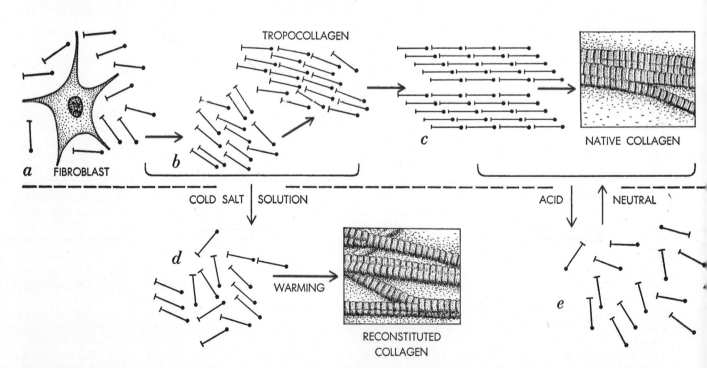

RECONSTITUTION OF COLLAGEN can take place in three basic ways. In an animal, tropocollagen molecules (*b*), manufactured by cells called fibroblasts (*a*), overlap to form native collagen (*c*). Newly formed molecules are soluble in cold salt solution (*d*); simple warming yields reconstituted fibrils duplicating the native form. Alternatively, native collagen can be disoriented in the same direction? And why should there be variations in the number of fibrils stacked one above the other within different layers? These puzzles might be explained if the secretion of collagen molecules from the cells were discontinuous in time, or "pulsed." In different sites at different times the duration of the pulse and the total amount of collagen secreted may differ. The thickness of the collagen layer would be determined by the duration of the "secretion pulse" and the amount of collagen secreted. The interval between pulses would allow time for fibrils to polymerize and would prevent intermingling of collagen in one layer with the next. The collagen molecules laid down during the single secretion pulse would be oriented in one direction only. Molecules secreted in the next pulse might then be oriented perpendicularly to the molecular rows of the preceding layer, being laid down along the rows of electric charge created by the periodicities lying in register.

We would also like to account for the remarkable uniformity in diameter of fibril bundles. Although the diameter varies from one type of tissue to another, it is often constant for each type. Again the idea of discontinuous secretion of collagen molecules seems useful. As a first step we can imagine that a single group of molecules is secreted at about the same time and that these molecules are rather evenly distributed as a loose liquid crystal in a particular layer of extracellular space. The next step calls for the appearance of nuclei of some sort around which the molecules can begin to condense. It seems reasonable that nuclei would appear at about the same time throughout a given volume of space. In the last step the fibrils grow in size until the original supply of collagen molecules is exhausted. We need only specify that the fibrils grow at the same rate to explain how they all end up having about the same diameter. It is possible that the timing and the rate at which fibrils form and grow are determined by noncollagenous substances in the matrix; the viscosity and charge distribution of this matrix might control the freedom of movement of the individual collagen molecules. Attempts at explanation such as these prove nothing in themselves. Their principal value is that they suggest experiments that may lead to better understanding.

One incentive for studying collagen

BLOCK OF COLLAGEN (*right*) forms when a cold neutral salt solution of collagen (*left*) is warmed to body temperature for 10 minutes. Gel redissolves if cooled promptly.

so intensively is that it provides a valuable model for investigating the way in which the body assembles complex, reproducible structures from simple molecular building blocks. Another incentive is the medical one. In its growth and development the organism is continuously remodeling its tissues, a process of exquisite precision with regard to place, time and degree. It is possible that disturbances in the precise sequences of remodeling are responsible, at least in part, for some congenital malformations, for the crippling end results of rheumatic diseases and perhaps even for some of the changes in aging. Whether or not collagen itself is

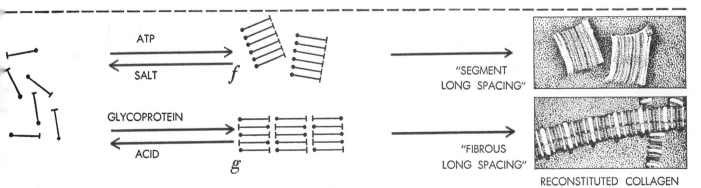

solved in acetic acid (*e*). Treating the resulting solution with adenosine triphosphoric acid produces the nonoverlapping, segment-long-spacing form of collagen (*f*). Treating the solution with glycoprotein produces the fibrous-long-spacing form (*g*), in which molecules face randomly in addition to not overlapping. The fine structure reflects the asymmetry of the tropocollagen molecule.

a target for the causative agent in diseases affecting connective tissue, such as rheumatoid arthritis, is still a matter of dispute. There is little doubt that severe crippling deformities of bones and joints, and the scarring of the heart, kidneys, blood vessels, lungs and other organs, are a manifestation of excessive production and aberrant arrangement of collagen in the affected tissues.

It is characteristic of medicine that frontal attacks on problems such as these seldom yield quick results. There is every reason to hope, however, that studying collagen at the fundamental level of molecular structure will lead ultimately to knowledge that can be applied in medicine.

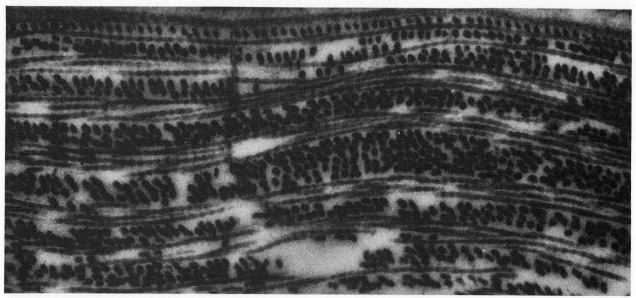

TADPOLE CORNEA consists of layers of collagen fibrils; the fibrils in one layer are at right angles to those in the next. The magnification is 56,000 diameters. This electron micrograph and the one below were made by Marie A. Jakus of the Retina Foundation.

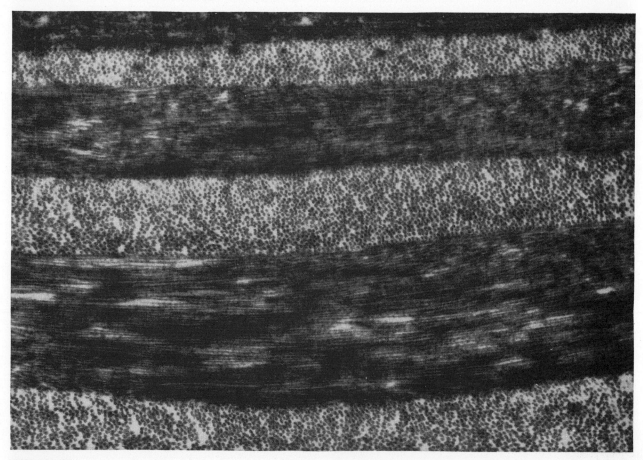

FISH CORNEA surpasses that of the tadpole in precision. In some unexplained manner a random collection of collagen molecules is converted into fibrils that are stacked neatly into layers lying at right angles to each other. The magnification is 28,000 diameters.

Part V

DIVISION AND DIFFERENTIATION

V Division and Differentiation

INTRODUCTION

Cells reproduce themselves by a process that entails the replication of their genetic material and the subsequent division of the rest of the cellular materials. If the program is carried out normally, the result is a pair of daughter cells, each with an identical genetic constitution and appearance. In microorganisms, this is the mechanism of population growth. Some bacteria may divide as frequently as every twenty minutes, and this enables them to exploit a transiently favorable set of conditions by quickly generating a large number of identical cells. The same kinds of divisions determine the cellular composition of a multicellular organism; but in this case the final population of cells is not an identical array but a heterogeneous, cooperating assembly. In such a system the process of growth by division is linked with the processes of differentiation, and of migration and association.

The mechanics of division, and the participation of cellular organelles in it, is discussed by Daniel Mazia in "How Cells Divide." Techniques for removing the spindle apparatus from cells undergoing mitosis have made possible some chemical characterizations of mitotic materials. In addition, the role of centrioles has been analyzed by using electron microscopy and various chemical means of arresting division. Though these are promising beginnings, we still need information about the supply of energy for, and the molecular basis of, the movements of spindle fibers and chromosomes.

The three other articles in this section concern the changes that occur in cells of multicellular organisms during successive divisions in development. One of the striking features of early embryonic development in animals is the degree to which changes in form are generated by the migration of individual cells or groups of cells from place to place. In order to account for the precision that individual cell movements exhibit, we must assume that the cells have means of "recognizing" one another and of thereby selecting the place in which they ultimately settle down in the embryo. A. A. Moscona describes the experimental analysis of such mechanisms in "How Cells Associate." It has been known for some time that dissociated sponge cells or tissue-cultured mammalian cells will "sort out" from mixtures. Moscona describes experiments showing that whereas in sponges the recognition is by species, in vertebrates it appears to be by tissue; indeed, cells from species as diverse as chicken and mouse will cooperate in the formation of mosaic kidneys or other tissues. Such experiments have exposed some of the ground rules that govern cell associations, but they have not yet been successful in identifying the surface-associated materials that bind compatible neighbors together or provide the signals that enable them to recognize one another. The importance of these materials is not restricted to these two functions; unspecified chemical signals are also involved in the influence that one tissue may exert over the course of differentiation of another. Such *inductions* occur in a variety of systems. They illustrate dramatically the variety of external influences to which the genetic equipment of the cell is subjected during the process of differentiation.

Michail Fischberg and Antonie W. Blackler discuss such influences in "How Cells Specialize." Experiments on nuclear transfer from the cells of early embryos into enucleated zygotes reveal, as expected, that the mitotically derived nuclei still possess an unrestricted genetic potential. Thus the cytoplasm exerts the major influence on subsequent differentiation; in fact, the distribution of various components in the unfertilized frog's egg enables one to predict the fates of cells that will arise in various regions during cleavage. Whether or not one accepts the somewhat controversial position, for which the authors offer some favorable evidence, that cytoplasmic influences ultimately produce genetically stable changes in the nuclei, the intimate nature of nuclear-cytoplasmic relationships is a critical problem in development.

In "Chromosome Puffs," Wolfgang Beermann and Ulrich Clever recount what has become one of the most promising approaches to this problem. As is described in the preceding article, the "puffs" observed on giant chromosomes in certain insect cells appear to represent selective synthetic activity of specific genetic loci. Beermann and Clever cite evidence that these puffs are produced only under restricted conditions: in a certain kind of cytoplasm, or at a specific stage of development, or under the influence of a hormone. The genome of each cell is presumed to contain the information required for all the syntheses performed by the organism during its life; but if so, this encyclopedia is read with great selectivity during establishment of the differentiated state.

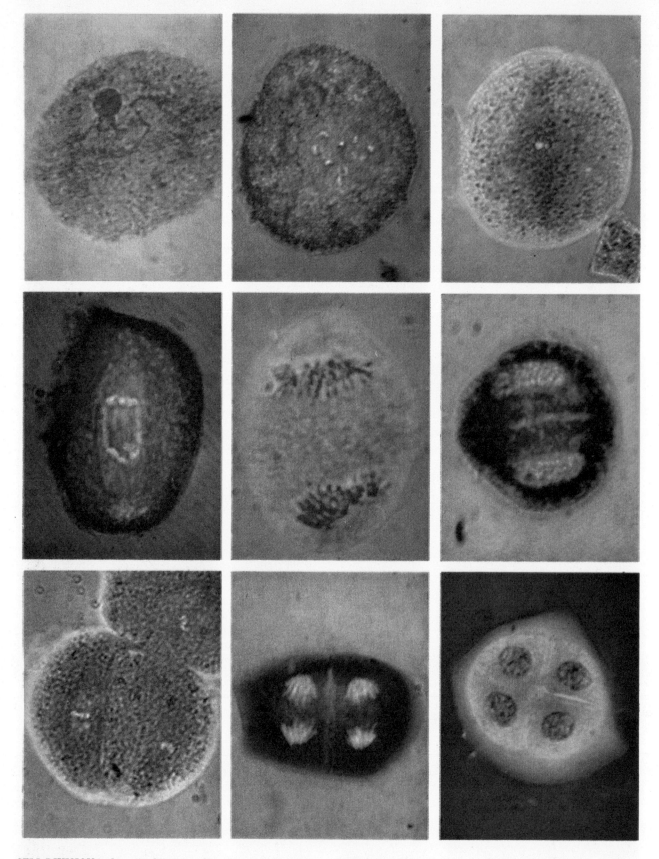

CELL DIVISION is shown in this series of interference-microscope photographs made by Arlene Longwell of the Children's Cancer Research Foundation in Boston. The first four photographs and the seventh show microsporocytes of corn; the remaining four, of wheat. The cells are seen undergoing meiosis, in which two successive cell divisions produce four germ cells. At start of first division (*top left*) chromosomes are seen as thin blue strands; they condense (*white spots at top center*), and spindle (*blue-green "diamond" at top right*) forms. After chromosomes move to spindle poles (*middle left and center*), cell wall forms between daughter cells (*middle right*). In second division chromosomes move to poles (*bottom left and center*), and germ cells form (*bottom right*).

17

HOW CELLS DIVIDE

by DANIEL MAZIA   September 1961

By the reproduction of cells, life thwarts time. Under the best circumstances the life span of individual cells is measured in days, weeks, months—at most in decades; the slope of time is the declivity of aging. But time can be reversed, with 100 per cent profit to boot, by the reproduction of a cell.

Each cell may begin its individual existence endowed with all the potentialities of its parent and may annihilate its individual existence in the production of two cells that inherit those potentialities unaged and undiluted. The daughters of these daughters may do the same and so on to immortality.

Although we shall be dealing here with the ideal case of the indefinite reproduction of cells, producing successive generations of identical individuals, it must be said that immortality need not be quite so monotonous in the real world. In organisms composed of many cells, some cells become different and subserve the needs of those special cells—the germ cells—which are responsible for the continuity from generation to generation. Such differentiated cells usually cease to reproduce and are therefore destined to age. Moreover, mistakes are made in the reproduction of cells; evolution turns such mistakes—mutations—into history.

The over-all reproductive cycle of a cell consists of the doubling of all the components of the cell, followed by a division that distributes the components to the daughter cells. The most fundamental part of the process, because it is the part responsible for the conservation of the character and potentialities of each kind of cell, is the replication of those molecules which carry the genetic code. The identification of the self-replicating molecule of deoxyribonucleic acid (DNA) as the agent of genetic continuity is one of the most impressive—and fateful—accomplishments of modern science [see "How Cells Make Molecules" on page 138].

But the reproduction of cells and organisms is not completely described by, although it is controlled by, molecular replication. Imagine a giraffe reproducing by what might be called the fission method: each molecule in the giraffe would replicate, and the products would sort themselves out into two giraffes—a clumsy process to say the least. In the normal generative scheme a giraffe gives rise to a giraffe egg, which is capable of generating another giraffe. (We ignore the male, whose function is only to provide a little variety.) The generation of the new giraffe depends on the reproduction of cells, which also follows a generative scheme. Only a limited number of molecules, the most important of which are the nuclear genes, are capable of genuine self-replication. These molecules not only reproduce themselves but also control, anew in each generation, the production and assembly of the rest of the materials and structure of the cell.

The essence of the plan (as it applies to the cells of plants and animals generally and to certain one-celled organisms) is that the genetic material is packaged into a small number of chromosomes. The behavior of the chromosomal packages can be observed and interpreted rather easily. In the period between cell divisions, the so-called interphase, the genetic material is contained within a nuclear envelope, but in a highly extended and attenuated form. Normally we cannot recognize individual chromosomes in their longest and thinnest state with the light microscope, and we have not yet characterized them with any certainty with the electron microscope. The duration of interphase in plant and animal cells varies between 10 and 20 hours.

During the period of division, which occupies about an hour (with wide variations, of course), the genetic apparatus goes through a complex but intelligible series of acts. The chromosomes condense to compact bodies. In most cases the nuclear envelope disintegrates. The chromosomes are now part of a mitotic apparatus, the structure of which defines the logic of the process of mitosis. On the cellular scale the mitotic apparatus is a large body. It possesses definite poles, which represent the destinations of the chromosomes, and its "equator" determines the plane through which the cell will divide. By means of the mitotic apparatus the chromosomes are deployed in an exact way. They are first moved to the equator. Then sister chromosomes, the products of the reproduction of each chromosome at an earlier time, split apart and move to opposite poles. The cell now divides through the equator of the mitotic apparatus, producing two cells, each with a full set of replicas of the chromosomes that the parent cell received at the division in which it was born.

The chromosomes of each daughter cell now uncoil. A new nuclear envelope is formed around them, and they are ready to begin careers that will end when each becomes two cells in the same way.

In an idealized version of the reproductive cycle of a plant or animal cell, we observe that it divides into halves, each daughter doubles in mass, seldom growing beyond the mass its parent had at the time of its division, then divides. Division creates the conditions for growth; growth culminates in division. It was quite logical, therefore, to assume that there was a causal connection between division and growth to some critical mass. Unfortunately we must reject this idea, because closer observation has shown that a cell can divide even if it has not doubled its entire mass. An alternative is that certain of the events taking place between divisions can be thought of as specific preparations for division. So long as the cell completes these preparations, it can divide even though it may not have accomplished the normal doubling of other constituents. If this is the case, we cannot limit our study of division to the period when the cell is visibly engaged in the act, because some of the most important events may have taken place beforehand. What are the prerequisites of division?

It is now well known that in plant and animal cells the actual replication of the genetic material—the doubling of DNA—takes place only between divisions. This can best be shown by experiments in which a population of cells is fed for a brief interval with some radioactively labeled substance (usually thymidine) that is built into the newly formed DNA. The newly synthesized DNA is found only in the nuclei of cells that are in interphase—never in cells that are going through mitosis. Refinements of such experiments show that DNA synthesis occupies only a certain part of the period between divisions.

If a given cell is not destined to divide again, as is the case with the cells of many specialized organs (muscles and brain, for example), DNA synthesis does not begin. If it does begin, the rule is that it goes to completion; that is, the original amount of DNA is doubled. A less rigid rule is: If a cell does undertake DNA synthesis, not only is the doubling completed but also the cell will usually enter division. Studies of the intestinal cells of the rat, made by Henry Quastler and Frederick Sherman at the Brookhaven National Laboratory, have shown that every cell makes a crucial decision within the first few hours after division; either it enters DNA synthesis and will divide again or it adopts the career of a differentiated cell and will never divide again. The mechanism controlling this decision is still unknown. This is unfortunate, because it is surely one of the keys to the normal balance of cell division and differentiation and to the disturbance of this balance that is malignant growth.

The replication of the chromosomes merely gives us one cell with a doubled set of chromosomes. To make two cells,

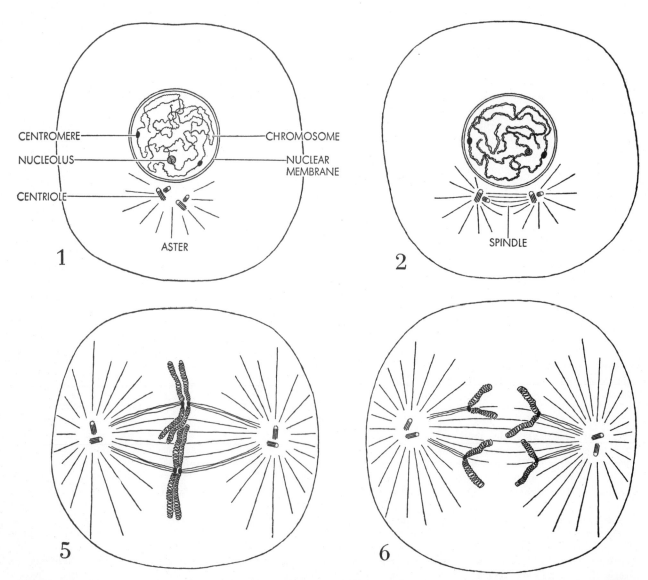

MITOTIC DIVISION of a cell is depicted in these eight drawings. During interphase, the period between divisions (1), the chromosomes are thin, extended threads. Each of two "parent" centrioles is paired with a smaller "daughter" centriole. At some point before division the chromosomes replicate (2), the centrioles begin to separate and the spindle starts to form. In prophase (3 and 4) the chromosomes coil, becoming highly condensed, the nuclear membrane and nucleolus break down and the centrioles move apart to

these chromosomes must move into an equator defined by poles; then sister chromosomes move to opposite poles. In many—perhaps all—cells, the poles that dictate the destinations of the chromosomes are not physical abstractions but definite, and most interesting, physical particles. What is more, the movements of the chromosomes depend on precise physical connections between the chromosomes and these particles.

In animal cells, where such particles can always be found, they were first aptly named "polar corpuscles," but they are generally called centrioles. The centrioles originally could be identified only as small dots that were made visible by staining techniques; further clarification of their structure came only with the electron microscope. In 1956 Wilhelm Bernhard and Etienne de Harven of the Institute for Cancer Research at Villejuif-sur-Seine, near Paris, described the centrioles of a cell in mitosis as cylindrical bodies about .3 to .5 micron long and about .15 micron in diameter, the walls of which consist of fine, parallel, tubular-appearing structures. Further work has shown that the cylinders are formed by nine groups of tubule-like bodies, each group often containing three of the tubules. The same particle can apparently serve in ways other than as a pole in mitosis; for example, the bodies found at the base of cilia and flagella have a fundamentally similar structure.

It must now be confessed that centriolar particles have not been seen in plant cells. Nonetheless the occurrence in plant cells of all the normal and abnormal features of mitosis that can be explained in terms of what we know about animal centrioles leads some of us to the opinion that equivalent particles will yet be discovered.

One of the prerequisites of division, then, is the production of centrioles, in animal cells at least. The most important statement we can make about this is that it is a reproductive process; centrioles are permanent and self-replicating structures. The centrioles generally are found in pairs, and it is a curious fact that the two centrioles of a pair commonly lie at right angles to each other. A cell inherits one set of them and makes two sets.

Something is known about the timing and the sequence of events. Experiments in our laboratory at the University of California had shown that mercap-

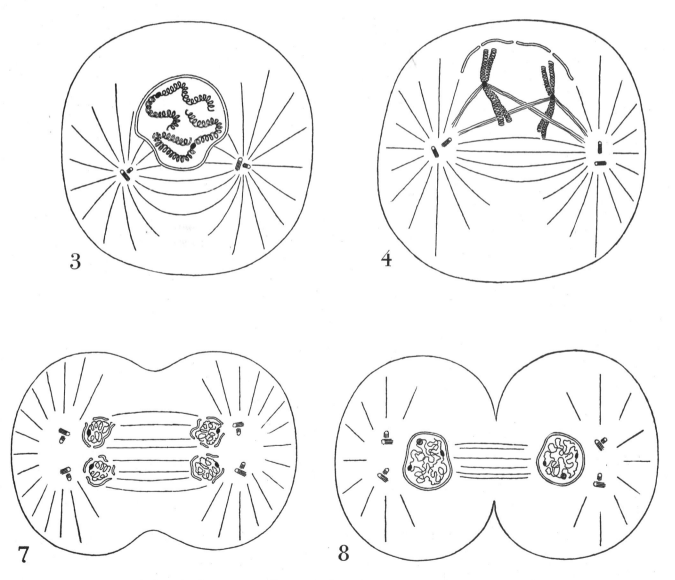

establish poles toward which the chromosomes will move. At the same time connections between the centromeres and the poles are being established. In metaphase (5) chromosomes move to the cell "equator." After splitting apart, sister chromosomes move toward poles during anaphase (6). In telophase (7 and 8) chromosomes uncoil, and the nuclear membranes and nucleoli of the daughter cells are formed. Each centriole has produced a new centriole. When division is completed, both cells will enter interphase (1).

toethanol, which is merely ethyl alcohol with a sulfur atom replacing a hydrogen atom, would block mitosis if applied before the time the chromosomes lined up and began to move. If cells were blocked for a sufficiently long time and then the block was removed, they consistently divided into four cells instead of two! Observation of what was going on inside the cells when they were blocked showed that the poles had split, each had given rise to two poles, and so the cell divided into four because it now had four poles. When the four daughter cells tried to divide, they could not do so at first because their mitotic apparatus had only one pole. (They then corrected the situation by going through an extra cycle of reproduction of centrioles, after which they could divide normally.) A simple interpretation of the observation is this: The poles of the mitotic apparatus are normally double; two actual poles contain four potential poles. While division is blocked by mercaptoethanol, the two units at each pole separate; the four potential poles become the actual poles of a four-way division.

The experiment also told us that the mercaptoethanol did not suppress the separation of existing centriolar units but did block the formation of new ones.

Using this fact, we were able to confirm the prevailing opinion that the replication of centrioles takes place long before division. If a four-way division following blockage by mercaptoethanol means that four potential poles are present, then the cells should be able to divide only in two if they are blocked earlier, before the centrioles have replicated. This turns out to be the case, and by systematic experiments we can determine at just what stage two potential poles give rise to four potential poles. By this test it was found that the decisive event in the generation of new centrioles takes place long before division; in fact, it seems to take place during the last part of the previous division. As the parent cell is dividing, it is conceiving centrioles for the next division.

From these experiments we conclude that the doubleness of the centriole depends on what may be called a generative mode of reproduction. On the molecular scale the centriole is a large three-dimensional body; it is difficult to imagine such a body making a copy of itself the way a strand of DNA does. But the first step could be the replication of a molecule carrying all the information for making a new centriole, just as the first step in the reproduction of a complex virus is the replication of the nucleic acid that will later assemble the other structures making up the complete virus. If some time must elapse between the conception of a new centriole and the completion of its development, then its fundamental doubleness can be taken as representing the coexistence of two generations. If we could see the production of new centrioles, we should expect to observe fully developed units with the new generation growing up beside them. This is exactly what has been seen in the electron microscope by Joseph G. Gall of the University of Minnesota [*see bottom illustration on opposite page*]. It is interesting enough that the reproduction of a particle takes place by the outgrowth of the daughter from the parent particle, but it is astounding that the new particle should invariably grow at a right angle to the old.

Once they have reproduced, the centrioles move apart. Their separation polarizes the cell for mitosis. When we have located the poles, we can tell where the chromosomes will go and through what plane the cell will divide. The separation takes place long before division in some kinds of animal cell; in others it occurs abruptly, just before the chromosomes begin their mitotic maneuvers. Superficially it has all the attributes of a repulsion; the polar particles move apart in a straight line. Measurements by Edwin W. Taylor of the University of Chicago indicate that this movement takes place in cells of newts at a constant velocity of about one micron per minute. We must not take the analogy of a repulsion literally. A more apt image is that the poles are pushed apart by the growth of fibers that continue to connect the poles and that together are called the central spindle. This is descriptively correct, but it remains to be explained how the growth of the central spindle is translated into an actual movement of the centriolar bodies.

The essence of the plan of mitosis is clear, and the precision is secured in a uniquely biological way. The centrioles double exactly, and the products separate to form two poles—no more and no less. The chromosomes reproduce exactly, and sister chromosomes are transported to sister poles. The rest of the story is a tale of complex molecular mechanics into which we are just beginning to gain some insight.

Since the act of mitosis involves the performance of work, it must also require the expenditure of energy [see "How Cells Transform Energy" on page 85]. Experiments suggest that the ener-

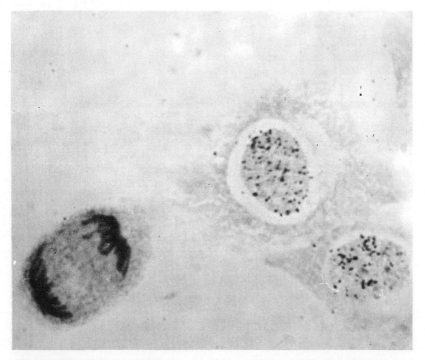

CHROMOSOME REPLICATION takes place during interphase, as this autoradiograph made by J. Herbert Taylor of Columbia University demonstrates. Two Chinese hamster cells (*right*) in interphase have incorporated radioactive cytidine (*numerous dark grains*) in their nuclei. A third cell undergoing division (*left*) contains none. Since cells incorporate cytidine only during synthesis of RNA and DNA, replication of chromosomes (containing DNA) must occur between cell divisions. Cells are magnified here approximately 1,200 diameters.

getic expenses of division are met by a paid-in-advance accumulation of energy. Up to a point, as a cell proceeds toward division, it can be brought to a halt by depriving the cell of oxygen or by poisoning its oxidative enzymes with carbon monoxide. But when the cell reaches a certain point—about the time the chromosomes are coiling up—it is no longer possible to stop the division by throttling the oxidations. Michael M. Swann of the University of Edinburgh concludes that the preparations for division include the filling of an "energy reservoir" that is adequate to meet the requirements of mitosis. The chemical identification of the energy reservoir may be one of the important problems of research on cell division.

Once these molecule-building preparations have been completed, the cell ordinarily is committed to enter mitosis. Whether a further special "trigger" is required or whether the completion of the last of the synthetic preparations is itself the trigger, we do not commonly encounter cells that are stalled on the brink of mitosis.

If the essence of mitosis is the movement of sister chromosomes to sister poles, the inauguration of the process calls for the establishment of connections between chromosomes and poles. But before the connections are established a radical rearrangement of the structure of the cell takes place, the prophase, which is intelligibly a mobilization for action.

The chromosomes condense into visible threads. What we see suggests that this condensation is largely a matter of packing the chromosome strands into tight coils and then imposing still another order of coiling. As a coiled coil, the chromosome exhibits handsomely the theme of helical design that pervades the study of molecular order. Although we do not know the inner mechanisms of this large-scale coiling, its significance is clear enough. It converts a tangle of long and tenuous threads into compact masses that can be moved freely and without entanglement. Fully extended, the DNA in a human nucleus could make a thin thread some 10 million microns—a meter—long. Packaged into chromosomes, it is deployed as two sets of 46 chromosomes, each a few microns long.

Toward the end of the period during which the chromosomes are coiling up, the nuclear envelope in many kinds of cell disintegrates. It is easy to understand this as the removal of a barrier between the chromosomes and the poles; it is more difficult to understand the

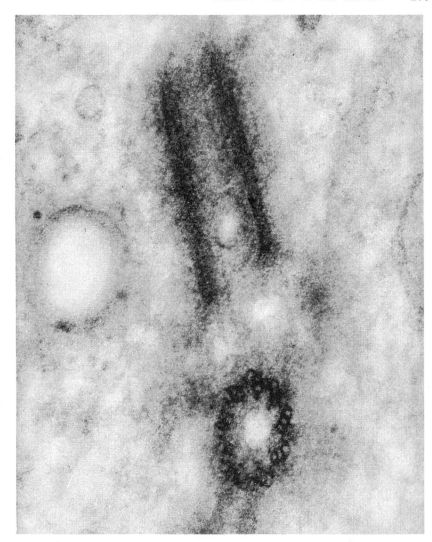

SISTER CENTRIOLES in a human tumor cell lie at right angles to each other. The cross section of one (*bottom*) shows that the centriole consists of nine groups, each containing three "tubules." Magnification of this electron micrograph, which was made by Wilhelm Bernhard of the Institute for Cancer Research at Villejuif-sur-Seine, is 160,000 diameters.

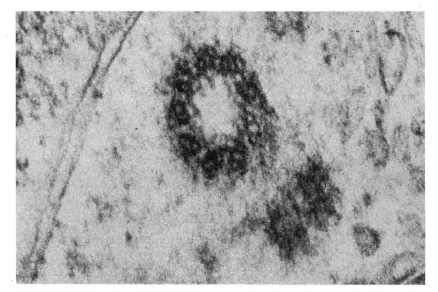

REPRODUCTION OF A CENTRIOLE in a cell from the snail *Viviparus* is magnified 175,000 diameters in this electron micrograph made by Joseph G. Gall of the University of Minnesota. The short daughter centriole, which is seen in longitudinal section near lower right, grows out at right angles from the parent centriole, seen in cross section at center.

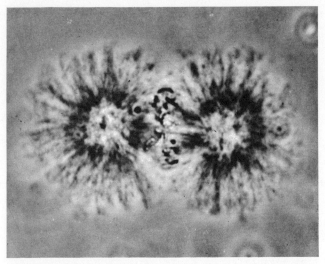
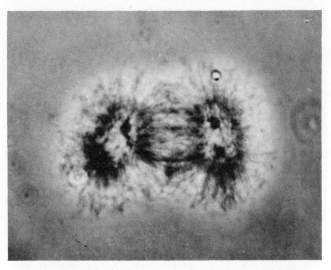

FOUR MITOTIC POLES are formed by centrioles in sea-urchin eggs when the movement of chromosomes during mitosis is blocked by the use of mercaptoethanol. The chromosomes (*seen in thin vertical grouping at center of photograph at left*) remain in the spindle as the centrioles split and separate (*second from left*). The centrioles then begin to move apart at right angles to the spindle

cases in which the envelope persists.

At the same time the body of the mitotic apparatus is assembling. We have described how the poles are established. The final destinations of the chromosomes are now fixed. Between the poles and around the nucleus we can often detect a gathering of a mass of material, still not well organized, that will be the mitotic apparatus. Descriptively we have every reason to say that the substances of the mitotic apparatus were originally spread through the whole cell but now are gathered in and concentrated by the centrioles, although we do not actually know how this comes about. In some kinds of cell we have the impression that the material of the future mitotic apparatus is collected inside the nucleus.

Only now—with the chromosomes neatly consolidated, the poles established and the substance of the body of the mitotic apparatus gathered—can the action begin. The chromosomes come under the control of the poles and begin to move. We describe this climax so cursorily because we know so little about it; actually it contains the essential mystery of mitosis.

The proper execution of the mitotic maneuvers demands strict obedience to a rule: All chromosomes must be engaged to a pole but it is prohibited that two sisters engage to the same pole. What we see suggests the establishment of physical connections—which we shall call fibers without invoking any special properties—between centrioles and chromosomes. We must also take into account another body: the centromere, or kinetochore. This is the anchor point on the chromosome at which its connection to its pole is made. The kinetochore has a constant position on each chromosome; and when we say that a chromosome in mitosis is V-shaped or J-shaped, it is because it consistently behaves as though it were being dragged by an attachment at its middle or toward one end. Clearly the kinetochore is that part of the chromosome which takes part in mitosis; the rest of the chromosome goes along for the ride. Yet we have no really detailed knowledge of this remarkable body.

Once the chromosomes are engaged by the poles and begin to move, the movement proceeds in two steps. First the still-paired sisters move into the

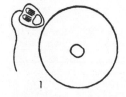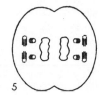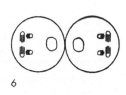

NORMAL CENTRIOLE REPRODUCTION is depicted here. After fertilization of egg (*1*) the parent-daughter pair of centrioles (*small cylinders in 2*) split, each producing a smaller daughter centriole (*3*). Centrioles form two mitotic poles (*4*). As cell di-

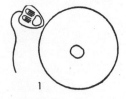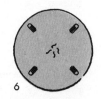

ABNORMAL CENTRIOLE REPRODUCTION was induced in sea-urchin eggs in an experiment performed by the author and his colleagues to show that a single mitotic pole (formed by a centriole) consists of two potential poles. In first part of experiment (*1, 2 and 3*) centrioles reproduced normally (*see first three drawings in illustration above*). Mercaptoethanol was then added (*4, 5 and 6*):

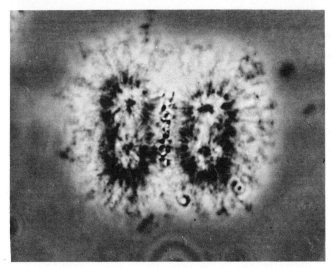
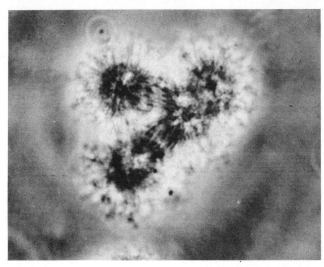

axis (*third from left*), moving farther apart to form a spindle with four poles (*right*). The shape of this mitotic figure is that of a pyramid with a triangular base. The magnification of these four phase-microscope photographs, which were taken in an experiment at the author's laboratory (*see illustrations at bottom of these two pages*), is approximately 1,100 diameters.

equatorial plane defined by the poles; then they split apart and move to the poles. We can understand this by the simple and instructive (but not necessarily faithful) image of a little puppet performance. The sister chromosomes, still paired and so connected by strings to both poles, are pulled into the equatorial—metaphase—position under equal tension from the two poles. When they split apart, the same tension carries them toward the two poles.

The separation of the sister chromosomes and their migration toward the poles—the anaphase movements—have been described and measured in great detail in recent years, thanks to better microscopes, the motion-picture camera, better methods of maintaining cells alive under the microscope and great expenditures of patience. On the cellular scale the distances traveled can be considerable: between five and 25 microns. The velocity of the movement is about one micron a minute; this works out to four hundred-millionths of a mile per hour, which is not a sensational speed. The chromosomes move in straight lines, usually converging on the poles. Often, as the chromosomes move to the poles, the poles themselves move farther apart, carrying the chromosomes with them. Commonly, but not always, the chromosome-to-pole movement precedes the further separation of the poles. The marveling observer has the impression that the chromosomes are pulled to the poles and then dragged by the poles as the poles push apart. The shapes of the chromosomes in movement, so often those expected of a flexible body being dragged through a liquid by a thread attached at one point, reinforce this impression.

Much of what has been said here is descriptive, but it is nevertheless intelligible. We perceive a consistent plan, followed in principle by a vast variety of cells, that does achieve the required end of genetic distribution. The meaning of each structure and step in relation to the others is clear, and the consequences of failure in any respect are predictable. Description is not necessarily "mere" description. Yet it is "mere" if we accept the objectives of contemporary cell biology, the ideal of which is a molecular (and submolecular and supermolecular) accounting for precisely those biological operations which, like heredity or mitosis, are already reasonably comprehensible in their own terms.

If mitosis is not desperately discouraging as a problem of molecular biology, it is because the complex operations are embodied in a definite structural assembly—the mitotic apparatus—that can be regarded as a gadget for perform-

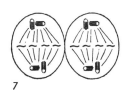
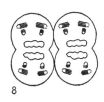
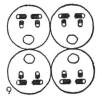

vides (*5 and 6*), parent-daughter pairs split and produce four new poles. When daughter cells divide (*7, 8 and 9*), this process is repeated.

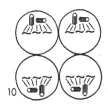

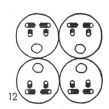

centrioles formed poles (*4*), separated without duplicating (*5 and 6*) and formed a tetrapolar spindle when the mercaptoethanol was removed (*7*). Four daughter cells formed and each centriole reproduced (*8 and 9*). Each cell, with half the normal number of centriolar units, formed a mitotic apparatus with one pole (*10*). Centrioles duplicated (*11*) and cells were ready to divide normally (*12*).

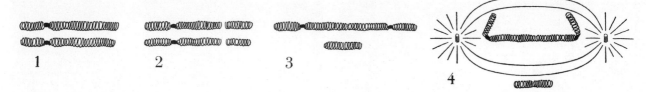

ROLE OF CENTROMERES, the sites (*small black ovals in 1*) at which chromosomes are connected to the poles of the mitotic apparatus, is illustrated by the experiment depicted here. If exposed to ionizing radiation, the chromosomes break (2). Broken ends always rejoin, but do so in various ways. As shown (3), the fragments with centromeres have joined together, as have those without centromeres. In mitosis (4) the former is pulled toward opposite poles; the latter does not take part in mitosis.

ing the operations. We can approach the physics and chemistry of mitosis through the study of the formation, structure and changes of the mitotic apparatus, without forgetting that mitosis is an operation of the whole cell.

Let us consider the fully formed mitotic apparatus at a crucial stage in mitosis: the metaphase, when the chromosomes have lined up on the equator but have not yet begun to move to the poles. The light microscope sees the chromosomes in a spindle, a body between the poles that has been thought to consist of fibers connecting pole to pole, fibers connecting chromosomes to poles, and a matrix of a rather undefined character. In animal cells the aptly named asters often radiate from the poles.

The mitotic spindle has been described as a gel, as a coherent body of limited rigidity and as a loose aggregate or network of chains or sheets of molecules. As J. Gordon Carlson of the University of Tennessee and others have shown, it can be pushed and poked about in the cell. That the spindle often appears as a transparent region against a more turbid background suggests that the assembly of materials to form the spindle excludes large particles such as mitochondria from that region, and this is confirmed by electron microscopy. The polarization microscope reveals that the molecular components of the spindle tend to be oriented along the pole-to-pole axis, in keeping with the impression of pole-to-pole and chromosome-to-pole "fibers." Recent advances in electron microscopy—especially in the preparation and fixing of cells for inspection in the microscope—go far toward confirming the impression. Pictures made by K. R. Porter of the Rockefeller Institute and by Bernhard and De Harven show fine straight filaments, usually double and sometimes occurring in bundles, running from the kinetochores to the vicinity of the centrioles. These are sometimes described as being tubules about 150 angstrom units in diameter. But this refers only to the image provided by the electron optics and does not mean that we actually are dealing with hollow pipes. It is these filaments that shorten as the chromosomes move to the poles and lengthen when the poles move apart. We are inclined to assign them an important part in the movement of the chromosomes. The picture remains, however, distressingly incomplete.

Obviously we shall not achieve a molecular analysis of mitosis before we know something about the molecules of the mitotic apparatus. The most straightforward way of going about this is to isolate the mitotic apparatus from dividing cells. For this purpose we require an abundant supply of cells in division, and this can be had. Marine organisms such as sea urchins produce great quantities of eggs. When such eggs are fertilized in the laboratory by mixing them with spermatozoa, they proceed to divide synchronously. Therefore one can obtain gram quantities of cells in division.

But the mitotic apparatus is notoriously evanescent. As a structure that assembles at the time of division and is dismantled when division has been accomplished, it is clearly not a permanent organ of the cell. Its chemical instability is revealed when one tries to isolate it; under most conditions it simply vanishes. In 1952 Katsuma Dan of the Tokyo Metropolitan University and I succeeded in isolating the mitotic apparatus [see "Cell Division," by Daniel Mazia, page 199, for further information]. It was evident from the first that a price of chemical damage had to be paid for the isolation of a stable mitotic apparatus; the aim of developing improved methods has been to reduce the price by seeking gentler procedures.

How could it be that the structure

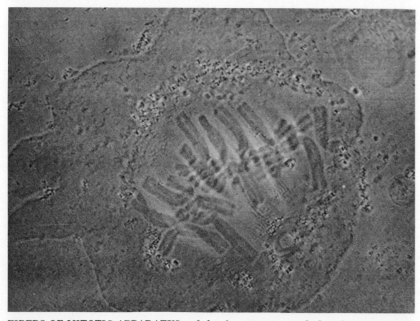

FIBERS OF MITOTIC APPARATUS and the chromosomes attached to them are seen in this polarization-microscope photograph made by Shinya Inoué of the Dartmouth Medical School and Andrew Bajer of the University of Cracow in Poland. The living endosperm cell containing them is that of the flowering plant *Haemanthus katherinae*. Sausage-like structures with faint lines running through their central axes are pairs of sister chromosomes. The thin bright lines perpendicular to the chromosome axes are the fibers, connected to specific chromosomal regions: the centromeres. Magnification is 1,500 diameters.

that held together within the cell was so incoherent once it was outside the cell? Obviously the cell was providing, in its internal environment, some protection for the structure. From evidence that sulfur bonds play an important role in holding the mitotic apparatus together in the cell and by a rather complex argument, I guessed that such protection might be given by a compound incorporating sulfur-to-sulfur bonds. The next compound tried was dithiodiglycol ($HOCH_2CH_2S$—SCH_2CH_2OH). J. M. Mitchison of the University of Edinburgh and I found that the addition of dithiodiglycol to a sucrose or dextrose medium did protect the stability of the mitotic apparatus, and the apparatus could be isolated merely by disrupting cells in such a medium. They could then be purified by further washing in the same medium. Such isolated preparations are being used in most of the current work in our laboratory at the University of California.

After eight years of study of the isolated mitotic apparatus, what have we learned? The reader who expects to be told how chromosomes are moved may skip the next few paragraphs. We have learned something about the kinds of molecule that are present in the apparatus and how they are put together; perhaps this is all that can be expected from studies of isolated parts of cells.

The mitotic apparatus contains a great deal of protein. John Dale Roslansky and I found that it represents an investment of at least 10 per cent of all the protein in the dividing sea-urchin egg. Is this protein synthesized at the time of division or is it preformed and then assembled at the time of division? Hans Went, now at Washington State University, attacked the question by an immunological method, asking whether the isolated mitotic apparatus contained any proteins—detected as antigens—that were not already present in the cell before division. Thus far no such antigens have been found, so we infer that the synthesis of the proteins of the mitotic apparatus is one of the prerequisites of division. A cell must anticipate division by providing these molecules.

Arthur M. Zimmerman, now at the Downstate Medical Center of the State University of New York, has made a careful study of these proteins. So far the picture has been surprisingly simple in the sense that most of the protein in the isolated mitotic apparatus seems to be of one kind, although there must be many other kinds that are present in smaller amounts.

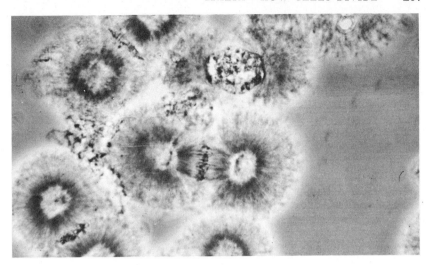

ISOLATED MITOTIC APPARATUSES from sea-urchin eggs are magnified 1,000 diameters in this phase-microscope photograph made by the author. In apparatus just below the exact center of the photograph the two light areas are mitotic poles; structure between them is the spindle. Chromosomes are seen as a thin, dark area at center of spindle.

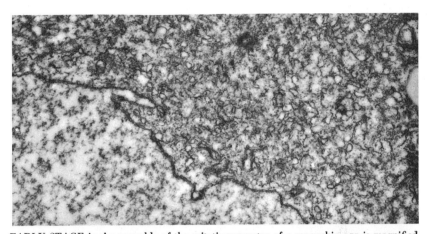

EARLY STAGE in the assembly of the mitotic apparatus of a sea-urchin egg is magnified 16,000 diameters in this electron micrograph made by Patricia Harris of the University of California. The apparatus is beginning to form around a centriole (*small dark C-shaped structure at top center*), which will be at one pole of the future apparatus. At such an early stage the nucleus (*large light gray area*) is still surrounded by the nuclear membrane.

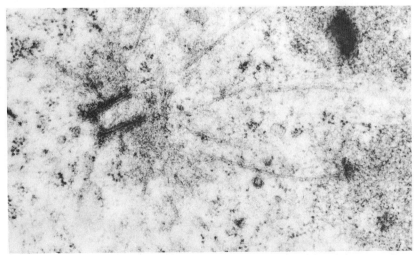

CENTRIOLE AND CHROMOSOMES, seen respectively at left and at right (*amorphous gray areas*), are connected by spindle fibers in a chicken-spleen cell. One fiber is attached to centromere (*small black area at lower right*). Micrograph was made by Jean André of the Institute for Cancer Research at Villejuif-sur-Seine. Magnification is 53,000 diameters.

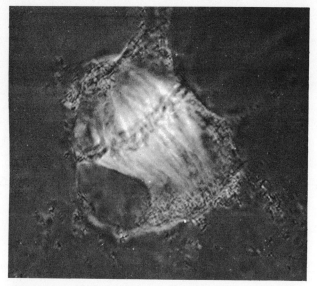 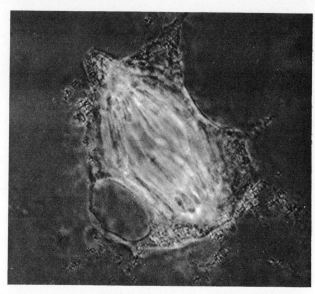

CHROMOSOME MOVEMENT in living endosperm cell of *Haemanthus katherinae* is seen in these three polarization-microscope photographs from a film sequence made by Inoué and Bajer. During metaphase (*left*) the chromosomes are aligned at the equator of the spindle. The chromosomes then move toward the mitotic poles (*second from left*) with their "arms" trailing, as though

The mitotic apparatus also contains ribonucleic acid (RNA), and much of the RNA seems to be associated with the major protein. The function of this RNA is a puzzle. RNA is usually identified with protein synthesis, but the mitotic apparatus does not seem to be manufacturing protein. The RNA associated with the mitotic apparatus may have something to do with the assembly of this structure and not merely of its molecules. It is tempting to imagine that genetic information is involved in the architectural activities of the cell as well as in the shaping of the bricks.

Our recent work has also shown the presence in the mitotic apparatus of a considerable amount of lipids, the fatty molecules that are so prominent in other kinds of structure, such as the external and internal membrane systems of the cell, the mitochondria and so on. Perhaps the lipids account for the presence in electron micrographs of so many vesicles, membranes and tubular-appearing structures in the mitotic apparatus.

Our early experiments were guided by the theory that the molecules in the mitotic apparatus were held together by disulfide bridges, chemical bonds between sulfur atoms on neighboring protein molecules. It was proposed that the assembly of the apparatus was essentially a process of establishing such bridges. More recent work, particularly in Dan's

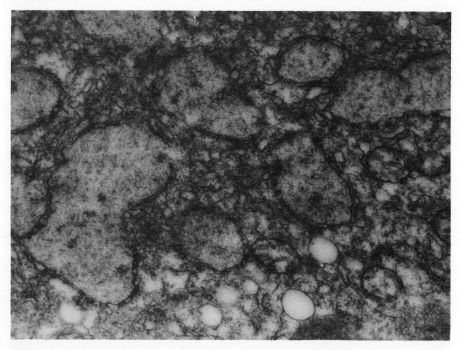 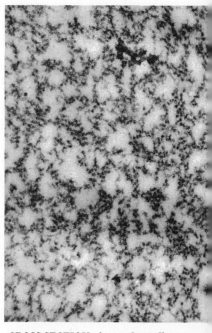

NUCLEAR MEMBRANE (*doubled lines surrounding small gray areas*) first re-forms around individual chromosomes as they approach mitotic pole at end of division. As chromosomes "flow" together, separate membranes form one. This electron micrograph of a cell from a sea-urchin embryo was made by Patricia Harris. Magnification is 20,000 diameters.

CROSS SECTION of central spindle (that part of the spindle connecting the mitotic poles) of *Barbalunympha*, a flagellate protozoon, is here magnified 18,000 diameters. The individ-

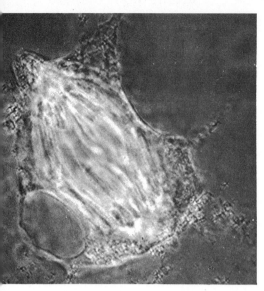

they were being pulled by their centromeres. The centromeres reach the poles (*right*). Magnification is 1,200 diameters.

laboratory in Tokyo, has shifted the emphasis from disulfide bonds to interactions that still involve sulfur-containing groups but which are not necessarily fully oxidized disulfide bonds such as are formed in such stable structures as hair and vulcanized rubber. Using a method that combines a colored compound specifically with the thiol groups (—SH groups) of proteins, Dan and a student of his, N. Kawamura, were able to show that the assembly of the mitotic apparatus is an ingathering toward the centrioles of proteins having a particularly high content of thiol groups, and that when the mitotic apparatus goes through the period of transport of the chromosomes to the poles (the anaphase), these thiol groups disappear, only to reappear at the next division. We would like to know how the congregation of proteins rich in thiol groups is related to the assembly of the mitotic apparatus, and whether the disappearance of these groups, perhaps by oxidation, is part of the chemistry of the chromosome movement itself. It is tantalizing to have so much evidence that the sulfur-containing groups are particularly important in mitosis without knowing just how or why.

The chemistry of movement in biological systems has provoked biologists throughout the recent history of the life sciences. The most influential idea is that movement somehow involves the reaction of the motile system with adenosine triphosphate (ATP) and the splitting of phosphate groups from the ATP. Proteins involved in movement—and not just proteins in muscle—are expected to react with and split ATP [see "How Cells Move" on page 268]. The early methods of isolating the mitotic apparatus did not, however, yield any material capable of this reaction. Using the newer sucrose-dithiodiglycol medium, Ray M. Iverson (now at the University of Miami), Rowland C. Chaffee (now at the University of California in Riverside) and I have discovered an active enzyme in the mitotic apparatus that splits ATP. So far as the observations go, they do favor the prediction that proteins of the mitotic apparatus would, like the contractile proteins of muscle, react with and split ATP.

We may let our love of unity take us still a step further, asking whether the mitotic apparatus is a system of contractile fibers—a little muscle. Fibers connecting chromosomes to the poles, and also fibers running from pole to pole, were seen long ago in microscope preparations of dead cells and are now seen with the electron microscope, but until they were observed in living cells in mitosis they could be discounted as artifacts of the preparation methods. The observations of live cells in polarized light by Shinya Inoué of the Dartmouth Medical School, who has designed advanced polarization microscopes for this purpose, leave little doubt that the fibers of the mitotic apparatus are real [*see bottom illustration on page 206 and illustration at top of pages 208 and 209*].

But a crude image of a system of contractile fibers will not take us far. The chromosome-to-pole fibers shorten to a fraction of their original length, if not to the point where they simply vanish. The pole-to-pole fibers grow longer, sometimes much longer. As the fibers be-

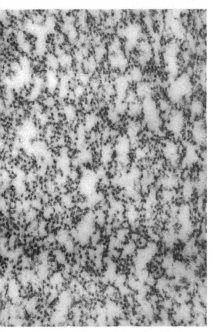

ual fibers appear to be tubular when they are viewed in cross section. This electron micrograph and that below were made by Joan Erickson Cook of the University of California.

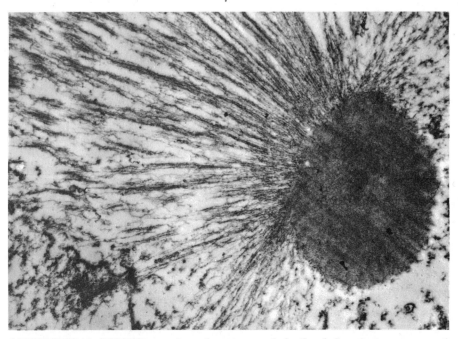

LONGITUDINAL SECTION seen here shows structural details of the mitotic apparatus of *Barbalunympha*. The main mass of fibers radiating from the end of the large centriole (*right*) runs to the other pole (*not shown*). A centromere (*small, dark crescent at lower left*) connects a chromosome (*small gray area farther to left*) to several fibers. Scalloped line passing between them is the nuclear membrane. Magnification is 12,000 diameters.

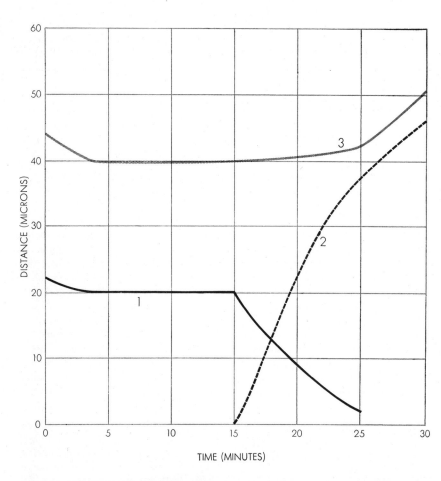

CELL MOVEMENTS IN MITOSIS are here plotted in three curves that show the changes in distance between chromosomes and poles (*1*), between sister chromosomes (*2*) and between poles (*3*). Curves are keyed to drawings at top, which indicate the distance plotted. On the time scale at bottom "0" marks beginning of anaphase, when chromosomes are aligned at the cell "equator." The sister chromosomes separate after 15 minutes (*1* and *2*).

come shorter or longer they do not become thicker or thinner, nor do they become less straight. The filaments seen in the electron microscope [*illustrations on page 209*], seem to retain the same diameter although they grow shorter or longer. In fact, we wonder if the "contraction" of fibers in the mitotic apparatus is not a shortening due to the actual removal of molecules and if the elongation is not a growth in one dimension, the addition of molecules. The question would be whether or not the removal or interpolation of substance could be carried out in such a way that the fibers could pull or push a mass. The growth itself can be accounted for by a model proposed by Inoué. He sees the molecular elements of the mitotic apparatus as being in two states: oriented (fibrous) and disoriented. Their transition from one state to the other is bound by an equilibrium such that the proportion of the material in the oriented state changes in response to the conditions in the cell as a whole.

Once the chromosomes have separated into two groups, the organization of two interphase nuclei, with their attenuated chromosomes contained by the characteristic envelope, begins. Only a few details of the reconstruction of nuclei are known; for instance, electron microscopy gives us the impression that the nuclear envelopes are not brand new but are made by assembling fragments of membranous material from the surroundings.

To the observer at the microscope the most remarkable event of cell division is the pinching of an animal cell into two, or the appearance in a plant cell, as though from nowhere, of a wall between the nuclei that have just gone through mitosis. The most ingenious theories have been proposed; for example, the idea that the cell surface forms a contractile belt around the equator, or that the cell surface can expand and push into the equatorial plane.

The requirement of a successful theory is that it explain how the poles of the mitotic apparatus can dictate the laying down of new cell membrane at the equator, whether the appearance is that of a furrowing of the original membrane or the building of a partition from within. If the mitotic apparatus is pushed to one side or rotated by 90 degrees, the plane of division will be displaced correspondingly, as has been shown recently by K. Kawamura at the University of Tennessee. Yet the completion of the act of division does not seem to require the immediate co-operation of the mitotic apparatus. Y. Hiramoto of the Misaki Marine Station in Japan has been able to remove the mitotic apparatus from dividing sea-urchin eggs, literally sucking them out of the cell by means of a very fine pipette controlled by a micromanipulator. If the apparatus was removed some time before the cell body began to divide, no division occurred. But if the operation took place just before division, at the time when the chromosomes were moving to the poles, then division proceeded. On the other hand, division does not depend on the chromosomes. We know this from various experiments in which the mitotic apparatus goes through its performance after the chromosomes have been removed.

An account of mitosis and cell division sounds more like the libretto of an Italian opera than like a page from Euclid. Cell reproduction is not a unit process, and it is not to be described by an equation. Its essence is the doubling of all the potentialities of a cell—the generation of twoness. The twoness is not only doubleness of quantity but the twoness of separation and independence. Indeed, we have seen that all the doubling of molecules takes place before mitosis, and only then is the material of one cell remolded into two cells.

Basically, biological increase is a scale-of-two process. A single cell can only attain a limited degree of growth. The limit seems to be a restriction of the domain of living mass that can be administered by a single nucleus. The limitation is not an exhaustion of the capacity for growth; if, after a cell has reached its maximum size, a piece is amputated, the cell will grow back to the maximum size but no larger. The genetic material of a plant or animal cell, measured as DNA, can only double, and it cannot increase further until the chromosomes have gone through the mitotic cycle. If the mitotic apparatus is damaged by drugs such as colchicine, the split chromosomes will not separate and will be retained in a single nucleus. The "polyploid" cell thus produced can grow to a size that is proportional to the number of sets of chromosomes present. If normal mitosis takes place but cell division fails, the cell with two nuclei can also grow to twice the normal size. If the whole division cycle is normal, each of the daughter cells can grow to the size attained by the parent cell.

Reproduction as the generation of twoness is not merely another interesting biological phenomenon. In it we can find reasons for everything else that happens in living things. Indeed, the biologist, unlike some other brands of scientist, can allow himself such a term as "reasons" because he does have a standard of judgment: the unambiguous criterion of survival.

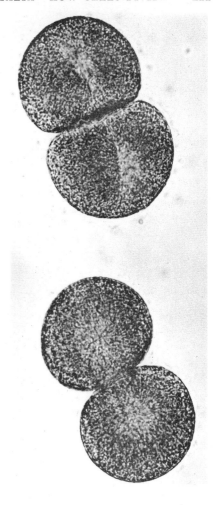

DIVISION FURROW, which pinches cells in two at the end of mitosis, is depicted in sea-urchin eggs. The furrow in the egg at top is almost complete; the furrow in the egg at bottom is at an earlier stage. The mitotic apparatus is still faintly visible. This photomicrograph, made in the author's laboratory, enlarges cells 800 diameters.

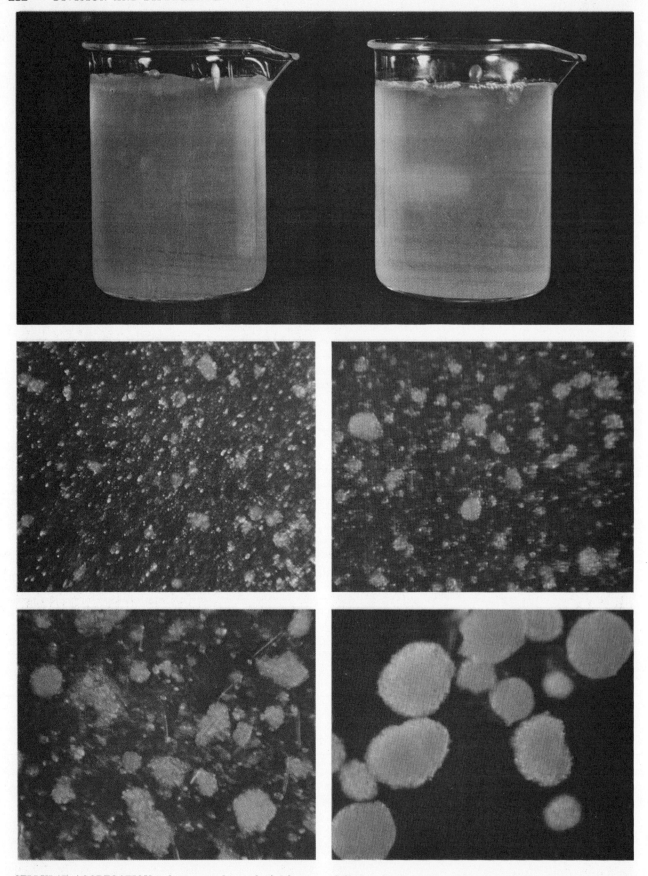

CELLULAR AGGREGATION is demonstrated in author's laboratory at the Marine Biological Laboratory in Woods Hole, Mass., using naturally orange sponge *Microciona* and yellow sponge *Cliona*. Solutions containing cells from each are in beakers at top. Cells are mixed together (*photomicrograph at middle left*). In the course of 12 hours they creep along the floor of the dish in which they have been placed and clump together by species, finally forming tiny orange sponges and yellow sponges (*bottom right*).

HOW CELLS ASSOCIATE

by A. A. MOSCONA   September 1961

To explain how cells join with one another to form the tissues and organs of multicellular organisms, the biologist must answer questions that are as basic and pressing in their way as those that surround the nature of the chemical bond. In the absence of the intercellular bonds that hold cells together, the human body would collapse in a heap of disconnected, individual cells, many of them quite indistinguishable from certain free-living protozoa. Were it not for the high specificity of these bonds and the selectivity with which cells interact with one another, there would be no tissues or organs, only nondescript clumps of cells. To devise an approach to these questions—to submit masses of cells to experimental test as they proceed to associate, interact and synthesize tissues—challenges the ingenuity of the investigator.

The study of cell association proceeds along the parallel paths of analysis and synthesis. Since the turn of the century workers in this field have been developing techniques of tissue culture that make it possible to study tissue cells in the simplified environment of laboratory glassware and, by one means or another, to cause the tissues to dissociate into cells. Biochemical analysis has sought to identify the substances involved in the bonding and interaction of cells; morphological analysis, facilitated by the electron microscope, has concentrated on the connection between function and structure. But it is the relatively novel and direct method of synthesis—the experimental synthesis of tissues from free cells under controlled conditions—that offers particular promise in this field. Only by such frontal approach can one put hypothesis to the test and find out how cells actually associate.

In nature, as the fertilized egg proliferates into a mass of rapidly dividing cells, the cells first bunch together in no clearly apparent order. But the lack of order is only superficial. The cells have fundamentally identical genetic endowments. Their initial diversification must arise, therefore, in large measure from their different positions in the embryo. There is an impressive body of evidence for this. In the early embryo, for example, one can graft cells from a skin-forming area to the eye-forming one. The grafted cells develop in harmony with their new site, acquiring their neighbors' "eyeness" as their persisting identity, recognized as such by their kind and by other cells. If they are thereafter transferred to other sites, they remain unchanged.

While the embryonic cell may thus "learn" a specific functional identity in response to influences in its environment, it also retains an intrinsic identity established by its genetic endowment. Oscar E. Schotté and Hans Spemann performed an experiment many years ago that strikingly demonstrates this principle. In amphibians the ectodermal tissue (the outer of the three primary embryonic layers) of the mouth forms the teeth. But it does so only if it is in contact with the mouth endoderm (the inner embryonic layer). A "signal" from the endodermal cells triggers a sequence of events in the ectoderm that leads to the formation of teeth. Actually the matter is more complex; the endodermal signal apparently reciprocates a prior stimulus from the ectoderm. So before any noticeable appearance of teeth several "messages" may have been exchanged by the cells in this region. In the early embryo it is possible to transfer ectoderm from any part of the body to the mouth region and make it form teeth by placing it in proper association with the endoderm. Schotté and Spemann took advantage of the fact that newts have bony teeth and frogs have horny teeth to see what would happen if they transplanted frog ectoderm cells to the mouth endoderm of the newt. It turned out that the frog cells get the "message" to form teeth but, being frog cells, they form horny teeth. The learned identity acquired in this experimental association is interpreted by the cells in accordance with their genetic endowment.

The movement of cells from one place to another in the embryo constitutes an essential and conspicuous feature of normal development. Singly and in groups, cells move to new sites where, in association with new neighbors, they form new structures. The mammalian kidney, for example, arises from two separate and initially distant components. A little pocket of cells on each side of the cloaca elongates into a finger-like process, destined to form the ureters, and extends into the body cavity toward a mass of mesodermal (middle layer) cells that at this stage shows no definite structure. As soon as the two groups of cells come into contact, however, they begin to change rapidly. The ureter branches and sends out secondary processes; the mesodermal cells with which these processes make contact are organized into kidney tubules. These changes come quite promptly, as if by an exchange of signals between the two groups of cells. Proximity and association are necessary to the interaction. If the cell groups are kept separate, they do not produce their typical responses. In one strain of mice a genetic defect keeps the two kidney components from making contact, and the kidney does not form.

Next to nothing is known about the

signals that are supposed to be involved in such "inductive" interactions. Jean Brachet and H. de Scoeux, working at the Catholic University of Louvain, found many years ago that the messages did not get across when they interposed a strip of cellophane between two prospectively reactant masses of cells. Cellophane allows the passage of only very small molecules. On the other hand, L. W. McKeehan of the University of Chicago used thin strips of agar, through which larger molecules can diffuse, and observed interaction between two tissues. Clifford Grobstein of Stanford University has performed similar experiments with the two components of mouse kidney isolated in tissue culture; he has found that cellophane blocks their interaction, whereas a filter that passes larger molecules permits the interaction to proceed.

The simplest and perhaps likeliest deduction from these experiments is that the tissues, as they associate, react toward one another through the medium of certain metabolic products. These products may provide both the signals and the means of linking the cells in a specific manner. It must be emphasized, however, that at present, with one possible exception, no such products have been isolated from the cells of any higher organism; moreover, there are acceptable alternative explanations for the experimental results.

But it would seem that some means of intercommunication between cells in a developing system must exist. The cells act as if they were capable of mutual recognition and of specific responses to messages conveyed by their neighbors. There is support, on general biological grounds, for the idea that the messenger is chemical in work on slime molds initiated by Kenneth B. Raper of the University of Wisconsin and continued by John Tyler Bonner of Princeton University, by Maurice and Raquel Sussman at Brandeis University and others. The slime molds live part of their life cycle as free amoebae; under certain conditions they come together and form aggregates that differentiate into "fruiting bodies." Their aggregation is directed by a substance (named acrasin) that has been isolated by Brian Shaffer of the University of Cambridge and that is being investigated in a number of laboratories. It emanates from the initial cluster of amoebae and attracts other cells to them. Here is an established case of chemical communication and guidance in the interaction of cells.

It is not too farfetched to assume that all cell contact implies interaction through the production of specific reaction products. The Australian biologist Sir Macfarlane Burnet suggested recently that production of antibodies by cells in adult organisms might present a

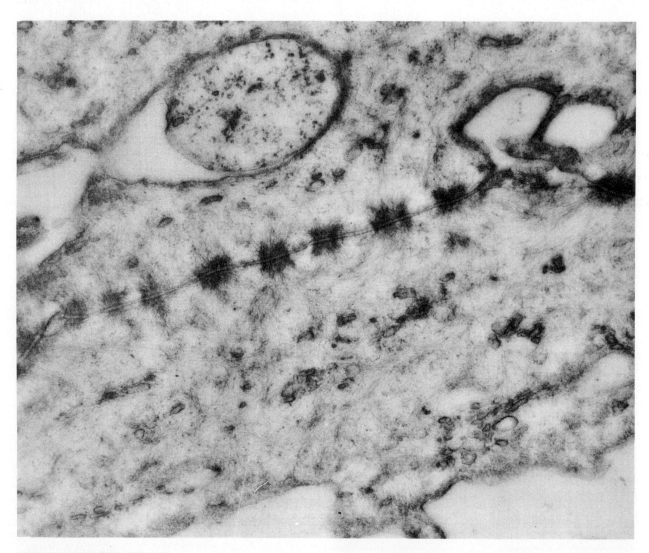

BRIDGES BETWEEN CELLS (desmosomes) are apparently special devices for mutual attachment of cells across their membranes. In this electron micrograph by K. R. Porter of the Rockefeller Institute more than a dozen such bridges (*dark, squarish areas*) connect two skin cells from a salamander larva. The cell membranes run horizontally across picture. Magnification is 35,000 diameters.

model, and perhaps an extreme case, of specific cellular response to chemical signals. The interactions among embryonic cells are, of course, different in detail from the true antibody reaction, and the subtlety and intricacy of these processes are probably of a different order. But it is precisely such subtle chemistry that could provide embryonic cells with the means of mutual communication and integration.

As for the intercellular bond, the term must not be taken as implying that the cells are firmly stuck together or even in direct contact with one another. Electron micrographs made by K. R. Porter of the Rockefeller Institute, by Don W. Fawcett of the Harvard Medical School and by others have suggested that cells may have special devices for mutual attachment on the outer surface of their membranes [see illustration on opposite page]. Furthermore, there is always some distance between cells in contact; this space may be extremely narrow or quite wide, and it seems to be filled with a cementing substance. Unlike brick-binding mortars, these intercellular cements have remarkably flexible and dynamic properties. Although they bind the cells, they permit them to move about and regroup without actual dissociation or loss of contiguity.

This dynamic linking is a cardinal feature of cell contact at all levels of multicellular organization. Consider the case of the everted hydra, described by R. L. Roudabush of Iowa State College in 1933. This tiny, vase-shaped animal can be made to turn itself inside out like the finger of a glove. Its internal digestive cells are then on the outside and the skin cells inside. The cells sense this change, and the hydra promptly proceeds to revert to normal. With the intercellular bonds destabilized, the cells migrate, gliding past each other from wrong side to right side. Throughout the process the hydra retains its over-all configuration, keeping its identity as an organization despite the flux of its constituent parts.

It is in terms of such flexibility of contact and such perception of position by the cells that one must visualize the nature of the intercellular bonds. Variations and changes in the stability of cell contacts are part and parcel of any organism—embryonic or adult. Pigment cells begin their embryonic development in the so-called neural crest; they soon lose their contact with this tissue and move out, singly and in groups, to find positions throughout the integument. Their migrations are clearly not random: they reach specific destinations and form

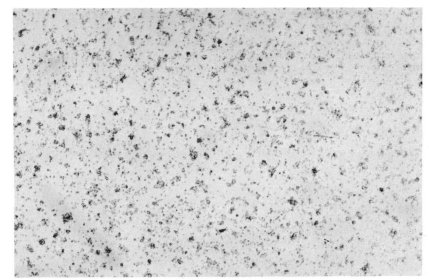

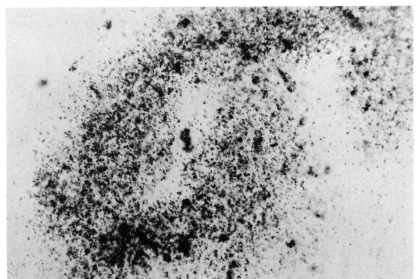

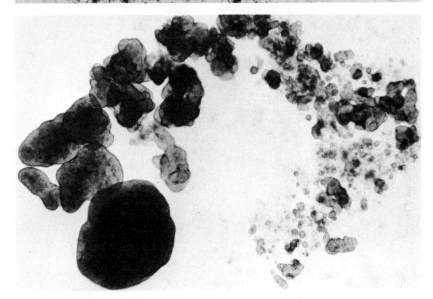

AGGREGATION IN ROTATING FLASK is illustrated in this series of photographs by the author. At top is a suspension of cells from the retina of a seven-day-old chick embryo. In middle is the initial stage of aggregation in a gyrating flask with cells and intercellular material accumulating in the vortex of the liquid culture medium. At bottom a later stage shows compact aggregations at the "head" of the spiral, with continuing aggregation toward the "tail." Magnification in these photographs is approximately 30 diameters.

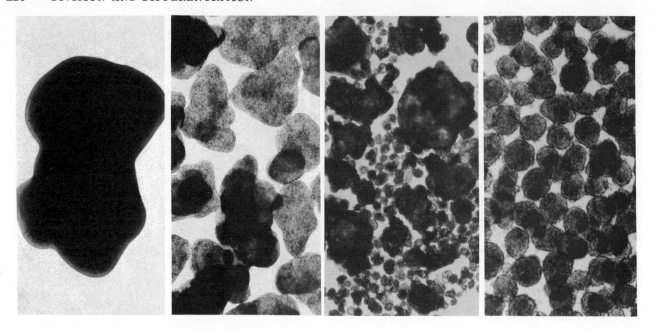

SPECIFIC AGGREGATION PATTERNS characterize each type of cell population. These aggregations were made by (*left to right*) liver cells, retina cells, kidney cells and limb-bud cells, all rotated for 24 hours at 70 revolutions per minute. The first three types came from seven-day-old chick embryos, the last from a four-day chick embryo. Enlargement is approximately 30 diameters.

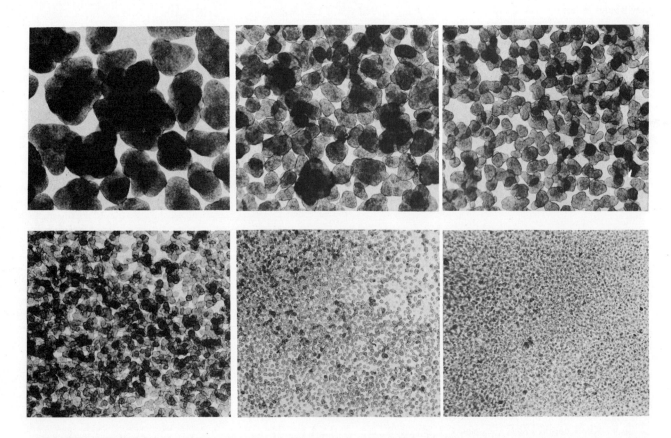

AGGREGATIONS ARE SMALLER when older cells are used. In these photomicrographs the concentration of dissociated retina cells was the same in every case; all were rotated for 24 hours at 70 r.p.m. The cells, however, were taken from chick embryos aged 7, 9, 11, 14, 17 and 19 days respectively. At 19 days the cells simply do not form aggregations. Lowering the temperature or increasing the rotation rate, while all other experimental conditions remain the same, has a similar effect on the size of the aggregations.

typical pigmentation patterns. Other cells leave the neural crest in loose swarms, "homing" toward certain sites in the head of the embryo, where, in conjunction with the cells of that region, they form the lower jaw.

Changes in cell-contact stability continue to play an important part in the life of the organism past the embryonic stage. The steady supply of blood cells involves the continuous disconnection of precursor cells from the bone marrow and their entry as free cells into the circulatory system. Similarly, sperm and egg cells free themselves, as they mature, from their tissues of origin. Elsewhere stability is greater, but definitely relative. Living cells cannot be disengaged from their places in the skin by mere pinching. But when the skin is cut, cells rapidly dissociate from the periphery of the wound, move into the gap, fill it and re-establish stable contacts.

Few questions about cell association yield to fruitful study in the intact organism. It is necessary to separate the cells and tissues from the complexity of the organism in order to control the conditions of observation and experiment. The first steps in this direction necessarily involved the tissues of lower organisms. At the turn of the century Curt Herbst, working at the Zoological Station in Naples, found that young sea-urchin embryos would fall apart and dissociate into single cells when placed in sea water from which he had removed the calcium. He then made the even more interesting discovery that the cells would coalesce and re-form into an embryo when calcium was restored to the water. Calcium has since proved to be an important element in the binding of cells, but not always so dramatically as in the sea-urchin embryo. In general calcium acts more directly as a cell binder in early embryos; later on it seems to operate in conjunction with organic materials to which the primary role seems to shift. There are, however, many invertebrates whose tissues fall apart in the adult state when deprived of calcium. In 1927 James Gray of the University of Cambridge isolated living ciliated cells from the mantle tissue of mussels by placing fragments of the mantle in calcium- and magnesium-free sea water.

An experiment by H. V. Wilson of the University of North Carolina in 1907 pointed to even deeper questions. By gently pressing a marine sponge through a fine sieve he found that he could dissociate it into free cells. He then noticed that as soon as the dispersed cells settled through the sea water onto the dish they started to coalesce. The resulting clumps, when suitably cultured, grew into small but complete sponges. At first it was thought that the sponges regenerated from cells called archeocytes, which, along with skin and digestive cells, make up the loosely associated tissues of the sponge. But further observation showed that all three types of cell persisted following dissociation and that they reassociated in the new aggregations.

Work by later investigators, particularly by Paul S. Galtsoff at the Marine Biological Laboratory in Woods Hole, Mass., and by Tom Humphreys of our laboratory at the University of Chicago, has added new dimensions to these early findings. When cells of different sponge species, preferably of different color for easy recognition, are dispersed and then mixed together, they separate and re-aggregate by species, forming separate clusters [*see illustration on page 212*]. The cells, in other words, are able to identify one another, to give out and register some kind of signal and so associate preferentially with their kin.

A sponge is in some respects a differentiated colony of cells rather than a true multicellular organism. One might question whether the capacity of sponge cells for mutual recognition and sorting out represents a phenomenon of general significance, found in other cellular systems and particularly in higher organisms. Certainly in the case of mammalian tissues it would be difficult to answer the question one way or another in the absence of techniques for dissociating them into individual cells. Some years ago, however, I found that the cementing substances in these tissues will yield to digestion by trypsin and certain other enzymes that break down proteins without serious injury to the cells. Practically any tissue of embryonic origin can now be dissociated and reduced to a suspension of its constituent cellular units. The cells may then be maintained in suitable nutrient media in a germ-free, temperature-controlled environment. It was now possible to conduct studies of the bonding and interaction of the tissue cells of mammals, birds and other higher organisms.

The next step—the resynthesis of complete systems from individual cells—also proved to be feasible. We found that, like sponge cells, the dispersed cells of mammalian or bird embryos will readily aggregate into clusters, migrating over the surface of the culture dish and forming stable connections. Cells from different kinds of tissue were even observed to sort themselves out by cell type in forming these clusters.

The technique lent itself to the study of many previously unanswerable questions, but it fell short of being an exact and adequately controlled procedure. For one thing, it depended primarily on active movement by the cells, a highly variable capacity susceptible to a host of poorly understood conditions. The results in consequence varied unpredictably from one experiment to another.

How could one harness cell aggregation and make it into a critical tool for the study of interactions among cells? The solution turned out to be extremely simple. Most of the irrelevant chance factors that dominate the situation in a stationary cell culture can be neutralized by setting the culture in motion and thereby suspending the cells in a controlled field of force. To do so we place the culture flasks on a horizontally gyrating platform that rotates the flasks 70 times a minute. In each flask the spinning liquid forms a vortex in which the cells concentrate rapidly. They soon link into clusters, within which they construct tissues.

The formation of these clusters depends on and reflects a dynamic equilibrium among the major factors in the system: a balance between the concentrating and shearing-flow forces in the liquid; the differential capacities of the cells to cohere; and the effects of the suspension medium on the cohesiveness of the cells. In this relatively simple system all the pertinent factors—the speed of rotation, the size of the flasks, the character and volume of the medium, the kind and concentration of cells and so on—can be effectively controlled. Thus if the rotation speed and the medium are made the constants of the experiment, the results will reflect the native cohesiveness of the cells in the population tested. The more cohesive they are, the larger and fewer will be their aggregates; the less their cohesiveness, the smaller and more numerous their aggregates. In experiments employing this system we have obtained strikingly consistent results. The rate of aggregation, the number, size distribution, shape and internal structure of the aggregates are always the same when cells of a given kind are aggregated under the same set of conditions.

Such experiments yield an aggregation pattern that is characteristic of the cells in question and of the particular set

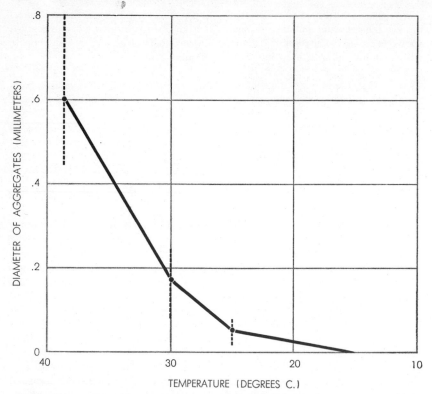

EFFECT OF LOWER TEMPERATURES on the size of aggregations of seven-day chick-embryo retina cells is plotted on this graph. The largest aggregations appear at 38 degrees centigrade; no aggregation occurs after 24 hours of rotation at 15 degrees C. The vertical broken lines show the range of size of the aggregations that build up at each temperature.

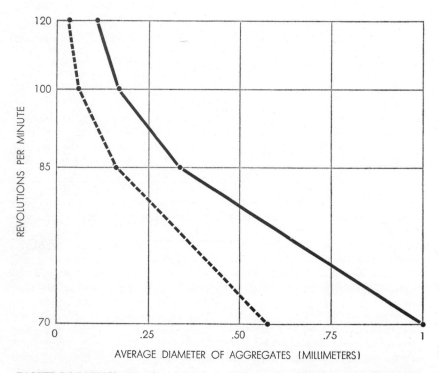

FASTER ROTATION makes the aggregations smaller, as shown by these curves. The broken curve represents chick-embryo retina cells; the solid curve, chick-embryo liver cells. The optimal aggregation size is achieved at 70 r.p.m. The vertical scale on this graph is logarithmic.

of conditions under which they are tested. These patterns can be readily described in terms of numbers, ratios and rates. The traditionally elusive subject of cell-bonding can now be reduced to laboratory prose. Moreover, since the patterns are reliably repeatable and sensitive to changes in conditions, they serve as useful base lines for the bioassay of the effects under study in a given experiment.

We soon found that aggregation patterns vary with different types and mixtures of cell. Under otherwise identical conditions, different kinds of cell "crystallize" into distinct and characteristic aggregates. Some kinds of cell consistently form a single mass; others produce numerous clusters of predictable shapes and sizes. Remarkably, those patterns that showed themselves to be characteristic of particular kinds of tissue proved to be similar for cells from different species. Whether from mouse or chick embryo, cells of the same tissue aggregate into very similar patterns. Their collective reactions seem to be guided by signals legible to both species.

For cells in general we also soon found that certain factors operate with uniform effect. The relationship of the embryonic age of the cells to their capacity for aggregation proved to be particularly striking. Under otherwise equivalent conditions, cells dissociated from tissues of older embryos are less cohesive than their counterparts from younger embryos. For each kind of cell, aggregation patterns provide a characteristic age profile. With increasing age in the donor embryo, the dissociated cells produce smaller and more numerous aggregates and eventually fail to aggregate. Cells dissociated from adult animals usually do not recohere at all.

The precise meaning of this effect of aging is not clear. There are grounds for believing that it reflects the loss by the cells of their ability to produce either the right kind or the right quantity of cell-linking substances. It may be that, as cells mature and acquire specialized functions within their stabilized associations, their metabolic machinery is gradually switched over from those processes that manufacture cell-linking materials to more pressing activities. As a result, when they have been isolated and denuded of their coatings, such cells can no longer recohere effectively. In contrast, embryonically young cells exhibit the capacity to manufacture those materials and to recohere.

If the recohesion of cells does depend

on metabolic processes, then it should be possible to inhibit it simply by lowering the temperature at which the experiment is performed, because metabolic processes are known to be dependent on temperature. This has proved to be the case. Cells that aggregate readily at the usual body temperature of 38 degrees centigrade cohere less effectively at lower temperatures; they remain separate indefinitely at 15 degrees C., even when brought together by rotation. Transferred back to 38 degrees C. after two or three days, such cooled cells aggregate well.

We do not know which of the many temperature-dependent metabolic activities that are depressed by cooling are involved in the production of cell-binding materials. But the answer, we are confident, is only a matter of time. The important point is that the problematical issue of cell-bonding can now be approached by means of concrete tests and experiments. Given the right temperature and otherwise favorable conditions, cells of suitable embryonic age construct tissues of the kind from which they have come. Aggregated liver cells make liver lobules; kidney cells reconstitute kidney tubules and corpuscles; intestinal cells produce digestive tissue; skeletal cells, cartilage and bone; retinal cells, sensory epithelium; heart cells, lumps of beating heart tissue; and so on. Although they are arbitrarily bunched by rotation, the cells rapidly organize orderly fabrics in the pattern of their original tissue. Like parts of an animated jigsaw puzzle, they re-establish a new whole in accordance with the original blueprint. At the Rockefeller Institute, Paul Weiss and Cecil A. Taylor recently grafted such aggregated cells back to embryos; the lumps became joined to the circulatory system of the embryo and developed into remarkable facsimiles of their original organs.

As in experiments with stationary cultures, mixtures of cells in rotating flasks sort themselves out by cell type. One can, for example, readily coaggregate intermingled skeletal and kidney cells. At first the cells are lumped by the spinning liquid into chaotic conglomerates, but soon they segregate by kind—skeletal cells congregating in the middle as nodules of cartilage, kidney cells lining up on the surface. Throughout these cellular maneuvers the aggregates maintain their over-all configuration. The situation obviously resembles the case of the everted hydra or of the embryo that retains its over-all configuration in spite of the extensive movement of its constituent cells.

As might be expected, the final patterning of such composite aggregates reflects their cellular composition. Depending on the nature of their partners in the common aggregate, cells of the same kind may settle inside or outside.

By testing various combinations of cells one discovers a kind of hierarchical order—a "who goes where" in aggregations of various kinds. Preference as to site and competition for physiological need obviously play an important role in the patterning of aggregates. The differen-

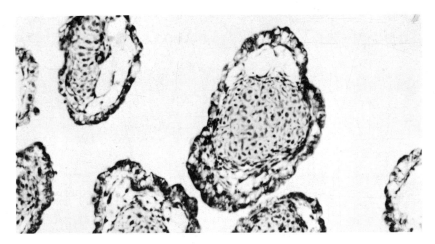

CHARACTERISTIC COMPOSITE AGGREGATIONS form when two different kinds of cell are mixed together in a rotating flask. In the resulting aggregations the cells sort out according to kind. Shown here are sections through such organized aggregations. They are composed of cartilage-forming cells surrounded by kidney cells taken from chick embryos.

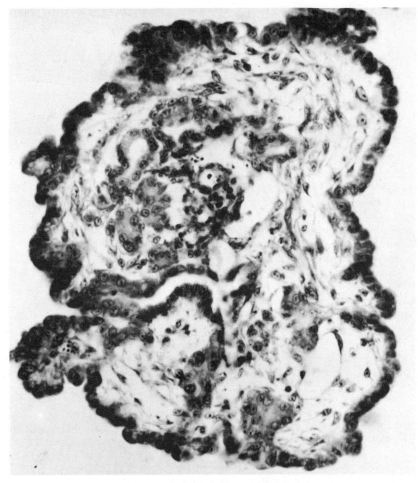

KIDNEY CELLS from the chick embryo form a complex organ-like aggregation after 24 hours of rotation in a flask. This is a highly enlarged section of such a kidney-cell aggregation.

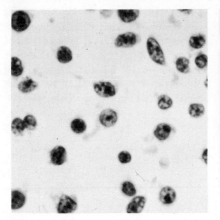 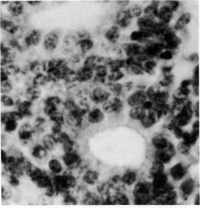 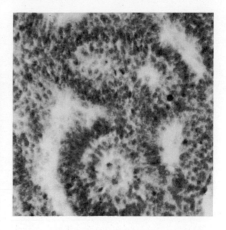

RETINA CELLS from 7-day chick embryo are already differentiated into types that will make up retina and its nerves. At left are stained dissociated cells. Center is section through aggregation formed in 24-hour rotation. 56-hour aggregation shows advanced reconstruction.

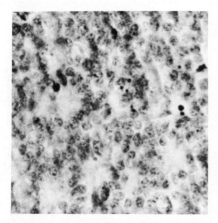 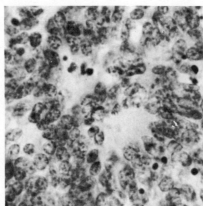 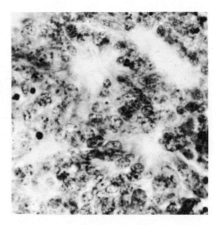

CELLS OF TWO SPECIES will form aggregations. These are sections through aggregations of chick-embryo retina cells (*left*), mouse-embryo retina cells (*center*) and mixed cells of both (*right*). Coming from the same kind of tissue, cells form a common fabric.

tial diffusion and availability of various constituents of the medium, of oxygen and carbon dioxide, also contribute to the outcome. But the patterning of aggregates also reflects the ability of the cells to "recognize" each other, to discriminate between self and nonself, to sort out and to associate in accordance with functional kinships.

One of the more remarkable aspects of such communication-by-contact in embryonic cells is that the signals characteristic for cells of a given tissue are not unique to a given species. One can coaggregate cells from mouse and chick embryos, either from different or from similar tissues. The cells from the dissimilar tissues aggregate separately, as might be expected. But cells of similar kind co-operate in the construction of chimeric fabrics, incorporating the cell of both species. Coaggregated kidney cells from the two species produce tubules of mouse and chicken cells. Liver, cartilage, retina and other cells likewise join in the formation of bi-specific tissues. The means by which these cells recognize each other and become effectively linked into tissues evidently transcend differences between species.

It occurred to us that one might learn more about communication among cells by trying to interfere with its specificity and effectiveness. We found that when dissociated cells are maintained in the dispersed state for some time (the time required varies from days to weeks, depending on the kind of cell and the conditions in which they are kept), they lose two significant capacities progressively and concurrently. Their cohesiveness decreases, and their precision in distinguishing between self and nonself in the organization of tissue drops markedly. These time-related changes raised the question of whether they are not also causally related; that is, whether the materials on the surfaces of cells and between them that link them together might not also play a key role in their interaction and communication.

It seemed not unlikely that these materials are bound to change in response to the novel conditions to which the cells are exposed when they are maintained in the dispersed state. The specificity of the materials could thereby become inactivated or blunted, and this would impair the ability of the cells to interact effectively. These were thin speculations, but we decided to test them by trying to reactivate modified cells by coaggregating them with freshly dissociated cells of their own kind. The fresh cells would presumably be effective producers of the cell-surface materials. When coaggregated by rotation with freshly obtained cells of the same kind, the modified cells did recover their ability to construct tissues. But when coaggregated with fresh cells of a different kind, the modified cells were largely left out of the aggregates. If they were included, they formed no clear structures.

Such findings lend themselves to different interpretations. Until we know more about the whole problem, they are at best suggestive. As such they focus attention on the possible role of cell-surface and intercellular materials in communication among cells and in their developmental association. Wherever such materials could be adequately examined they have been found to contain protein-bound carbohydrates. This fact is of considerable interest since it would seem to place them in the same chemical family with certain other cell products that have highly specific functions: the substances that determine blood groups, that compose antibodies, that are involved in the mating of microorganisms and that influence the selective susceptibility of tissue cells and bacteria to viruses. Could it be that the chemically similar materials that bind cells together also equip them with the means for mutual recognition and specific association? If so, then these cell-binding materials would be mortars of an extraordinary kind. Produced by the bricks themselves, they would serve also to co-ordinate the construction of the tissue, the organ and the organism.

HOW CELLS SPECIALIZE

by MICHAIL FISCHBERG and ANTONIE W. BLACKLER
September 1961

Long before men knew anything about cells, much less molecules, they were familiar with one of the most tangible mysteries in nature: out of a simple-looking egg emerges a living organism, complete and perfect in every detail and unimaginably complex. Each organ is normally just the right size and in the right place and contains the right kinds of cell to carry out its specialized function. Today we are scarcely less mystified. How does the undifferentiated cell of a cleaving egg turn into the specialized cell of heart, liver, nerve, bone or muscle?

Although the complex riddle of differentiation yields its secrets most unwillingly, great progress is now being made. This progress is mainly due to rapid advancements in biochemistry, the development of new techniques and the choice of organisms particularly suited for the study of the problem of differentiation. But perhaps most important of all is a change in the philosophical approach to the problem. One formerly thought too much in terms of the isolated role of the cell nucleus, or the role of the cell cytoplasm, or the role of the environment of the cell. Today we have become much more aware of the dynamic interplay among the three variables and we have learned to observe all three as the embryonic organism develops.

Having said this, however, we will limit our discussion to examples and experiments that demonstrate the roles and mutual interactions of the cell nucleus and cytoplasm in process of differentiation. The nature of the nucleus and the cytoplasm has been presented elsewhere in this book [see "The Living Cell," on page 5]. We are not suggesting that the role of the cell environment is negligible in the cases that we shall present; only that it appears to be secondary to that of nucleus and cytoplasm. Its role, in any case, is not dominant, as it often is in the development of slime molds or in the differentiation of cells in tissue culture.

We shall start by describing how an egg cell develops, for we now believe that the foundation of the future embryo is already laid down while the egg is growing and before the mature egg is even fertilized.

Before a future egg cell begins to grow it looks like any other undifferentiated cell in that it lacks the characteristics by which it would be assigned to a particular specialized cell type. In the frog it measures about 17 microns in diameter, or about twice the diameter of specialized cells. By the time the frog egg has matured its diameter has increased to about 2,000 microns, or two millimeters. This means the volume has increased 1,600,000 times. This tremendous increase in volume is due to the uptake of raw material from the ovarian environment and the use of this material in the synthesis of egg substrate. In some animals highly complicated molecules, the product of synthesis of other cells (usually the "nurse" cells or follicle cells), are taken up into the egg cell and incorporated. In frogs and other amphibia, however, it appears that the incoming material is largely in the form of simple molecules and that these are actively synthesized into more complex substances by the nucleus and cytoplasm of the egg cell itself.

Cytoplasm, nucleus, nucleoli and the nuclear membrane all increase in mass during egg development, so evidently the simple precursors from the environment are taken up by all these cell components and at least stored in them. Research with radioactive precursors has shown that strong synthesis occurs in the nucleus as well as in the cytoplasm. Within the nucleus very active synthesis takes place on the chromosomes, in the nuclear sap and probably also in the nucleoli.

The substances synthesized in the growing egg cell are mainly glycogen, lipids, proteins and nucleoproteins,

CHROMOSOME LOOPS at locus of a lampbrush chromosome are formed by filaments (with surrounding matrix) that connect a pair of chromomeres. The chromosome axis, chromomere pair and filaments consist of DNA; the matrix, of RNA and protein.

which are proteins combined with the nucleic acids, DNA and RNA [see "How Cells Make Molecules" on page 138]. The proteins are partly in the form of clear cytoplasm rich in RNA and partly in the form of yolk particles called platelets, which come in various sizes and whose role is obscure. Mitochondria and many enzymes are found in abundance. In the eggs of some species DNA has also been found in the cytoplasm. It is probable that the main groups of chemicals in growing eggs are themselves tremendously heterogeneous since much of the complex chemical synthesis takes place at the sites of the 10,000 to 20,000 genes, each of which may give rise to a different substance. Let us, therefore, take a closer look at the chromosomes, where the genes reside.

The chromosomes of egg cells that have entered the growth period are not the densely spiralized, rodlike struc-

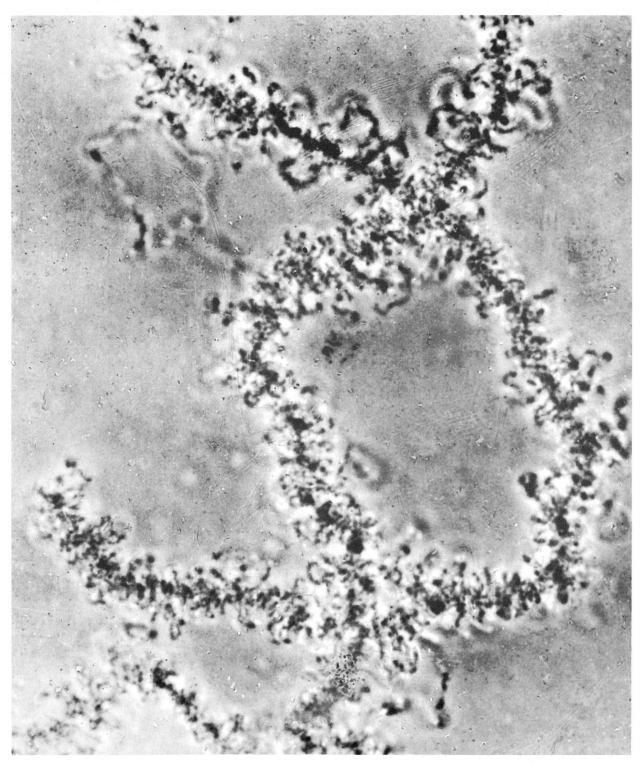

LAMPBRUSH CHROMOSOMES are named for their brushlike appearance, which results from the presence of numerous chromosomal loops (*small, dark wavy lines*) along the chromosome axes. This homologous pair of chromosomes, enlarged some 20,000 diameters, is from the oöcyte nucleus of a newt. This phase-microscope photograph was made by H. G. Callan of St. Andrews University.

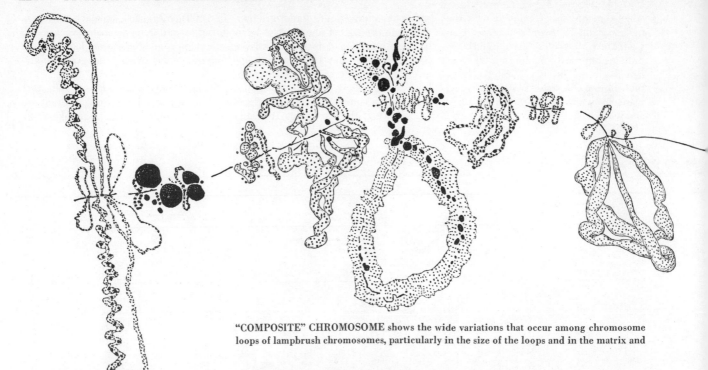

"COMPOSITE" CHROMOSOME shows the wide variations that occur among chromosome loops of lampbrush chromosomes, particularly in the size of the loops and in the matrix and

tures commonly seen at the time of mitotic division. Rather, they are largely despiralized and therefore very long and thin. Early in egg growth they consist of a single axial filament and later of a double filament, which contains at short intervals pairs of thicker and denser swellings: the chromomeres. The chromomeres are so plentiful that they agree roughly with the expected number of genes and they may even be the genes; they seem to be densely spiralized parts of the chromosome axis. A pair of thin filaments (probably themselves despiralized parts of the chromosome axis) run out and form a loop on each side of the chromosome and then return to it, entering the second chromomere of each pair [see illustration on page 222]. The over-all appearance of the chromosomes in this stage of egg growth has led to their being called "lampbrush" chromosomes.

The vast majority of the chromomeres possess lateral loops, which come in many lengths and varieties. H. G. Callan of St. Andrews University in Scotland and Joseph G. Gall of the University of Minnesota have found that the shape of the loop and the nature of the matrix surrounding a particular loop are characteristic for a given chromomere, which is always found in the same position along the axis of a particular chromosome. The chromosome axis, the chromomeres and the lateral loops consist of DNA, now regarded as the substance of which genes are composed. The matrix surrounding the lateral loops, however, consists of RNA and protein. These two substances, but not DNA, are abundantly synthesized at these loops during the whole period of egg cell growth.

Callan and Gall have shown that RNA (and probably protein) produced at the lateral loops of lampbrush chromosomes detaches itself from the loops and comes first to lie free in the nuclear sap and later passes through the porous nuclear membrane into the cytoplasm. The RNA in the cytoplasm increases during the early and middle stages of

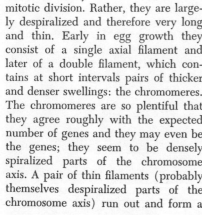

GROWTH OF FROG EGG CELL is marked by increased heterogeneity of cell components (keyed to legend at right). Size and number of RNA globules are greatest in "Later growth" and "Yolk formation" phases. The cell cytoplasm is packed with RNA granules in "Early growth"; granules are fairly evenly distributed in "Later growth," except for a peripheral ring that disappears in "Yolk formation." At "End of growth" the granules are distributed in a smooth and weakening gradient from the upper to the lower pole.

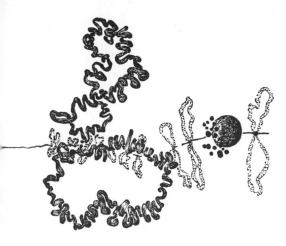

matrix inclusions. The loops depicted here do not all belong to the same chromosome.

egg growth, and much protein is synthesized at the sites of the cytoplasmic RNA granules. It is possible, therefore, that the number of different species of RNA and protein molecules in an egg cell equals the number of synthesizing genes present. Until methods for distinguishing among different species of RNA and proteins have improved one can only guess at the heterogeneity of the egg contents.

It is not difficult to see, in any case, that the mature egg has polarity, which means that its contents are distributed nonuniformly and in an orderly and specific way. The nucleus lies near what can be called the upper pole and the clear cytoplasm, the RNA granules and mitochondria are concentrated near this pole and decline toward the lower pole in quantity. The pigment on the surface shows a similar distribution, whereas the yolk granules are larger and more densely packed near the lower pole, becoming smaller and more widely spaced toward the upper pole. We can therefore speak of gradients in the cytoplasmic substances. These gradients run parallel to the egg axis and are radially symmetrical about it. In other words, all the points on a particular plane parallel to the "equator" of the egg have the same cytoplasmic composition but differ in composition from points on all other planes. As a consequence all the meridional slices going from the upper pole of the egg to the lower one (like segments of an orange) contain the same substances distributed in the same way.

This radial symmetry, observable in the eggs of many species, is eventually changed to bilateral symmetry. Depending on the species of animal producing the egg, this happens either shortly before or shortly after fertilization [see illustration on next page]. The change in symmetry is caused by a change in the distribution of the cytoplasmic components and is often linked to a change in the outermost (cortical) layer of the egg. Certain cytoplasmic substances sometimes accumulate in the form of a crescent on one side of the egg. Further development of these eggs shows that the plane that divides the body of the developing embryo into a right and left half passes through the broadest region of this crescent, roughly parallel to the equator, and through the two poles. In amphibia the broadest region of the crescent always becomes the dorsal or upper part of the body, while the region where the thin tips of the crescent come together develops into the ventral part, or underbelly. The head of the animal develops on the ventral side of the upper pole, and the tail comes to lie in the region between the lower pole and the lower edge of the crescent.

We see, then, that the future embryo is fully predetermined in the uncleaved egg by the distribution and peculiarities

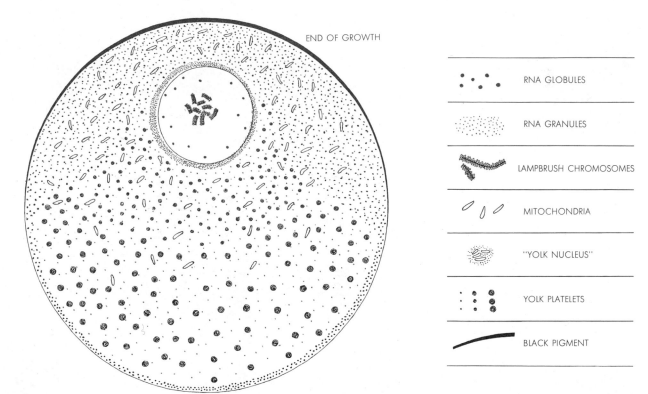

Throughout cell growth lampbrush chromosomes become increasingly condensed in center of nucleus. Mitochondria are equally distributed in "Early growth" except for concentration in the yolk nucleus; at "End of growth" they are distributed in a smooth and weakening gradient from the upper to the lower pole. Smallest yolk platelets appear in "Yolk formation." At "End of growth" platelets are distributed throughout the cell, increasing in size from top to bottom. Black pigment appears only at "End of growth."

of its cytoplasmic components. This predetermination is so clear in the eggs of many species that certain areas in the uncleaved egg can be easily recognized as those which are going to give rise to the brain, the intestine, the muscles or the future germ cells. And, as we have noted, the predetermination in some species is clearly evident even before the egg is fertilized. Fertilization is normally required, of course, before cleavage will take place.

During subsequent development, after fertilization, the chromosomes of the egg and sperm are incorporated into a single nucleus, which then divides by mitosis, leading to the first cleavage. As cleavage progresses, each daughter nucleus is accordingly surrounded by a distinctively different matrix of cytoplasmic substances as the spatially organized heterogeneity of the egg cytoplasm is stabilized by the appearance of cell membranes.

However, all the nuclei derived by mitosis from the one nucleus of the fertilized egg are initially, at least, identical. Being identical, they all have to act in the same manner. They can act differently only if they are ordered or stimulated to do so by a variable factor, namely their environment, particularly the immediate cytoplasmic environment. Unless one is willing to believe that the cytoplasm has full responsibility for differentiation, one has to believe that the cytoplasm, which varies from one part of the egg to another, influences identical nuclei to act in different ways. This assumption is not only a logical necessity but has found strong experimental support, as we shall now see.

Some insect larvae (mainly the larvae of the order Diptera) have salivary glands consisting of a small number of large cells with enormous chromosomes. This is one of those cases in which

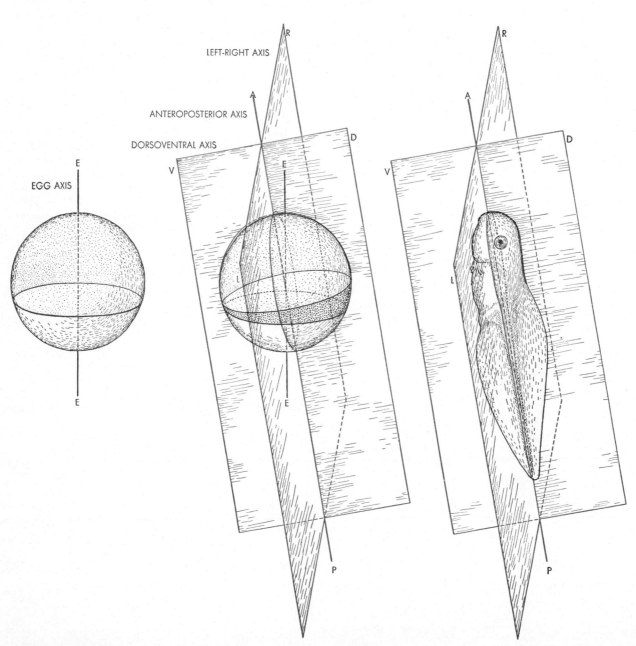

RADIAL SYMMETRY of a frog egg (*left*) changes to bilateral symmetry (*middle*) with the development of a "crescent" (*one side of which is defined by the heavily stippled area*). The tadpole (*right*) that will develop when such an egg is fertilized will also show a corresponding symmetry, as can be seen by comparing the main axes of both the egg and the tadpole. As depicted here, the "Left-right axis" (*L-R*) and "Dorsoventral axis" (*D-V*) have been extended to form planes that cut through egg and tadpole.

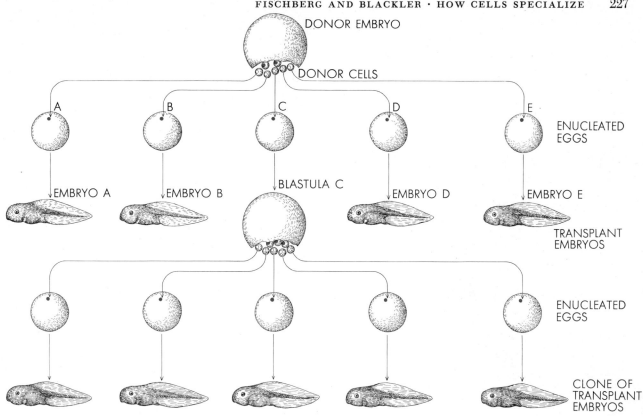

PRODUCTION OF NUCLEAR CLONES makes it possible to test various assumptions concerning the specialization of cells (see, for example, the experiment illustrated on the following page). Nuclei (colored dots) from donor cells of a partially developed frog embryo are injected into "Enucleated eggs," i.e., eggs from which the nuclei have been removed. These produce five normal "Transplant embryos." Nuclei from cells of "Blastula C" are injected into another set of enucleated eggs. The resulting "Clone of transplant embryos" is five genetically identical individuals whose chromosomes are all descended from chromosomes of one nucleus.

growth is achieved not by an increase in cell number but by an increase in cell size. Very large cells apparently cannot function with the usual number of chromosomes, and the chromosome number is therefore increased in proportion to cell size. The giant salivary gland chromosomes are really bundles of about 500 to 1,000 despiralized chromosomes that stick together in such a way that the identical parts—the chromomeres—of all the homologous chromosomes lie side by side. As a result all the chromomeres of a particular gene site form a disk, as do the chromomeres at other sites. These disks are separated by disks of the non-chromomeric parts of the sausage-like chromosomes, so that the giant chromosome looks as if it were made up of a large number of dark and light bands of varying thickness [see illustration on page 230].

Careful studies by M. E. Breuer and Clodowaldo Pavan of the University of São Paulo and by W. Beermann and his colleagues at the University of Tübingen have shown that chromomeres can swell up into so-called Balbiani rings, or puffs. They have also found that the puffs are the sites of strong synthesis of RNA and proteins. These observations are particularly exciting because the occurrence of the puffs follows a specific pattern. In mature larvae of a particular species only certain recognizable chromomeres will show puffs in all the specimens. But in young larvae other chromomeres, or gene sites, will be puffed up. It is reasonable to conclude that different genes puff up and become active at different stages of an organism's development.

This activity of chromomeres is not only specific with respect to age but also with respect to cell type. Salivary glands, at least in chironomid Diptera (midges), possess two types of secretory cell, and each type has its own pattern of swellings, which also undergoes changes during development. It appears, therefore, that specific factors in the cytoplasm call forth the activity of particular genes. Furthermore, it is likely that the cytoplasm varies from one cell type to another and changes progressively (probably as a result of specific gene activity) during development.

Some support for this idea is provided by new and original experiments conducted by H. Kroeger at Oak Ridge National Laboratory. He changed the environment of salivary gland chromosomes by transferring the nuclei from glands of advanced larvae into a preparation containing the cytoplasm characteristic of developing eggs. He found that the swellings of certain chromomeres of the salivary gland chromosomes disappear in the new environment and that other chromomeres are induced to produce puffs. There is even a certain amount of correlation between the puffing and the developmental stage of the eggs providing the new environment [see illustration on page 230].

The work on the salivary gland chromosomes is of the greatest importance to our views on how cells specialize. It provides a rare insight into the relationship of cytoplasmic environment and nucleus. The fact that the work deals mainly with advanced stages of development does not prevent it from serving as a model suggesting how differentiation can arise in early development.

Let us now see what happens if the experiment is turned about so that the cytoplasmic environment is held constant and nuclei are made the variable. It has long been accepted, with little direct evidence, that the chromosomes in all the cells of an organism are identical, regardless of how the cell itself is specialized. To test this assumption one can, by means of delicate techniques, extract the nucleus (containing all the chromosomes) from an unfertilized frog egg and replace it with a nucleus obtained from one of the partially specialized cells of a developing frog embryo. The cytoplasm of the egg and the injected nucleus will then undergo development and give rise to what is called a transplant embryo. Fertilization is not necessary because the injected nucleus is already the descendant of a fused egg and sperm nucleus. Since the cytoplasm contains all the qualities necessary for normal development, the actual development will be a measure of the quality or developmental potential of the injected nucleus.

Robert W. Briggs of Indiana University and Thomas J. King of the Institute for Cancer Research in Philadelphia, and the authors in collaboration with J. B. Gurdon of the University of Oxford, carried out such experiments with different species of frog and obtained in principle the same results. They found that if the nuclei were obtained from embryos in the early (blastula) stage of development, the new transplant embryos in most cases produced normal tadpoles [*see illustration on this page*]. If, however, nuclei of advanced embryonic stages are transplanted, fewer nuclei are able to participate in the production of normal embryos. Most of the transplant embryos either do not cleave normally and die or are arrested at later and abnormal developmental stages. The conclusion from this last experiment is that nuclei change during development and differentiation. They seem to lose their totipotentiality and become more limited in their ability to promote normal development.

One of the most striking characteristics of these experiments is that nuclei taken from a single future organ—say, the gut—of a single embryo do not give rise to similar transplant embryos but to a great variety of embryos. This includes arrested blastulae, a wide range of abnormal embryos and a few normal tadpoles. Accordingly, one can conclude not only that nuclei differ from one organ to another but also that even single organs are made up of cells containing quite different nuclei. In other words, a developing organ of an embryo is made up of a heterogeneous population of nuclei. If nuclei of newly hatched tadpoles are transplanted in the way described, the resulting transplant embryos show greater uniformity. This decline in variation of the transplant embryos could indicate that the nuclei become more similar because most of them have by now undergone the same kind of changes, whereas in experiments with embryos the nuclei were caught at different phases of change. On the other hand, the nuclei of tadpoles may be so limited in their potentialities that they lead usually only to arrest at the earliest

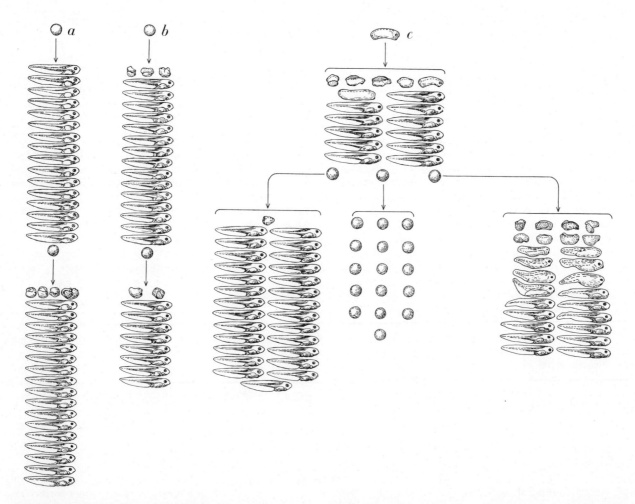

NUCLEAR-TRANSPLANTATION EXPERIMENT performed by J. B. Gurdon on South African clawed frogs involved the production of nuclear clones (*see illustration on preceding two pages*). Nuclei from two blastulae ("*a*" and "*b*") promoted predominantly normal development in two groups of transplant embryos (*top left*). Nuclear clones from a second transplant (*bottom left*) were also predominantly normal. Nuclei from an embryo at a later stage ("*c*") promoted a variety of normal and abnormal embryos from which three nuclear clones were produced: predominantly normal (*at left, under "c"*), undeveloped (*middle*) and predominantly abnormal (*at right*).

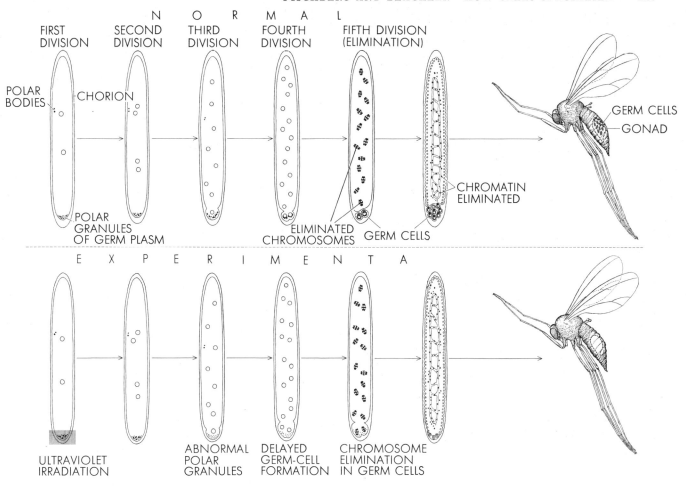

GERM-CELL SPECIALIZATION in gall midge *Mayetiola destructor* is contrasted under normal and experimental conditions. In normal development (*top*) nucleus of fertilized egg divides into two nuclei (*small circles in "First division"*), then four and so on. In "Third division" and "Fourth division" two nuclei move to posterior pole of egg; the cytoplasm contracts so as to separate the nuclei from the rest of the egg. In "Fifth division" the remaining 14 nuclei undergo a mitotic division in which only eight sets of chromosomes in each nucleus separate to the poles; the other 32 pairs eventually dissolve in the cytoplasm of the cell. The germ cells develop normally and adult gall midge is fertile (*top right*). In an experiment (*bottom*) discussed in the text the germ plasm was irradiated with ultraviolet light (*bottom left*), thus retarding germ-cell formation. As a result the two posterior nuclei also underwent mitotic division and lost 32 of their chromosome sets. The gall midge that developed (*bottom right*) was consequently sterile.

stages of development.

Another question is: Are the observed nuclear changes of a reversible nature or are they stable, irreversible and hereditary? This question can be answered by the production of "nuclear clones" [*see illustration on page 227*]. A nucleus of a donor embryo is injected into an unfertilized, enucleated egg. This develops without further nuclear differentiation into a blastula. The cells of this blastula are then dissociated and their nuclei are injected singly into unfertilized eggs. The embryos developing from this second transfer form a clone of embryos, each containing nuclei that are derived from the one initial nucleus used in the first transfer, and are therefore of identical genetic constitution. The experiment is repeated to produce a number of such clones.

The similarity within a clone is in remarkable contrast to the variation found among different clones. Each clone represents the quality of only one tested nucleus and the differences among clones represent differences among individual nuclei. The experiment indicates that the nuclear changes due to natural differentiation are relatively stable and of a heritable nature. It also strengthens the belief that the variation observed after the first transfer is a true one and not a result of damaging the nuclei during handling.

The stable, heritable and apparently irreversible nature of these nuclear changes poses, of course, a number of new questions. For example, one would like to find out which of the nuclear components is the site of the change. Is it the nuclear membrane, the nuclear sap, the nucleolus or the chromosomes embodying the genes? The chromosomes would seem the likeliest site, because so far as we know they are the only nuclear structure showing continuity throughout both mitosis and heredity. But a definite answer cannot be given until experiments now in progress shed some light on the problem.

Some years ago it would have been almost unthinkable to consider that nuclei might, during differentiation, change in their genetic qualities. Recently more and more cases of irreversible nuclear changes induced by cytoplasmic factors have been brought to our attention. They include examples from protozoa, ascarides, frogs and gall midges. In the last the changes occur at the level of chromosomes and are clearly visible under the light microscope.

Adult gall midges are unusual in that they have a large number of chromosomes in the germ cells and a small number in the somatic cells, which make up the whole body except the germ line.

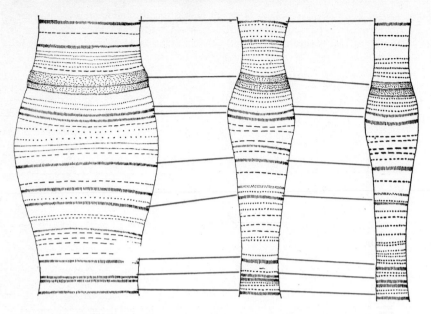

GIANT CHROMOSOME NUMBER 3 of larva of the midge *Chironomus* varies in shape and slightly in banding pattern depending on the type of cell in which it is found. Identical sections here are from the salivary gland (*left*), Malpighian tubule (*middle*) and rectum (*right*).

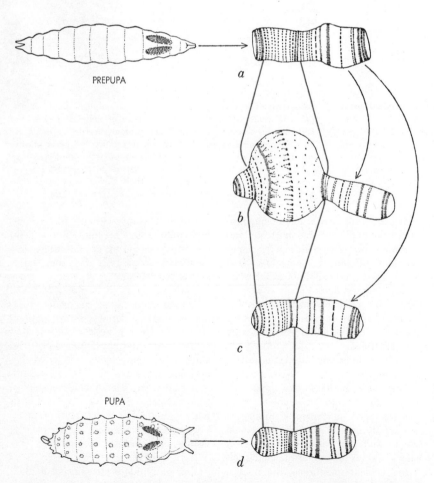

EXPERIMENT performed by H. Kroeger involved chromosome Number 2 from the salivary gland of the fruit fly *Drosophila busckii*. The chromosome is depicted here as it appears at the prepupal (*a*) and pupal (*d*) stages of the fly's development; and as it appeared after transferal to the preblastoderm (*b*) and blastoderm (*c*) egg contents of *D. melanogaster*.

We shall describe how this difference in chromosome number is established in one species, *Mayetiola destructor* [*see illustration on page 229*].

As in other insects, no cell membranes are formed during the earliest development. The zygote nucleus, containing about 40 chromosomes, divides into two, then four, then eight nuclei. Because of the lack of cell membranes, the nuclei are free to distribute themselves evenly through the cytoplasm of the egg. At the eight-nucleus stage one of the nuclei moves all the way down to the posterior pole of the egg and comes to lie in the vicinity of a particular cytoplasm, the germ plasm, which always accumulates at this pole.

During the next mitosis all eight nuclei divide again, so that 16 are present. At the same time the cytoplasm near the posterior pole constricts in such a way that two cells are formed, containing the two most posterior nuclei and all the germ plasm. This cleavage cuts off the two newly formed cells from the rest of the egg. These two cells are destined to become the primary germ cells, and all the future eggs or spermatozoa, as well as the nurse cells, will derive from them.

At this moment the peculiar fifth division begins in all 14 nuclei lying in the somatic part of the egg, but the two primary germ cells are exempt from this mitosis. When it is time for the duplicated chromosomes to start moving apart (at the anaphase stage), it becomes clear that the mitosis is most unusual. Only eight of the 40 chromosomes of each nucleus separate along their whole length into two chromosomes and move toward the opposite poles of the mitotic spindle. The remaining 32 fail to separate at their ends and remain immobile in the equatorial plane of the spindle. Meanwhile the eight chromosomes at each pole of the spindle form two small daughter nuclei. The 32 chromosomes left behind soon begin to dissolve and the material derived from them gradually disperses throughout the cytoplasm and is not seen again. In this way 32 chromosomes are eliminated from somatic cells.

The primary germ cells are exempt not only from this mitosis but also from chromosome elimination. During subsequent divisions of the germ cells the full chromosome number is maintained and the germ line is formed. The small nuclei (which contain eight chromosomes) go on dividing and give rise to the small nuclei of all the somatic cells.

It is evident that the presence or absence of a cytoplasmic factor determines the behavior of the nuclei during the fifth mitosis. Close observation of the future germ-cell nuclei shows that the germ plasm, rich in RNA and mitochondria, wraps itself intimately around the nuclear membranes of these nuclei after the fourth mitosis. One has the impression that the germ plasm protects the nuclei from the influence of the neighboring cytoplasm, an impression that is strengthened by the formation of cell membranes cutting this pair of cells off from the rest of the egg.

To study the cause of chromosome elimination, C. R. Bantock of the University of Oxford irradiated with ultraviolet light the extreme posterior ends of gall-midge eggs, the ends containing the germ plasm, before the future germ-cell nuclei had migrated into them. The remainder of each egg, containing all the nuclei, was carefully shielded from irradiation. In eggs so treated the posterior nuclei migrated normally into the germ plasm, but now, instead of being protected in some fashion, they shared the same fate as the other nuclei—that is, their chromosome number was reduced from 40 to eight. The primary germ cells, in spite of losing 32 chromosomes, still gave rise to the beginning of a germ line. If the embryos were allowed to develop, they produced adult male and female midges that looked normal but were actually sterile. Histological examination of the gonads of gall midges obtained from ultraviolet-treated eggs revealed that the reproductive cells failed to develop.

It appears, therefore, that the germ plasm normally prevents chromosome elimination. This might be due to an inhibition of the fifth mitosis in the future germ-line nuclei or to another protective mechanism merely coinciding in time with this inhibition. The absence of reproductive cells in gall midges hatched from irradiated eggs indicates that the missing chromosomes are necessary for the development of such cells. The strong effect of the irradiation suggests furthermore that the protective factor of the germ plasm is composed of RNA, because it is particularly sensitive to ultraviolet light.

Bantock's work on the gall midge provides a clear example of nuclear-cytoplasmic interactions. It shows that egg formation can proceed normally only with the participation of the whole chromosome complement of the species, whereas somatic differentiation can take its normal course with a severely reduced chromosome number (eight) provided that the egg cytoplasm contains gene-products of the genes active during egg formation.

Just as the germ plasm prevents chromosome elimination in the gall midge, it probably plays a comparable role in protecting the germ cells in other and more complex organisms from specialization and loss of their totipotentiality. In frogs a germ plasm of similar chemical composition has been discovered and here too ultraviolet irradiation of the germ plasm causes partial or total sterility of frogs developing from such treated eggs.

In summary, differentiation is most likely to result from nuclear-cytoplasmic interactions that cause progressive individuation of the cytoplasm and increasing specialization of the nuclei of particular cells. The original question of how cells specialize has not been solved, but we hope we have shown that complexity of the cytoplasm in the mature egg cell cannot fail to lead to specialization.

CHROMOSOME PUFFS

by WOLFGANG BEERMANN and ULRICH CLEVER

April 1964

The genetic material performs two functions that are basic to life: it replicates itself and it ultimately directs all the manifold chemical activities of every living cell. The first function is expressed at the time of cell division in the manufacture of more of the genetic material: deoxyribonucleic acid (DNA). The second is accomplished during the "interphase" between cell divisions; DNA directs the synthesis of ribonucleic acid (RNA), which in turn directs the synthesis of proteins, which as enzymes in turn catalyze the other reactions of the cell. In this way RNA translates the genetic information of DNA into the language of physiology and growth, into the everyday processes of synthesis and metabolism.

As readers of SCIENTIFIC AMERICAN are aware, the work of elucidating the genetic code is now being carried out by investigators in laboratories throughout the world, largely by the breeding and statistical study of certain bacteria and the viruses that infect them. In recent years our laboratory at the Max Planck Institute for Biology in Tübingen and several other laboratories have adopted somewhat different techniques for investigating the relation between DNA and RNA in the genetic material of higher organisms—those belonging to the insect order Diptera, such as the fruit fly *Drosophila* and the midge *Chironomus*. In these insects, as in all higher organisms, the DNA resides in the structures called chromosomes. In certain exceptionally large cells of *Drosophila* and *Chironomus* we have found that we can actually see the ultimate units of heredity—the genes—at work. These active genes take the form of "puffs" scattered here and there along the giant chromosomes of the giant cells. We have found that the puffs produce RNA and that the RNA made in one puff differs from the RNA made in another. Observations of the puffs have also enabled us to trace the time patterns of gene activity in several tissues of developing insect larvae. Furthermore, by administering hormones and other substances we can start, stop and prevent some of these activities.

The giant chromosomes were first observed late in the last century, but it was not until 1933 that Emil Heitz and Hans Bauer of the University of Hamburg recognized them as chromosomes. By 1933 breeding studies of the fruit fly had resulted in detailed "maps" on which genes were placed in relation to each other along the chromosomes. The genes, however, were still conceptions rather than physical entities, and the chromosomes had been recognized only during cell division, when they are coiled like a spring and present a condensed, rodlike appearance. During interphase, when they are directing cellular activity, the chromosomes in typical cells are virtually invisible because, although they are long, they are so thin that they can be seen only at the extremely high magnifications provided by the electron microscope, a comparatively recent invention.

Heitz and Bauer realized that giant chromosomes, which are clearly visible in the light microscope, are the equivalent of the interphase chromosomes of typical cells. In the words of T. S. Painter of the University of Texas, the giant salivary-gland chromosomes of fruit fly larvae were "the material of which every geneticist had been dreaming. The way led to the lair of the gene." Intensive work by Painter and others in the U.S., including H. J. Muller, Calvin B. Bridges and Milislav Demerec, soon identified specific characteristics of flies with particular loci, or bands, on the giant chromosomes. Since then the bands have been considered the material equivalent of the conceptual Mendelian genes.

The giant chromosomes are found primarily in well-differentiated organs that are engaged in vigorous metabolic activity, such as salivary glands, intestines and the Malpighian tubules (excretory

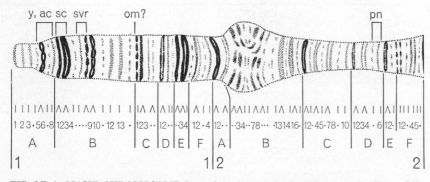

TIP OF A GIANT CHROMOSOME from the salivary gland of the fruit fly *Drosophila melanogaster* is shown in this diagram. The reference system below it was devised by Calvin B. Bridges of the California Institute of Technology. The letters and brackets above it mark certain sites known to be associated with specific bodily characteristics. For example, the "y" at left denotes the band or gene responsible for yellow body color.

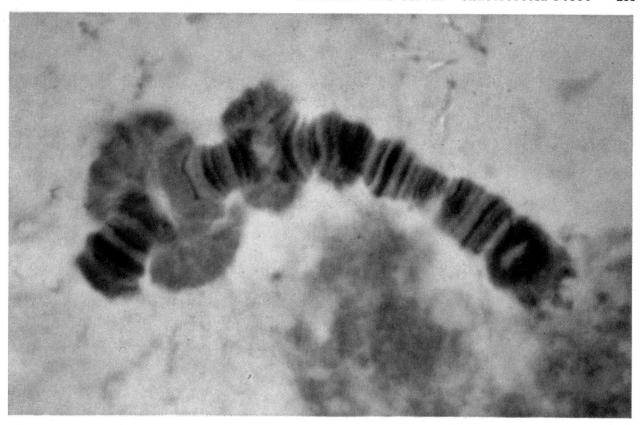

CHROMOSOME PUFFS are the protuberances on the left-hand portion of the giant chromosome in this photomicrograph. Very large puffs, of which two are seen, are called Balbiani rings. Protein has been stained green, deoxyribonucleic acid (DNA) brown.

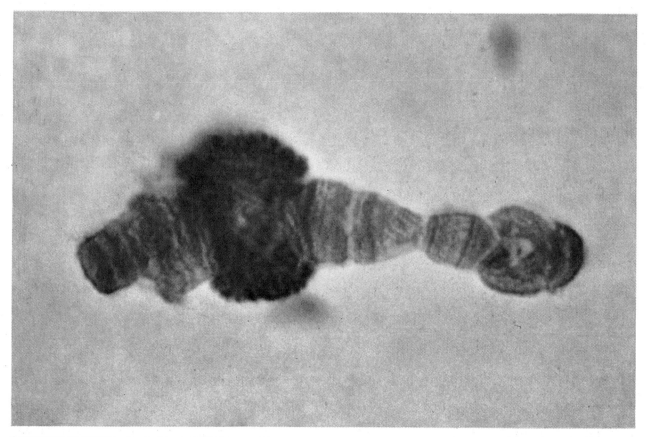

PRODUCT OF PUFFS, ribonucleic acid (RNA), is reddish-violet when dyed with toluidine blue. Here the DNA is blue. The photomicrographs on this page show two different specimens of the giant chromosome IV from the salivary gland of the midge *Chironomus tentans*. Both were made at the Max Planck Institute for Biology in Tübingen. The magnification in each is some 2,500 diameters.

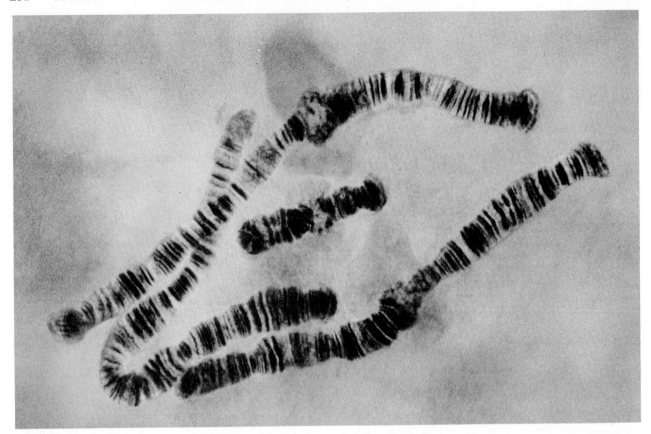

SET OF FOUR GIANT CHROMOSOMES from a cell in the salivary gland of *Ch. tentans* is here magnified some 700 diameters. The enlarged regions on two of the long chromosomes are nucleoli. Chromosome IV is the shortest of the four; it has a Balbiani ring. The banding pattern on each of the four chromosomes is visible in corresponding giant chromosomes from entirely different tissues.

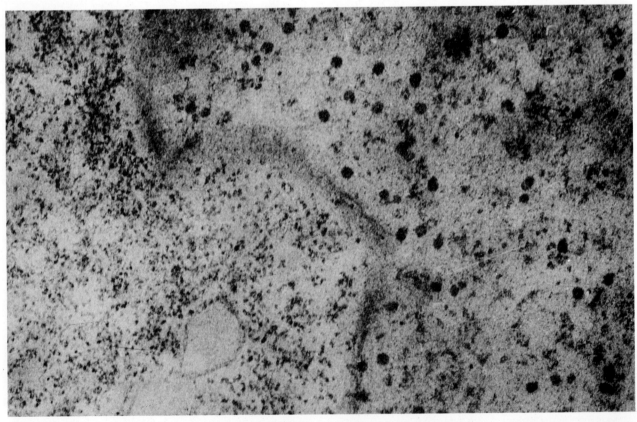

MIGRATING GRANULES, consisting of ribonucleic acid and protein (*right*), are penetrating pores in the nuclear membrane (*bottom center*) in this electron micrograph. Cytoplasm of cell is to left of membrane. The small particles in it are ribosomes, the sites of protein synthesis. The RNA in the large particles may be on its way to the ribosomes to act as a template for proteins.

organs). These tissues grow by an increase in cell size rather than in cell number. Apparently the giant cells require more genetic material than typical cells do; as they expand, the chromosomes replicate again and again and also increase in length. Along individual chromosome fibers there are numerous dense spots where presumably the structure is drawn into tight folds. These locations are called chromomeres.

As the chromosome filaments in giant cells increase in number, those of a particular chromosome remain tightly bound together; each chromomere is fastened to the homologous, or matching, chromomere of the neighboring filaments. Such locations become the bands, which are also known as chromomeres. The chromosome that results from this growth process is said to be "polytene": it has a multistrand structure resembling a rope. At full size the giant chromosomes are almost 100 times thicker and more than 10 times longer than the chromosomes of typical cells at cell division.

The bands, which vary in thickness, contain a high concentration of DNA and histone, a protein associated with DNA. The spaces between the bands, known as interbands, contain a very low concentration of these substances. It was discovered in 1933 that each giant chromosome in a set within a cell has its own characteristic sequence, or pattern, of banding and that, even more striking, every detail of the pattern recurs with the utmost precision in the homologous giant chromosome of every individual of the species.

In the past most cell geneticists were so occupied with localizing the genes in the salivary-gland chromosomes that they did not investigate the giant chromosomes in other tissues. Yet the presence of such chromosomes in cells with quite different functions poses an obvious challenge to the biologist interested in development and differentiation. It had long been held that every cell of an individual possesses exactly the same set of chromosomes and the same pattern of genes. Giant chromosomes in a variety of tissues provided an opportunity for testing this idea, that is, for determining if the special metabolic condition or function of a cell influences in any way the state of its chromosomes and genes. For example, in spite of the constancy of the banding pattern found in salivary-gland chromosomes, the same chromosomes in other organs of the same species might present a different banding pattern. If this were true, the localization of genes in specific bands would lose all general meaning.

Assertions that different tissues have different banding patterns were actually made 15 years ago by Curt Kosswig and Atif Şengün of the University of Istanbul. One of us (Beermann, then working in the laboratory of Hans Bauer at the Max Planck Institute for Marine Biology in Wilhelmshaven) checked these claims by a detailed comparative study of the banding of giant chromosomes from four different tissues of the midge *Chironomus tentans*. Independently Clodowaldo Pavan and Martha E. Breuer of the University of São Paulo carried out similar investigations on the fly *Rhynchosciara angelae*. We could not find any detectable variation in the arrangement and sequence of bands along the chromosomes in different tissues. The uniformity of chromosome banding lends strong support to the basic concept that the linear arrangement of the genes as mapped in breeding experiments corresponds to the pattern of the bands on giant chromosomes.

At the same time, however, we found that chromosomal differentiation of a very interesting kind does exist. The fine structure of individual bands can differ with respect to puffs that are in one location on a chromosome in one tissue and in another location on the same chromosome at another time or in another tissue. These localized modifications in chromosome structure of various Diptera had been noted many years earlier, but their possible significance was overlooked.

The coherence of the chromosome filaments is loosened at the puffed regions. The loosening always starts at a single band. In small puffs a particular band simply loses its sharp contour and presents a diffuse, out-of-focus appearance in the microscope. At other loci or at other times a band may look as though it had "exploded" into a large ring of loops around the chromosome [*see top illustrations on next two pages*]. Such doughnut-like structures are called Balbiani rings, after E. G. Balbiani of the Collège de France, who first described them in 1881. Puffing is thought to be due to the unfolding or uncoiling of individual chromomeres in a band. On observing that specific tissues and stages of development are characterized by

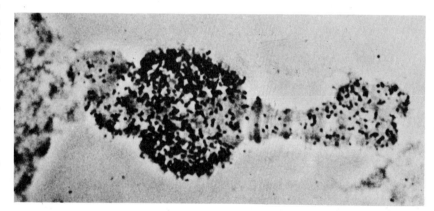

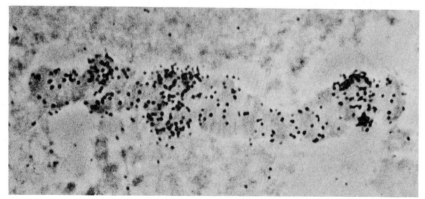

INHIBITION OF PUFFING and of RNA synthesis is accomplished by treatment with the antibiotic actinomycin D. At top an autoradiogram of a chromosome IV of *Ch. tentans* shows the incorporation of much radioactive uridine (*black spots*), which takes place during the production of RNA, as explained in the text. Another chromosome IV (*bottom*) that had been puffing shows puff regression and little radioactivity after half an hour of treatment with minute amounts of actinomycin D, which inhibits RNA synthesis by DNA.

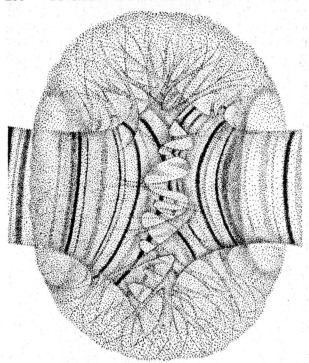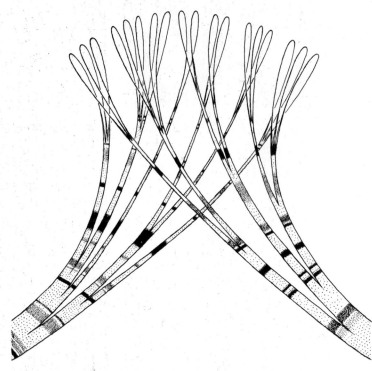

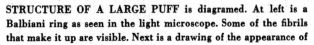

STRUCTURE OF A LARGE PUFF is diagramed. At left is a Balbiani ring as seen in the light microscope. Some of the fibrils that make it up are visible. Next is a drawing of the appearance of a few of the fibrils at very high magnification in the light microscope. The much greater magnification provided by the electron microscope (*third from left*) shows two puff fibrils with granules

definite puff patterns, one of us (Beermann) postulated in 1952 that a particular sequence of puffs represents a corresponding pattern of gene activity. At about the same time, Pavan and Breuer arrived at a comparable conclusion based on their experiments with *Rhynchosciara*.

If differential gene activation does in fact occur, one would predict that genes in a specific type of cell will regularly puff whereas the same gene in another type of cell will not. A gene of exactly this kind has been discovered in *Chironomus*. A group of four cells near the duct of the salivary gland of the species *Chironomus pallidivittatus* produces a granular secretion. The same cells in the closely related species *Ch. tentans* give off a clear, nongranular fluid. In hybrids of the two species this characteristic follows simple Mendelian laws of heredity. We have been able to localize the difference in a group of fewer than 10 bands in one of *Chironomus'* four chromosomes; the chromosome is designated IV. The granule-producing cells of *Ch. pallidivittatus* have a puff associated with this group of bands, a puff that is entirely absent at the corresponding loci of chromosome IV in *Ch. tentans*. In hybrids the puff appears only on the chromosome coming from the *Ch. pallidivittatus* parent; the hybrid produces a far smaller number of granules than

that parent. Moreover, the size of the puff is positively correlated with the number of granules. This reveals quite clearly the association between a puff and a specific cellular product.

Such analysis can demonstrate only that a specific relation exists between certain puffed genes and certain cell functions. We therefore sought to find a biochemical method for showing that puffing patterns along chromosomes are in fact patterns of gene activity. According to the current hypothesis the sequence of the four bases that characterize DNA—guanine, adenine, thymine and cytosine—represents a code for the sequence of the 20 kinds of amino acid unit that make up a protein. Most, if not all, protein synthesis takes place not in the nucleus of the cell but in the surrounding cytoplasm. The DNA always remains in the nucleus. As a result the instructions supplied by DNA must be carried to the cytoplasm, where the translation is made. The carrier and translator of the DNA information is thought to be the special form of RNA called messenger RNA. Each DNA molecule serves as a template for a specific messenger RNA molecule, which then acts as a template in the synthesis of a particular protein. Hence what we have termed gene activity becomes equivalent to the rate of production of messenger RNA at each gene.

It has been known for some time that chromosome puffs contain significant amounts of RNA. As we have noted, the normal, unpuffed bands chiefly contain DNA and histone. In general the amount of these compounds remains unchanged in the transition from a band to a puff, whereas the amount of RNA increases considerably. The presence of RNA is beautifully demonstrated by metachromatic dyes such as toluidine blue, which simultaneously stains RNA red-violet and DNA a shade of blue. A great increase in the amount of RNA, however, is not sufficient to demonstrate that RNA synthesis is the main function of puffs. For one thing, some dyes show that a protein other than histone accumulates in the puffs along with RNA. Perhaps it too is made there.

In order to find out if RNA is the major puff product, Claus Pelling of our laboratory employed the technique of autoradiography. His "tracer" was uridine, a substance the cell tends to use to make RNA rather than DNA, that had been labeled with the radioactive isotope hydrogen 3 (tritium). He injected the uridine into *Chironomus* larvae, which he later killed. When giant cells from the larvae were placed in contact with a photographic emulsion, the radioactive loci in their chromosomes darkened the emulsion [*see illustration*

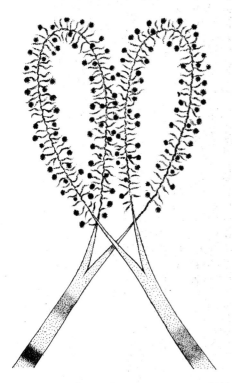

that are believed to be messenger RNA produced by the genes. In particularly small puffs the loops cannot actually be observed.

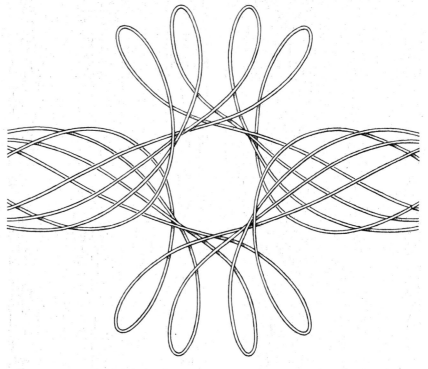

SCHEMATIC REPRESENTATION of how a large puff is formed shows fibrils untwisted and "popped out" of the cable-like structure. A giant chromosome in reality contains thousands of fibrils. Those untwisted here are tightly coiled when in the form of bands.

on page 235]. In every case in which Pelling killed the larvae soon after injection, sometimes as quickly as two minutes afterward, only the puffs, the Balbiani rings and the nucleoli were labeled. (Nucleoli are large deposits of RNA and protein that are formed in all types of cells by chromosomal regions known as nucleolar organizers.

Presumably they are involved in the formation of ribosomes, which are the sites of protein synthesis in the cytoplasm.) The rest of the chromosomal material and the cytoplasm showed very little radioactive label until long after the injection.

When the preparations were treated with an enzyme that decomposes RNA

before placing them in contact with the emulsion, the label was absent. Pelling demonstrated further that the rate of RNA synthesis is closely correlated with the relative size of the puffs. The administration of the antibiotic actinomycin D, a specific inhibitor of any RNA synthesis that depends directly on DNA, stopped the formation of RNA.

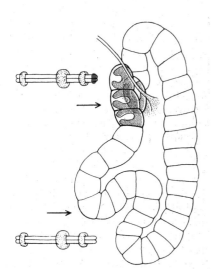
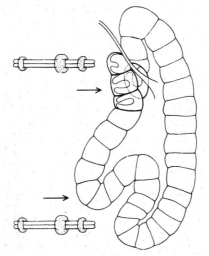
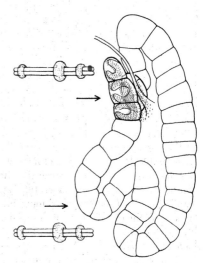

DIFFERENTIAL GENE ACTIVATION occurs on homologous, or matching, chromosomes of *Chironomus*. Four salivary-gland cells in the species *Ch. pallidivittatus* (left) produce granules (colored stippling). The species *Ch. tentans* (center) makes no granules. Chromosome IV from the four granule-producing cells (at left of cells) has a puff at one end (color), whereas the same chromosome from other cells (lower left) of the same gland and from all salivary-gland cells of *Ch. tentans* have no puff there. (In each case the chromosome inherited from both parents is shown.) Hybrids of the two species (right) have a puff only on the chromosome from the *Ch. pallidivittatus* parent in the four granule-producing cells. As indicated in the drawing, they make far fewer granules.

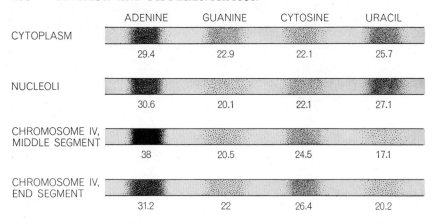

RNA'S FROM VARIOUS REGIONS of salivary-gland cells of *Chironomus* differ from one another in percentages of the four RNA bases. The RNA from each place was decomposed by an enzyme and the bases were then separated by electrophoresis on rayon threads. Samples from chromosome IV differ widely from nucleolar and cytoplasmic RNA's.

This proved that the synthesis was taking place at the site of the DNA in the chromosome. All these results agree with the assumption that the pattern of puffing along chromosomes is a quantitative reflection of the pattern of synthetic activities from gene to gene.

The protein in the puffs, in contrast to RNA, takes up little or no radioactive material if we inject the larvae with radioactively labeled leucine or another labeled amino acid. The labeled protein always appears first in the cytoplasm and does not reach the chromosomes for at least an hour. One of us (Clever) obtained the same result when he injected labeled leucine together with the hormone ecdysone, which elicits puffing at several sites in a short time: We have concluded, therefore, that the puff protein is made elsewhere than in the chromosome. Probably some of this protein is the enzyme RNA polymerase, which presides at the synthesis of RNA.

In order to learn if the RNA made in the puffs is messenger RNA, we collaborated with Jan-Erik Edström of the University of Göteborg in Sweden. He has developed an elegant microelectrophoretic technique that makes it possible to determine the base composition of very small amounts of RNA. He applies RNA that has been decomposed by an enzyme to moist rayon threads, which are then laid between two electric poles. Thereafter the different bases move different distances along the threads in a given time. Their quantities can be determined by photometry and their relative proportions established. We made separate analyses of the proportions of bases in RNA's from various parts of the salivary-gland cells of *Chironomus*, including the cytoplasm, the nucleoli, the entire chromosome I and the three large Balbiani rings of chromosome IV. This involved, among other things, cutting several hundred IV chromosomes into three pieces. The base compositions of all these RNA's differ from one another. The RNA's of the cytoplasm and the nucleoli appear to be nearly identical, but both differ from the RNA of the entire chromosome I and particularly from the RNA of puffs. In addition, there are slight but significant differences among the RNA's of the three Balbiani rings. One conclusion is that puff RNA certainly represents a special type of RNA. Is it therefore messenger RNA? An unusual feature of its base composition suggests that it is.

The RNA of salivary-gland chromosome puffs consistently contains more adenine than uracil—twice as much in the case of one Balbiani ring. (RNA contains uracil in place of the thymine in DNA.) Deviations from a one-to-one ratio are also found with respect to guanine and cytosine. In typical DNA the ratios of adenine to thymine and of guanine to cytosine invariably equal one because the bases are paired in the double-strand helix of the DNA molecule. RNA, being single-stranded, is not subject to this rule. In the case of messenger RNA, however, if one assumes that both strands of DNA make complementary copies of RNA, the ratios of adenine to uracil and guanine to cytosine should also be one. Most investigators confirm this expectation. Our data, on the other hand, strongly suggest that puff RNA is a copy of only one DNA strand. This appears to us to be a more reasonable way to make messenger RNA, since in protein synthesis only one of the two putative RNA copies of double-strand DNA could serve as a template. Messenger RNA fractions similar in composition to ours have now been discovered in other organisms.

Evidence for the physical movement of our messenger RNA has been found recently in electron micrographs of sections through the Balbiani rings that reveal the presence of ribonucleoprotein (RNA and protein) particles. In other electron micrographs such particles are seen floating freely in the nuclear sap and through pores in the nuclear membrane [*see bottom illustration on page 234*]. They break up in the cytoplasm. We believe these particles carry the messenger RNA to the ribosomes, where it would serve as the template for the synthesis of proteins.

In the hope of delineating at least some of the forces that control the behavior of genes, one of us (Clever) set out to learn about the conditions under which puffs are produced or changed. Since insect metamorphosis has been studied rather fully, a good starting point seemed to be the changes of the puff pattern in the course of metamorphosis.

Insect metamorphosis is the transformation from the larva to the adult. In the higher insects, to which the Diptera belong, it begins with the molting of the larva into the pupa and ends with the molting of the pupa into the imago, or adult. The moltings are caused by the hormone ecdysone, which is produced by the prothorax gland located in the thorax. So far this is the only insect hormone that has been purified. Because ecdysone affects single cells directly, injection of it induces changes related to molting in all cells of the insect body.

First we examined the time relation between the changes in puffing of individual loci and the metamorphic processes in the larvae. In the great majority of the puffed loci in the salivary glands, phases in which a puff is produced alternate with phases in which a puff is absent [*see illustration on opposite page*]. Some of the phases of puff formation have no recognizable connection with the molting process. Other puffs, however, appear regularly only after the molting of the larva has begun; some at the start of molting, others later. Apparently these chromosomal sites participate in metabolic processes that take place in the cell only during the molting stage. Finally, a third group of puffs, which are found in larvae of all ages, always become particularly large during metamorphosis. This indicates that some components of the metabolic process not specific to molting are intensified at that time.

Further experiments and observations have given some indication of how ecdysone regulates the activity of single sites during molting. In the first place, the hormone not only initiates the process; it must also be present continuously in the hemolymph, or blood, of the insect if molting is to continue. The secretion of ecdysone may stop for a time in *Chironomus* larvae that had begun to molt. In such larvae all the puffs characteristic of molting are absent, which shows that the hormone controls the pattern of gene activity specific to molting. Hans-Joachim Becker of the University of Marburg confirmed this by knotting a thread around *Drosophila* larvae at the start of metamorphosis so that the prothorax gland and part of the salivary gland were in front of the knot and another part of the salivary gland was behind it, cut off from the prothorax secretions. After a time he killed the larvae and found that the puff pattern of metamorphosis was absent in the salivary-gland cells behind the knot but present in cells in front of it.

In detail ecdysone affects the puffs in a variety of ways. If we inject the hormone into *Chironomus* larvae, most of the puffs do not react until long afterward. For some the interval is a few hours, for others one or more days, and this is independent of the quantity of ecdysone. Two puffs, on the other hand, appear quite soon after the injection of ecdysone into larvae that have not begun to molt. One puff arises in 15 to 30 minutes at locus 18-C of chromosome I, the other in 30 to 60 minutes at locus 2-B of chromosome IV. These are the earliest observable gene activations produced so far by the administration of ecdysone. At both loci the higher the dosage of hormone, the longer the puffs last. The injection of more hormone slows the regression of the puffs at both loci, and if ecdysone is injected after the puffs have regressed, they swell up again. From this we conclude that the cause of puff regression is the elimination of the hormone.

The two loci exhibit different reaction thresholds. At locus 18-C on chromosome I a minimum ecdysone concentration of about 10^{-7} microgram (one ten-trillionth of a gram) per milligram of larval weight is required to induce puffing. The locus 2-B on chromosome IV reacts only to about 10^{-6} microgram per milligram of larval weight. In these concentrations there can be no more than 100 ecdysone molecules at each of the chromosome strands in a puff, assuming that each giant chromosome

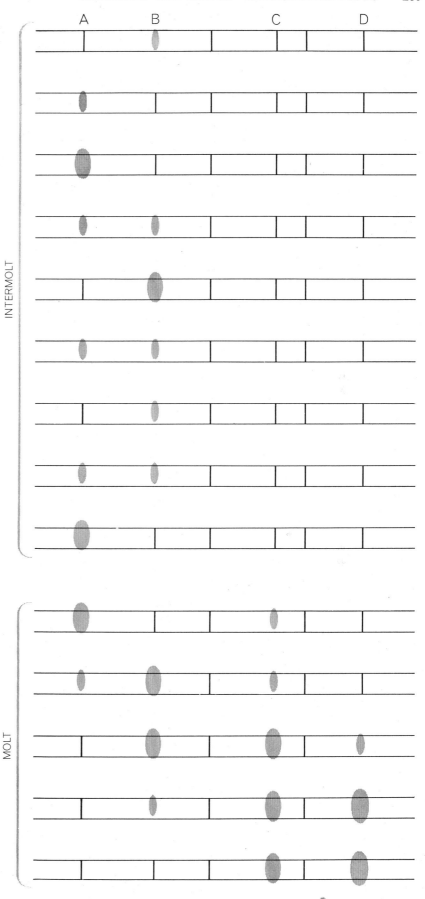

SEQUENCE OF PUFFING at four sites (*A, B, C, D*) of one chromosome in the salivary gland of *Ch. tentans* is diagramed. Some bands that do not puff are also shown. Starting from the top, the changes occur before and during the molt that begins pupation.

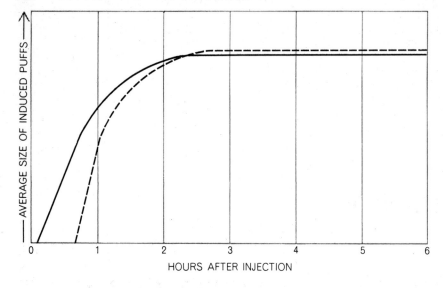

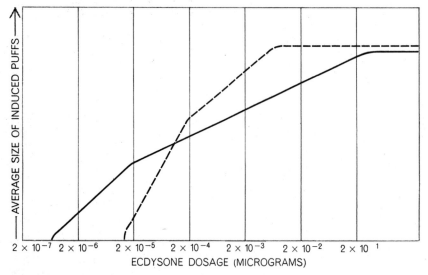

INDUCED PUFFING follows injection of the hormone ecdysone at locus 18-C of chromosome I in *Ch. tentans* (*solid curves*) and locus 2-B of chromosome IV (*broken curves*). Upper diagram shows time schedule of puffs, lower diagram the relation of puff size to quantity of hormone. Dosage is in micrograms per milligram of total weight of larva.

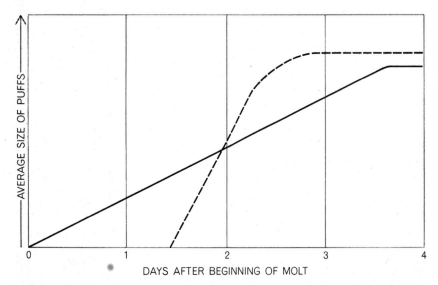

NORMAL PUFFING SEQUENCE at loci 18-C of chromosome I (*solid curve*) and 2-B of chromosome IV (*broken curve*) follows a schedule different from that of induced puffing.

consists of 10,000 to 20,000 single strands and that the hormone is distributed evenly throughout the larva. The puff at locus 2-B attains maximum size at a lower concentration than that at locus 18-C.

By applying these findings to a record of the growth and change of the two sites during normal molting, we find that the hormone level apparently increases gradually during metamorphosis. Thus the different activity patterns of these two loci can be explained as responses to the same factor: the changing hormone concentration.

In the case of locus 2-B, however, ecdysone is not the only active agent. The puff at this locus begins to regress during the second half of the prepupal phase (the last larval stage before pupation) even though the puff at locus 18-C persists to the end of the phase. In larvae that are ready to pupate, the puff at locus 2-B has usually regressed altogether. Yet when we inject hemolymph from these prepupae into young larvae, puffing is induced both at locus 18-C and locus 2-B. The former puff is quite large, which indicates that the hemolymph from the older larvae still contains ecdysone in high concentration. Evidently in these larvae, although not in their hemolymph, there is an antagonistic factor that actively represses puffing at locus 2-B in spite of the presence of ecdysone. This demonstrates that in higher organisms the activity of genes falls under the control of more than one factor.

Whereas gene activity at the loci 18-C and 2-B is subject to stringent regulation by very specific factors, other tests show that later puffs elsewhere are not the result of changes in the concentration of ecdysone. Rather they behave in an all-or-nothing manner that appears to depend on the duration and size of puffing at the two sites of earliest reaction.

We know nothing as yet about the mechanism by which the hormone regulates the genes. We are not even certain of the exact point of action of the hormone, although we would like to believe that it is the gene itself. The induction of puffing can be prevented with inhibitors of nucleic acid metabolism such as actinomycin or mitomycin, but inhibitors of protein synthesis, such as chloramphenicol and puromycin, have no apparent effect on the puffing. Thus ecdysone does not seem to act through the stimulation of protein synthesis in the cytoplasm, or to depend for its action on this synthesis. Only further investigation will solve such problems.

Part VI

SPECIAL ACTIVITIES

VI

Special Activities

INTRODUCTION

The outcome of the processes of differentiation described in the last section is an assemblage of cells, each member of which has, to some degree, a special activity. For some, this may be a synthetic task—the manufacture of hemoglobin by a red blood cell, of keratin by an epidermal cell, of yolk by the cells surrounding a developing egg. For others, the specialty may entail a certain *property* not shared by other cells. Of these, no cell types are more diagnostic of the basic character of complex animals than those composing nerves, muscles, and sense organs. The integration of all other systems depends upon their activities; not only are they collectively responsible for the ordered movement of the animal in his environment, but they also preside over a host of more routine and internal coordinations.

The property of impulse conduction, discussed by Bernhard Katz in "How Cells Communicate," depends upon the ability of the membrane to alter some yet unknown component of its physical structure quickly and reversibly, in order to allow sodium to enter, and then—just as quickly—to restore selective permeability to potassium ions. Propagated down the nerve, this sequence of ionic shifts provides the all-or-nothing electrical signal that releases small amounts of a certain chemical at the ending and so excites another nerve, muscle fiber, or other effector organ. Though we still lack information about the precise nature of this change in membrane state, the problems of how the transmitter is released and how the impulse is propagated have been very successfully attacked. Most of the steps that remain are concerned with the interactions between units.

In sensory cells, the electrical events that generate trains of impulses are similar to those occurring in the synaptic regions of nerve cells. In "How Cells Receive Stimuli," William H. Miller, Floyd Ratliff, and H. K. Hartline describe the origin of such generator potentials in different types of receptor elements—in particular, mechanoreceptors and photoreceptors— and relate them to the special equipment of these cells that enables them to act as transducers of stimulus energy.

Like the reception and conduction of signals, the ability to respond mechanically to such signals is also a remarkable cellular achievement. In "How Cells Move," Teru Hayashi discusses the generation of biological motion from protoplasmic streaming and ciliary action to the contraction of muscle. All such phenomena seemingly involve reversible changes in the state of proteins. In cilia and flagella, where a well-documented pattern of nine peripheral and two central proteinaceous filaments exists, the problem of energy transmission along the chain is posed with special force. In a way, the situation is the reverse of that in the many sensory cells that are derived from cilia and therefore have the same structure; it is conceivable that signal migration in the opposite direction could be involved here.

The final article, "The Contraction of Muscle" by H. E. Huxley, deals with the most familiar movement-generating devices we know of in biology. The superb analysis, by Huxley himself and his collaborators, of the ultrastructure of muscle and the contribution of specific protein components to it, has made possible a new set of explanations for some old and perplexing data. It is appropriate that the analysis of a specialized cellular system that ends this volume should have much the same theme as the article that introduced it: the unity between structure and function at the level of molecular architecture.

HOW CELLS COMMUNICATE

by BERNHARD KATZ — September 1961

In the animal kingdom, the "higher" the organism, the more important becomes the system of cells set aside for co-ordinating its activities. Nature has developed two distinct co-ordinating mechanisms. One depends on the release and circulation of "chemical messengers," the hormones that are manufactured by certain specialized cells and that are capable of regulating the activity of cells in other parts of the body. The second mechanism, which is in general far superior in speed and selectivity, depends on a specialized system of nerve cells, or neurons, whose function is to receive and to give instructions by means of electrical impulses directed over specific pathways. Both co-ordinating mechanisms are ancient from the viewpoint of evolution, but it is the second—the nervous system—that has lent itself to the greater evolutionary development, culminating in that wonderful and mysterious structure, the human brain.

Man's understanding of the working of his millions of brain cells is still at a primitive stage. But our knowledge is reasonably adequate to a more restricted task, which is to describe and partially explain how individual cells—the neurons—generate and transmit the electrical impulses that form the basic code element of our internal communication system.

A large fraction of the neuronal cell population can be divided into two classes: sensory and motor. The sensory neurons collect and relay to higher centers in the nervous system the impulses that arise at special receptor sites [see "How Cells Receive Stimuli" on page

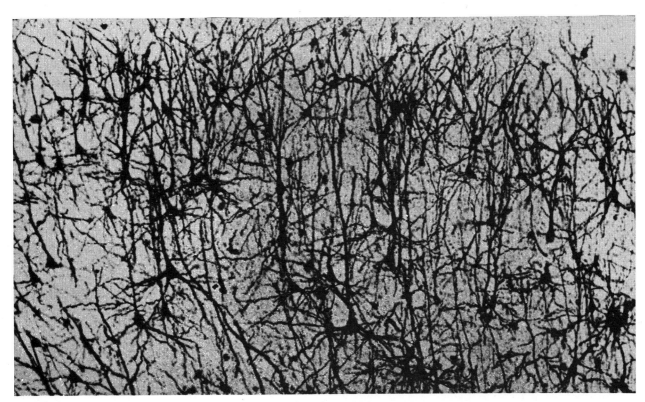

CEREBRAL CORTEX is densely packed with the bodies of nerve cells and the fibers called dendrites that branch from the cell body. This section through the sensory-motor cortex of a cat is enlarged some 150 diameters. Only about 1.5 per cent of the cells and dendrites actually present are stained and show here. The nerve axons, the fibers that carry impulses away from the cell body, are not usually shown at all by this staining method. The photomicrograph was made by the late D. A. Sholl of University College London.

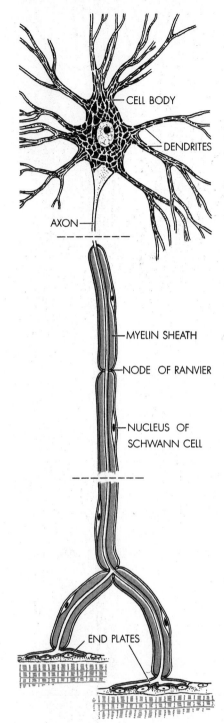

MOTOR NEURON is the nerve cell that carries electrical impulses to activate muscle fibers. The cell body (*top*) fans out into a number of twigs, the dendrites, which make synaptic contact with other nerve fibers (*see top illustration on opposite page*). Nerve impulses arising at the cell body travel through the axon to the motor-plate endings, which are embedded in muscle fibers. Myelin sheath is formed by Schwann cells as shown at bottom of opposite page. By insulating the axon the myelin wrapping increases the speed of signal transmission.

257], whose function is to monitor the organism's external and internal environments. The motor neurons carry impulses from the higher centers to the "working" cells, usually muscle cells, which provide the organism's response to changes in the two environments. In simple reflex reactions the transfer of signals from sensory to motor neurons is automatic and involves relatively simple synaptic mechanisms, which are fairly well understood.

When a nerve cell, either motor or sensory, begins to differentiate in the embryo, the cell body sends out a long fiber—the axon—which in some unknown way grows toward its proper peripheral station to make contact with muscle or skin. In man the adult axon may be several feet long, although it is less than .001 inch thick. It forms a kind of miniature cable for conducting messages between the periphery and the central terminus, which lies protected together with the nerve-cell body inside the spinal canal or the skull. Isolated peripheral nerve fibers probably have been subjected to more intense experimental study than any other tissue, in spite of the fact that they are only fragments of cells severed from their central nuclei as well as their terminal connections. Even so, isolated axons are capable of conducting tens of thousands of impulses before they fail to work. This fact and other observations make it clear that the nucleated body of the nerve cell is concerned with long-term maintenance of the nerve fibers—with growth and repair rather than with the immediate signaling mechanism.

For years there was controversy as to whether or not our fundamental concept of the existence of individual cell units could be applied at all to the nervous system and to its functional connections. Some investigators believed that the developing nerve cell literally grows into the cytoplasm of all cells with which it establishes a functional relationship. The matter could not be settled convincingly until the advent of high-resolution electron microscopy. It turns out that most of the surface of a nerve cell, including all its extensions, is indeed closely invested and enveloped with other cells, but that the cytoplasm of adjacent cells remains separated by distinct membranes. Moreover, there is a small extracellular gap, usually of 100 to 200 angstrom units, between adjoining cell membranes.

A fraction of these cell contacts are functional synapses: the points at which signals are transferred from one cell to the next link in the chain. But synapses are found only at and near the cell body of the neuron or at the terminals of the axon. Most of the investing cells, particularly those clinging to the axon, are not nerve cells at all. Their function is still a puzzle. Some of these satellite cells are called Schwann cells, others glia cells; they do not appear to take any part in the immediate process of impulse transmission except perhaps indirectly to modify the pathway of electric current flow around the axon. It is significant, for example, that very few scattered satellites are to be found on the exposed cell surfaces of muscle fibers, which closely resemble nerve fibers in their ability to conduct electrical impulses from one end to the other.

One of the known functions of the axon satellites is the formation of the so-called myelin sheath, a segmented insulating jacket that improves the signaling efficiency of peripheral nerve fibers in vertebrate animals. Thanks to the electron microscope studies of Betty Ben Geren-Uzman and Francis O. Schmitt of the Massachusetts Institute of Technology, we now know that each myelin segment is produced by a nucleated Schwann cell that winds its cytoplasm tightly around the surface of the axon, forming a spiral envelope of many turns [*see bottom illustration on opposite page*]. The segments are separated by gaps—the nodes of Ranvier—which mark the points along the axon where the electrical signal is regenerated.

There are other types of nerve fiber that do not have a myelin sheath, but even these are covered by simple layers of Schwann cells. Perhaps because the axon extends so far from the nucleus of the nerve cell it requires close association with nucleated satellite cells all along its length. Muscle fibers, unlike the isolated axons, are self-contained cells with nuclei distributed along their cytoplasm, which may explain why these fibers can manage to exist without an investing layer of satellite cells. Whatever the function of the satellites, they cannot maintain the life of an axon for long once it has been severed from the main cell body; after a number of days the peripheral segment of the nerve cell disintegrates. How the nerve cell nucleus acts as a lifelong center of repair and brings its influence to bear on the distant parts of the axon—which in terms of ordinary diffusion would be years away—remains a mystery.

The experimental methods of physiology have been much more successful in dealing with the immediate processes of nerve communication than with the equally important but much more intractable long-term events. We know very little about the chemical interactions between nerve and satellite, or about the forces that guide and attract growing nerves along specific pathways and that induce the formation of synaptic contacts with other cells. Nor do we know how cells store information and provide us with memory. The rest of this article will therefore be concerned almost solely with nerve signals and the method by which they pass across the narrow synaptic gaps separating one nerve cell from another.

Much of our knowledge of the nerve cell has been obtained from the giant axon of the squid, which is nearly a millimeter in diameter. It is fairly easy to probe this useful fiber with microelectrodes and to follow the movement of radioactively labeled substances into it and out of it. The axon membrane separates two aqueous solutions that are almost equally electroconductive and that contain approximately the same number of electrically charged particles, or ions. But the chemical composition of the two solutions is quite different. In the external solution more than 90 per cent of the charged particles are sodium ions (positively charged) and chloride ions (negatively charged). Inside the cell these ions together account for less than 10 per cent of the solutes; there the principal positive ion is potassium and the negative ions are a variety of organic particles (doubtless synthesized within the cell itself) that are too large to diffuse easily through the axon membrane. Therefore the concentration of sodium is about 10 times higher *outside* the axon, and the concentration of potassium is about 30 times higher *inside* the axon. Although the permeability of the membrane to ions is low, it is not indiscriminate; potassium and chloride ions can move through the membrane much more easily than sodium and the large organic ions can. This gives rise to a voltage drop of some 60 to 90 millivolts across the membrane, with the inside of the cell being negative with respect to the outside.

To maintain these differences in ion concentration the nerve cell contains a kind of pump that forces sodium ions "uphill" and outward through the cell membrane as fast as they leak into the cell in the direction of the electrochemical gradient [*see illustration on page 252*]. The permeability of the resting cell surface to sodium is normally so low that the rate of leakage remains very small, and the work required of the

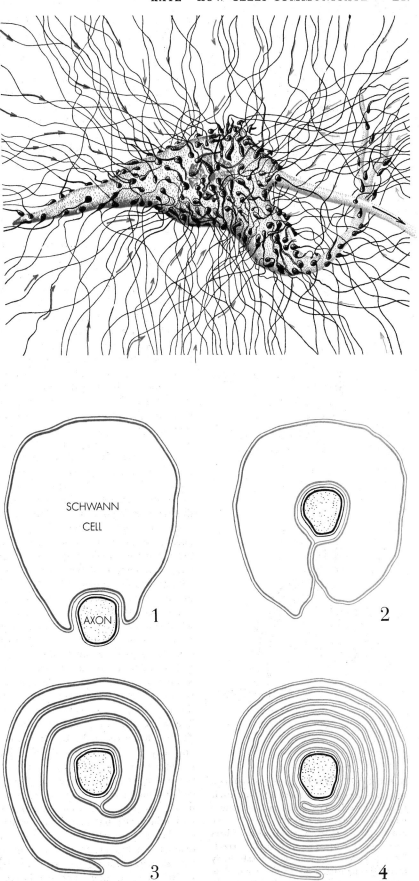

MYELIN SHEATH is created when a Schwann cell wraps itself around the nerve axon. After the enfolding is complete, the cytoplasm of the Schwann cell is expelled and the cell's folded membranes fuse into a tough, compact wrapping. Diagrams are based on studies of chick-embryo neurons by Betty Ben Geren-Uzman of Children's Medical Center in Boston.

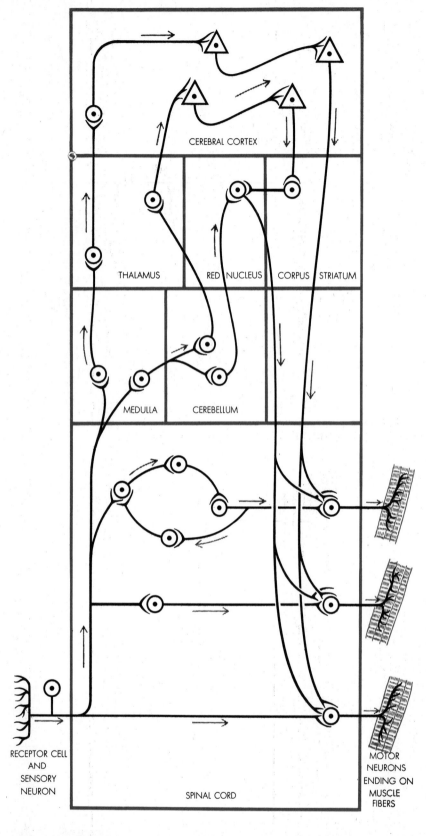

SIMPLIFIED FLOW DIAGRAM OF NERVOUS SYSTEM barely hints at the many possible pathways open to an impulse entering the spinal cord from a receptor cell and its sensory fiber. Rarely does the incoming signal directly activate a motor neuron leading to a muscle fiber. Typically it travels upward through the spinal cord and through several relay centers before arriving at the cerebral cortex. There (if not elsewhere) a "command" may be given (or withheld) that sends nerve impulses back down the spinal cord to fire a motor neuron.

pumping process amounts to only a fraction of the energy that is continuously being made available by the metabolism of the cell. We do not know in detail how this pump works, but it appears to trade sodium and potassium ions; that is, for each sodium ion ejected through the membrane it accepts one potassium ion. Once transported inside the axon the potassium ions move about as freely as the ions in any simple salt solution. When the cell is resting, they tend to leak "downhill" and outward through the membrane, but at a slow rate.

The axon membrane resembles the membrane of other cells. It is about 50 to 100 angstroms thick and incorporates a thin layer of fatty insulating material. Its specific resistance to the passage of an electric current is at least 10 million times greater than that of the salt solutions bathing it on each side. On the other hand, the axon would be quite worthless if it were employed simply as the equivalent of an electric cable. The electrical resistance of the axon's fluid core is about 100 million times greater than that of copper wire, and the axon membrane is about a million times leakier to electric current than the sheath of a good cable. If an electric pulse too weak to trigger a nerve impulse is fed into an axon, the pulse fades out and becomes badly blunted after traveling only a few millimeters.

How, then, can the axon transmit a nerve impulse for several feet without decrement and without distortion?

As one steps up the intensity of a voltage signal impressed on the membrane of a nerve cell a point is reached where the signal no longer fades and dies. Instead (if the voltage is of the right sign), a threshold is crossed and the cell becomes "excited" [*see illustrations on page 250*]. The axon of the cell no longer behaves like a passive cable but produces an extra current pulse of its own that amplifies the original input pulse. The amplified pulse, or "spike," regenerates itself from point to point without loss of amplitude and travels at constant speed down the whole length of the axon. The speed of transmission in vertebrate nerve fibers ranges from a few meters per second, for thin nonmyelinated fibers, to about 100 meters per second in the thickest myelinated fibers. The highest speeds, equivalent to some 200 miles per hour, are found in the sensory and motor fibers concerned with body balance and fast reflex movements. After transmitting an impulse the nerve is left briefly in a refractory, or inexcitable,

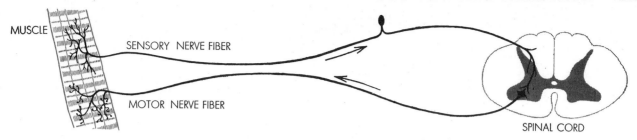

REFLEX ARC illustrates the minimum nerve circuit between stimulus and response. A sensory fiber arising in a muscle spindle enters the spinal cord, where it makes synaptic contact with a motor neuron whose axon returns to the muscle containing the spindle.

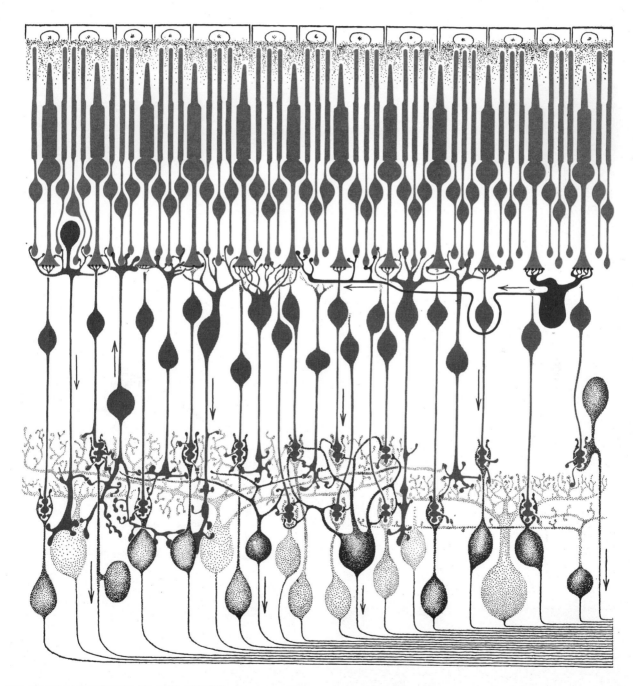

NERVE-CELL NETWORK IN THE RETINA, here magnified about 600 diameters, exemplifies the retinal complexity in man and apes. The photoreceptors are the densely packed cells shown in color; the thinner ones are rods, the thicker ones cones. To reach them the incoming light must traverse a dense but transparent layer of neurons *(dark shapes)* that have rich interconnections with the photoreceptors and with each other. The output of these neurons finally feeds into the optic nerve shown at the bottom of the diagram.

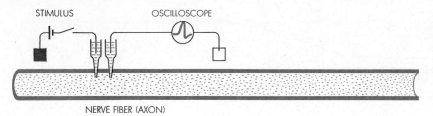

INVESTIGATION OF NERVE FIBER is carried out with two microelectrodes. One provides a stimulating pulse, the other measures changes in membrane potential (*see below*).

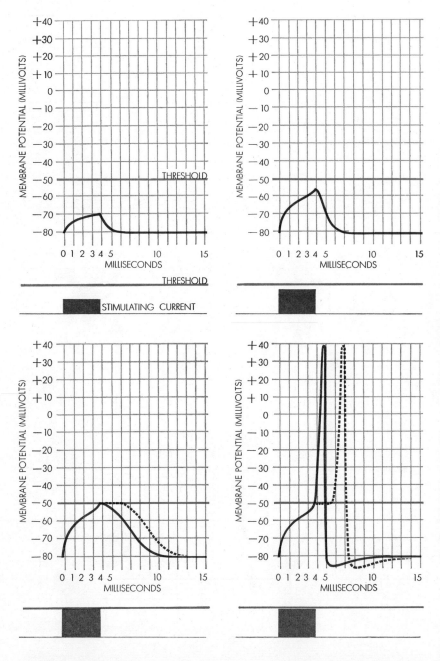

ELECTRICAL PROPERTIES OF NERVE FIBER are elucidated by measuring voltage changes across the axon membrane when stimulating pulses of varying size are applied. In the resting state the interior of the axon is about 80 millivolts negative. Subthreshold stimulating pulses (*top left and top right*) shift the potential upward momentarily. Larger pulses push the potential to its threshold, where it becomes unstable, either subsiding (*bottom left*) or flaring up into an "action potential" (*bottom right*) with a variable delay (*broken curve*).

state, but within one or two milliseconds it is ready to fire again.

The electrochemical events that underlie the nerve impulse—or action potential, as it is called—have been greatly clarified within the past 15 years. As we have seen, the voltage difference across the membrane is determined largely by the membrane's differential permeability to sodium and potassium ions. Many kinds of selective membrane, natural and artificial, show such differences. What makes the nerve membrane distinctive is that its permeability is in turn regulated by the voltage difference across the membrane, and this peculiar mutual influence is in fact the basis of the signaling process.

It was shown by A. L. Hodgkin and A. F. Huxley of the University of Cambridge that when the voltage difference across the membrane is artificially lowered, the immediate effect is to increase its sodium permeability. We do not know why the ionic insulation of the membrane is altered in this specific way, but the consequences are far-reaching. As sodium ions, with their positive charges, leak through the membrane they cancel out locally a portion of the excess negative charge inside the axon, thereby further reducing the voltage drop across the membrane. This is a regenerative process that leads to automatic self-reinforcement; the flow of some sodium ions through the membrane makes it easier for others to follow. When the voltage drop across the membrane has been reduced to the threshold level, sodium ions enter in such numbers that they change the internal potential of the membrane from negative to positive; the process "ignites" and flares up to create the nerve impulse, or action potential. The impulse, which shows up as a spike on the oscilloscope, changes the permeability of the axon membrane immediately ahead of it and sets up the conditions for sodium to flow into the axon, repeating the whole regenerative process in a progressive wave until the spike has traveled the length of the axon [*see illustration on opposite page*].

Immediately after the peak of the wave other events are taking place. The "sodium gates," which had opened during the rise of the spike, are closed again, and the "potassium gates" are opened briefly. This causes a rapid outflow of the positive potassium ions, which restores the original negative charge of the interior of the axon. For a few milliseconds after the membrane voltage has been driven toward its initial level it is difficult to displace the voltage and

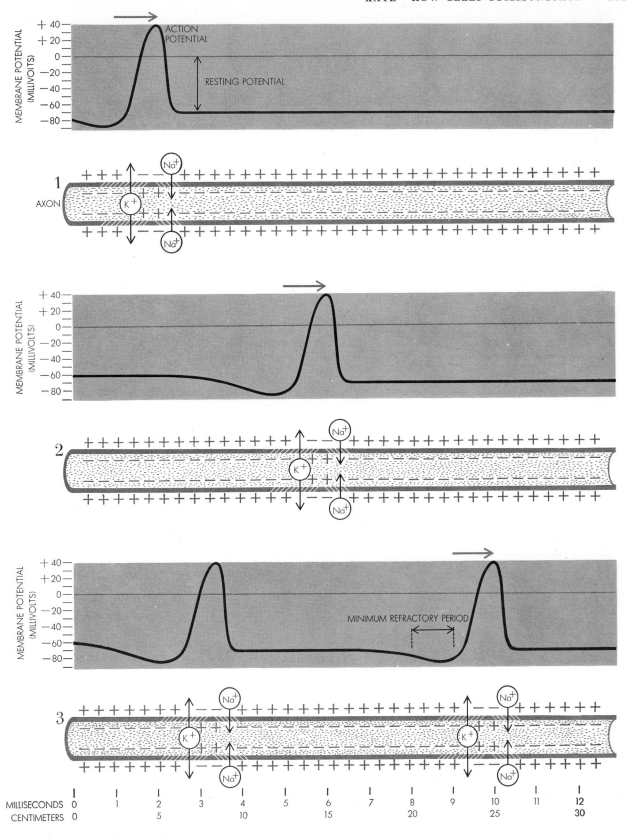

PROPAGATION OF NERVE IMPULSE coincides with changes in the permeability of the axon membrane. Normally the axon interior is rich in potassium ions and poor in sodium ions; the fluid outside has a reverse composition. When a nerve impulse arises, having been triggered in some fashion, a "gate" opens and lets sodium ions pour into the axon in advance of the impulse, making the axon interior locally positive. In the wake of the impulse the sodium gate closes and a potassium gate opens, allowing potassium ions to flow out, restoring the normal negative potential. As the nerve impulse moves along the axon (*1 and 2*) it leaves the axon in a refractory state briefly, after which a second impulse can follow (*3*). The impulse propagation speed is that of a squid axon.

set up another impulse. But the ionic permeabilities quickly return to their initial condition and the cell is ready to fire another impulse.

The inflow of sodium ions and subsequent outflow of potassium ions is so brief and involves so few particles that the over-all internal composition of the axon is scarcely affected. Even without replenishment the store of potassium ions inside the axon is sufficient to provide tens of thousands of impulses. In the living organism the cellular enzyme system that runs the sodium pump has no difficulty keeping nerves in continuous firing condition.

This intricate process—signal conduction through a leaky cable coupled with repeated automatic boosting along the transmission path—provides the long-distance communication needs of our nervous system. It imposes a certain stereotyped form of "coding" on our signaling channels: brief pulses of almost constant amplitude following each other at variable intervals, limited only by the refractory period of the nerve cell. To make up for the limitations of this simple coding system, large numbers of axon channels, each a separate nerve cell, are provided and arranged in parallel. For example, in the optic nerve trunk emerging from the eye there are more than a million channels running close together, all capable of transmitting separate signals to the higher centers of the brain.

Let us now turn to the question of what happens at a synapse, the point at which the impulse reaches the end of one cell and encounters another nerve cell. The self-amplifying cable process that serves within the borders of any one cell is not designed to jump automatically across the border to adjacent cells. Indeed, if there were such "cross talk" between adjacent channels, for instance among the fibers closely packed together in our nerve bundles, the system would become quite useless. It is true that at functional synaptic contacts the separation between the cell membranes is only 100 to a few hundred angstroms. But from what we know of the dimensions of the contact area, and of the insulating properties of cell membranes, it is unlikely that an effective cable connection could exist between the terminal of one nerve cell and the interior of its neighbor. This can easily be demonstrated by trying to pass a subthreshold pulse—that is, one that does not trigger a spike—across the synapse that separates a motor nerve from a muscle fiber. A recording probe located just inside the muscle detects no signal when a weak pulse is applied to the motor nerve close to the synapse. Clearly the cable linkage is broken at the synapse and some other process must take its place.

The nature of this process was discovered some 25 years ago by Sir Henry Dale and his collaborators at the National Institute for Medical Research in London. In some ways it resembles the hormonal mechanism mentioned at the beginning of this article. The motor nerve terminals act rather like glands secreting a chemical messenger. Upon arrival of an impulse, the terminals release a special substance, acetylcholine, that quickly and efficiently diffuses across the short synaptic gap. Acetylcholine molecules combine with receptor molecules in the contact area of the muscle fiber and somehow open its ionic gates, allowing sodium to flow in and trigger an impulse. The same result can be obtained by artificially applying ace-

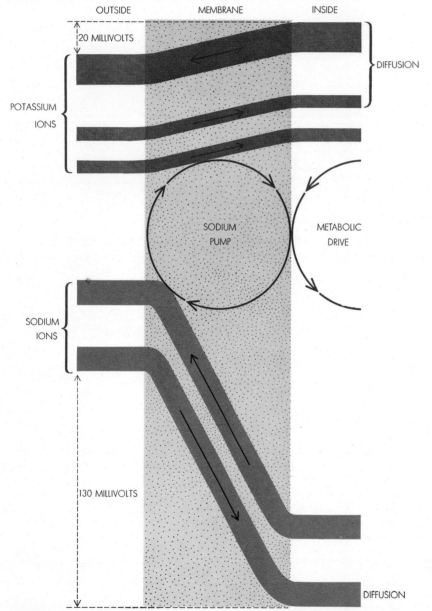

"SODIUM PUMP," details unknown, is required to expel sodium ions from the interior of the nerve axon so that the interior sodium-ion concentration is held to about 10 per cent that of the exterior fluid. At the same time the pump drives potassium ions "uphill" from a low external concentration to a 30-times-higher internal concentration. The pumping rate must keep up with the "downhill" leakage of the two kinds of ion. Since both are positively charged, sodium ions have the higher leakage rate (expressed in terms of millivolts of driving force) because they are attracted to the negatively charged interior of the axon, whereas potassium ions tend to be retained. But there is still a net outward leakage of potassium.

tylcholine to the contact region of the muscle fiber. It is probable that similar processes of chemical mediation take place at the majority of cell contacts in our central nervous system. But it is most unlikely that acetylcholine is the universal mediator at all these points, and an intensive search is being made by many workers for other naturally occurring transmitter substances.

Synaptic transmission presents two quite distinct sets of problems. First, exactly how does a nerve impulse manage to cause the secretion of the chemical mediator? Second, what are the physicochemical factors that decide whether a mediator will stimulate the next cell to fire in some cases or inhibit it from firing in others? So far we have said nothing about inhibition, even though it occurs throughout the nervous system and is one of the most curious modes of nervous activity. Inhibition takes place when a nerve impulse acts as a brake on the next cell, preventing it from becoming activated by excitatory messages that may be arriving along other channels at the same time. The impulse that travels along an inhibitory axon cannot be distinguished electrically from an impulse traveling in an excitatory axon. But the physicochemical effect that it induces at a synapse must be different in kind. Presumably inhibition results from a process that in some way stabilizes the membrane potential (degree of electrification) of the receiving cell and prevents it from being driven to its unstable threshold, or "ignition" point.

There are several processes by which such a stabilization could be achieved. One of them has already been mentioned; it occurs in the refractory period immediately after a spike has been generated. In this period the membrane potential is driven to a high stable level (some 80 to 90 millivolts negative inside the membrane) because, to put it somewhat crudely, the potassium gates are wide open and the sodium gates are firmly shut. If the transmitter substance can produce one or both of these states of ionic permeability, it will undoubtedly act as an inhibitor. There are good reasons for believing that this is the way impulses from the vagus nerve slow down and inhibit the heartbeat; incidentally, the transmitter substance released from the vagus nerve is again acetylcholine, as was discovered by Otto Loewi 40 years ago. Similar effects occur at various inhibitory synapses in the spinal cord, but there the chemical nature of the transmitter has so far eluded identification.

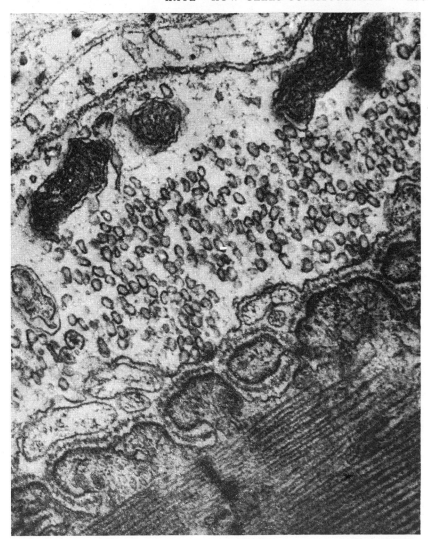

NERVE-MUSCLE SYNAPSE is the site at which a nerve impulse activates the contraction of a muscle fiber. In this electron micrograph (made by R. Birks, H. E. Huxley and the author) the region of the synapse is enlarged 53,000 diameters. Motor nerve terminal runs diagonally from lower left to upper right, being bounded at upper left by a Schwann cell. Muscle fiber is the dark striated area at lower right, with a folded membrane. Nerve terminal is populated with "synaptic vesicles" that may contain acetylcholine, which is released into the synaptic cleft by a nerve impulse and evokes electrical activity in the muscle.

Inhibition would also result if two "antagonistic" axons converged on the same spot of a third nerve cell and released chemically competing molecules. Although a natural example of this kind has not yet been demonstrated, the chemical and pharmacological use of competitive inhibitors is well established. (For example, the paralyzing effect of the drug curare arises from its competitive attachment to the region of the muscle fiber that is normally free to react with acetylcholine.) Alternatively, a substance released by an inhibitory nerve ending could act on the excitatory nerve terminal in such a way as to reduce its secretory power, thereby causing less of the excitatory transmitter substance to be released.

This brings us back to the question:

How does a nerve impulse lead to the secretion of transmitter substances? Recent experiments on the nerve-muscle junction have shown that the effect of the nerve impulse is not to initiate a process of secretion but rather, by altering the membrane potential, to change the rate of a secretory process that goes on all the time. Even in the absence of any form of stimulation, packets of acetylcholine are released from discrete spots of the nerve terminals at random intervals, each packet containing a large number—probably thousands—of molecules.

Each time one of these quanta of transmitter molecules is liberated spontaneously, it is possible to detect a sudden minute local response in the muscle fiber on the other side of the synapse.

Within a millisecond there is a drop of .5 millivolt in the potential of the muscle membrane, which takes about 20 milliseconds to recover. By systematically altering the potential of the membrane of the nerve ending it has been possible to work out the characteristic relation between the membrane potential of the axon terminal and the rate of secretion of transmitter packets. It appears that the rate of release increases by a factor of about 100 times for each 30-millivolt lowering of membrane potential. In the resting condition there is a random discharge of about one packet per second at each nerve-muscle junction. But during the brief 120-millivolt change associated with the nerve impulse the frequency rises momentarily by a factor of nearly a million, providing a synchronous release of a few hundred packets within a fraction of a millisecond.

It is significant that the transmitter is released not in independent molecular doses but always in multimolecular parcels of standard size. The explanation of this feature is probably to be found in the microstructural make-up of the nerve terminals. They contain a characteristic accumulation of so-called vesicles, each about 500 angstroms in diameter, which may contain the transmitter substance parceled and ready for release [*see illustration on page 253*]. Conceivably when the vesicles collide with the axon membrane, as they often must, the collision may sometimes cause the vesicular content to spill into the synaptic cleft. Such ideas have yet to be proved by direct evidence, but they provide a reasonable explanation of all that is known about the quantal spontaneous release of acetylcholine and its accelerated release under various natural and experimental conditions. At any rate, the ideas provide an interesting meeting point between the functional and structural approaches to a common problem.

Because of the sparseness of existing knowledge, we have left out of this discussion many fascinating problems of the long-term interactions and adaptive modifications that must certainly take place in nerve pathways. For handling such problems investigators will probably have to develop very different methods from those followed in the past. It may be that our preoccupation with the techniques that have been so successful in illuminating the brief reactions of excitable cells has prevented us from making inroads on the problems of learning, of memory, of conditioning and of the structural and operating relations between nerve cells and their neighbors.

HOW CELLS RECEIVE STIMULI

by WILLIAM H. MILLER, FLOYD RATLIFF, and
H. K. HARTLINE   September 1961

The survival of every living thing depends ultimately on its ability to respond to the world around it and to regulate its own internal environment. In most multicellular animals this response and regulation is made possible by specialized receptor cells that are sensitive to a wide variety of physical, chemical and mechanical stimuli.

In many animals, including man, these receptors provide information that far exceeds that furnished by the traditional five senses (sight, hearing, smell, taste and touch). Sense organs of which we are less aware include equally important receptors that monitor the internal environment. Receptor organs in the muscles, called muscle spindles, provide a continuous measure of muscle stretch, and other receptors sense the movement of joints. Without such receptors it would be difficult to move or talk. Receptor cells in the hypothalamus, a part of the brain, are sensitive to the temperature of the blood; pressure-sensitive cells in the carotid sinus measure the blood pressure. Still other internal receptors monitor carbon dioxide in special regions of the large arteries. Pain receptors, widely distributed throughout the body, respond to noxious stimuli of almost any nature that are likely to cause tissue damage.

Receptor cells not only have diverse functions and structures but also connect in various ways with the nerve fibers channeling into the central nervous system. Some receptor cells give rise directly to nerve fibers of their own; others make contact with nerve fibers originating elsewhere. All receptors, however, share a common function: the generation of nerve impulses. This does not imply that impulses necessarily occur in the receptor cells themselves. For example, in the eyes of vertebrates no one has yet been able to detect impulses in the photoreceptor cells: the rods and cones. Nevertheless, the rods and cones, when struck by light, set up the physicochemical conditions that trigger impulses in nerve cells lying behind them. Typical nerve impulses are readily detected in the optic nerve itself, which is composed of fibers of ganglion cells separated from the rods and cones by at least one intervening group of nerve cells.

Eventually physiologists hope to unravel the detailed train of events by which a receptor cell gives rise to a discharge of nerve impulses following mechanical deformation, absorption of light or heat, or stimulation by a particular molecule. In no case have all the events been traced out. In our discussion we will begin with the one final event common to all sensory reception—the generation of nerve impulses. We will then examine in some detail the events occurring in one particular receptor: the photoreceptor of *Limulus*, the horseshoe crab. Finally, we will describe some characteristics of the output of receptors acting singly and in concert with others.

The nerve fiber, or axon, is a thread-like extension of the nerve-cell body. The entire surface membrane of the cell, including that of the axon, is electrically polarized; the inside of the cell is some 70 millivolts negative with respect to the outside. This potential difference is called the membrane potential. In response to a suitable triggering event the membrane potential is momentarily and locally altered, giving rise to a nerve impulse, which is then propagated the whole length of the axon [see "How Cells Communicate" on page 245].

In any particular nerve fiber the impulses are always of essentially the same magnitude and form and they travel with the same speed. This has been known for some 30 years, since the pioneering studies of E. D. Adrian at the University of Cambridge. He and his colleagues found that varying the intensity of the stimulus applied to a receptor cell affects not the size of the impulses but the frequency with which they are discharged; the greater the intensity, the greater the frequency of nerve impulses generated by the receptor. Thus all sensory messages—concerning light, sound, muscle position and so on—are conveyed in the same code of individual nerve impulses. The animal is able to decode the various messages because each type of receptor communicates to the higher nerve centers only through its own private set of nerve channels.

Adrian and others have investigated the problem of how the receptor cell triggers sensory nerve impulses. Adrian suggested that the receptor must somehow diminish the resting membrane potential of its nerve fiber; that is, it must locally depolarize the axon membrane. The existence of local potentials in the eye has been known since 1865, and much later similar potentials were recorded in other sense organs. But the

---

VISUAL RECEPTOR of the horseshoe crab (*Limulus*) is enlarged 19,000 diameters in the electron micrograph on the opposite page. Called an ommatidium, it is one of about 1,000 photoreceptor units in the compound eye of *Limulus*. Here the ommatidium is seen in cross section; the individual receptor cells are arranged radially like segments of a tangerine around a nerve filament (dendrite) arising from an associated nerve cell *(see illustration on page 261)*. The dark ring around the dendrite and spokelike areas may contain photosensitive pigment. The electron micrograph was made by William H. Miller, one of the authors.

relationship of these gross electrical changes to the discharge of nerve impulses was not clear. For some simple eyes, however, the polarity of the local potential changes in the receptors is such that they appear to depolarize the sensory nerve fibers. This led Ragnar Granit of the Royal Caroline Institute in Stockholm to propose that they be called "generator" potentials. The present view is that stimulation of the receptor cell gives rise to a sustained local depolarization of the sensory nerve fiber, which thereupon generates a train of impulses.

Some of the first direct evidence for generator potentials at the cellular level was produced in 1935 by one of the authors of this article (Hartline), then working at the Johnson Research Foundation of the University of Pennsylvania. He found what appeared to be a generator potential when he recorded the activity of a single optic nerve fiber and its receptor in the compound eye of *Limulus*. Superimposed on the potential was a train of nerve impulses [see illustration on page 262].

In 1950 Bernhard Katz of University College London obtained unmistakable evidence for a generator potential in a somewhat simpler receptor: the vertebrate muscle spindle. When the spindle was stretched, a small, steady depolarization could be recorded in the nerve fiber coming from the spindle. As viewed on the oscilloscope, it appeared that the base line of the recorded signal had been shifted slightly upward. Superimposed on the shifted signal, or local potential, was a series of "spikes" representing individual nerve impulses. The stronger or the more rapid the stretch, the greater the magnitude of the potential shift and the greater the frequency of the impulses [see illustration on page 260]. Analysis of many such records showed that in the steady state the frequency of nerve impulses depends directly on the magnitude of the altered potential. If a local anesthetic is applied to the spindle, the impulses are abolished but the potential shift remains. Katz concluded that this potential shift is an essential link between the stretching of the spindle and the discharge of nerve impulses; indeed, that it is the generator potential. Moreover, the potential can be detected only very close to the spindle, showing that it is conducted passively—which is to say poorly—along the nerve fiber.

Important confirmation of the role of the generator potential was provided by the work of Stephen W. Kuffler and

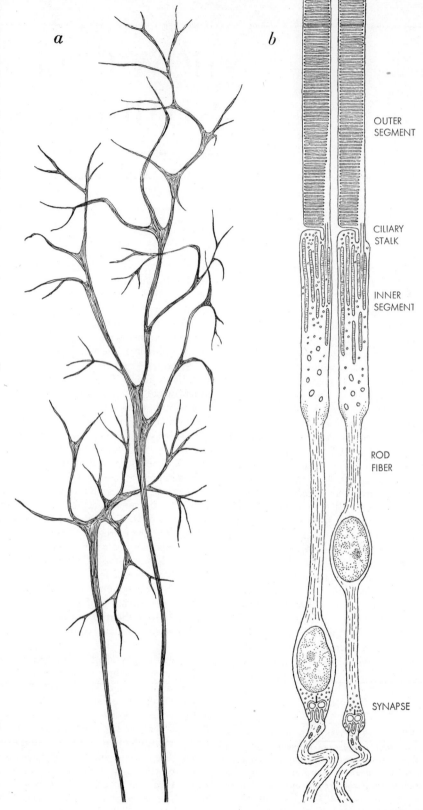

RECEPTOR CELLS, typical of those found in vertebrates, respond to a variety of stimuli: heat, light, chemicals and mechanical deformation. The "pit" on the head of the pit viper contains a network of free nerve endings (*a*) that are sensitive to heat and help the viper locate its prey. Rods (*b*) are light-sensitive cells in the retina of the eye; photosensitive pigment is in the laminar structure at top of drawing. Taste buds (*c*) are chemoreceptor cells embedded in the tongue. The cochlea, a spiral tube in the inner ear, contains thousands of

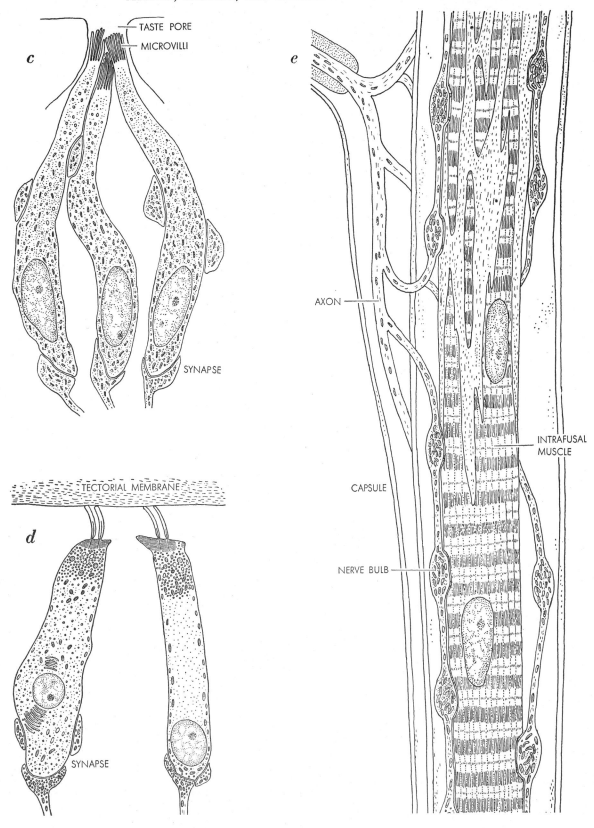

sensitive cells (*d*) in the so-called organs of Corti. When the hairlike bristles of these cells are mechanically deformed by sound vibrations, impulses are generated in the auditory nerve fibers leading to brain. Muscle spindle (*e*) contains a number of nerve endings that respond sensitively to stretching of muscle fibers surrounding them. These illustrations of receptor cells are based on the work of the following investigators: Theodore H. Bullock of the University of California at Los Angeles (*a*), Fritiof Sjöstrand of the same institution (formerly of the Karolinska Institute, Stockholm) (*b*), A. J. de Lorenzo of the Johns Hopkins School of Medicine, Baltimore, Md. (*c*), Salvatore Iurato of the University of Milan (*d*) and Bernhard Katz of University College London (*e*).

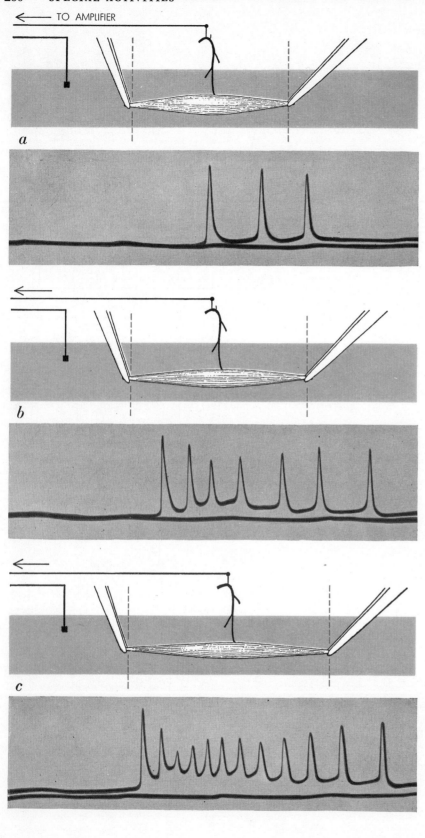

MUSCLE SPINDLE responds to stretch by firing nerve impulses at a rate proportional to the degree and speed of stretching. These recordings made by Bernhard Katz of University College London were the first to show that stretching causes depolarization of the nerve near the spindle (*base line shifted upward in the traces*) and that this depolarization is the precondition for the firing of nerve impulses. The shift is called the generator potential.

Carlos Eyzaguirre, then at Johns Hopkins University, using the so-called Alexandrowicz stretch-receptor cells in crustaceans. These are large single receptor cells with dendrites (short fibers) that are embedded in specialized receptor muscles. Kuffler was able to insert a microelectrode within the cell and record its membrane potential as well as the nerve impulses in its axon. He found that when he distorted the cell's dendrites by stretching the receptor muscle, the cell body became depolarized and the depolarization spread passively to the site of impulse generation, which is probably in the axon close to where it emerges from the cell body. When this generator potential reached a critical level, the cell fired a train of nerve impulses; the greater the depolarization of the axon above this critical level, the higher the frequency of the discharge.

There is now abundant evidence that a receptor cell triggers a train of nerve impulses by locally depolarizing the adjacent nerve fiber—either its own fiber or one provided by another cell. With few exceptions, a fiber of a nerve trunk will not respond repetitively if one passes a sustained depolarizing current through it; it responds only briefly with one to several impulses and then accommodates to the stimulus and responds no more. Evidently that part of the sensory nerve fiber close to the receptor must be specialized so that it does not speedily accommodate to the generator potential. It is nonetheless true that a certain amount of accommodation, or adaptation, almost always takes place when a receptor cell is exposed to a sustained stimulus. In any event, the initiation of nerve impulses in the axons of receptor cells by means of a generator potential appears to be a general phenomenon.

The question still remains: How does the external stimulus produce the generator potential? In most of the receptors studied there is no evidence whatever on this point. Only in the photoreceptor do we have precise knowledge of the first step in the excitation of the sense cell. Yet the study of the photoreceptor is beset by special difficulties. In most eyes the receptors are small and densely packed, and their associated neural structures are complex and highly organized. A fortunate exception is the compound eye of *Limulus,* which provided early evidence for the generator potential. In this eye the receptor cells are large and the neural organization is relatively simple.

The coarsely faceted compound eye

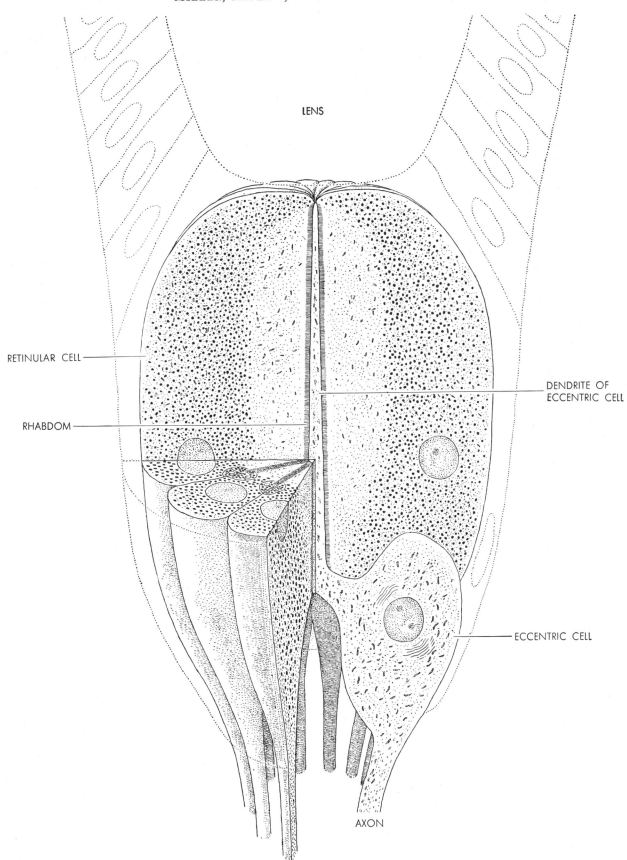

OMMATIDIUM OF LIMULUS is a remarkable structure roughly the size of a pencil lead. About 1,000 form the crab's compound eye. The ommatidium consists of about 12 wedge-shaped retinular cells clustered around a central fiber, which is the dendrite (sensitive process) of a nerve cell, the eccentric cell shown at lower right. When light strikes the ommatidium (*at the top*), the eccentric cell gives rise to nerve impulses (*see illustration on next page*). Photosensitive pigment rhodopsin is believed to be in the rhabdom.

of *Limulus* has some 1,000 ommatidia ("little eyes"), each of which contains about a dozen cells. The cells in each ommatidium have a regular arrangement. The retinular cells—the receptors—are arranged radially like the segments of a tangerine around the dendrite of an associated neuron: a single eccentric cell within each ommatidium [*see illustration on preceding page*].

Hartline, H. G. Wagner and E. F. MacNichol, Jr., working at Johns Hopkins University, found by the use of microelectrodes that the eccentric cell gives rise to the nerve impulses that can be recorded farther down in the nerve strand leaving the ommatidium. The microelectrode also records the generator potential of the ommatidium. Because of the anatomical complexity of the ommatidium, the site of origin of the generator potential has not been identified with certainty. Nor has activity yet been detected in the axons of the retinular cells. As in the vertebrate and invertebrate stretch receptors, local anesthetics extinguish the nerve impulses without destroying the generator potential. Moreover, as in the stretch receptors, there is a proportional relationship between the degree of depolarization and the frequency of nerve impulses.

Recently M. G. F. Fuortes of the National Institute of Neurological Diseases and Blindness has shown that illumination increases the conductance of the eccentric cell. He postulates that the increase is produced by a chemical transmitter substance that is released by the action of light and acts on the eccentric cell's dendrite. Presumably the increased conductance of the dendrite results in a depolarization that spreads passively to the site of impulse generation, where it acts as the generator potential.

In photosensory cells—alone among all receptors—there exists direct experimental evidence of the initial molecular events in the receptor process. It has been known for about a century that visual receptor cells in both vertebrates and invertebrates have specially differentiated organelles containing a photosensitive pigment. In vertebrates this reddish pigment, called rhodopsin, can be clearly seen in the outer segments of rods. The absorption spectrum of human rhodopsin corresponds closely to the light-sensitivity curve for human vision under conditions of dim illumination, when only the rods of the retina are operative. This is strong evidence that rhodopsin brings about the first active event in rod vision: the absorption of light by the photoreceptor structure. (There is evidence for similar pigments in the outer segments of cones, but they have proved more difficult to isolate and study.)

The visual pigments are known to be complex proteins, but the light-absorbing part of the pigment, called the chromophore, has been found to be a relatively simple substance: vitamin A aldehyde. Because it contains a number of double chemical bonds in its make-up, vitamin A aldehyde can exist in various molecular configurations known as "*cis*" and "*trans*" isomers. We know from the work of Ruth Hubbard, George Wald and their colleagues at Harvard University that the absorption of light changes the chromophore from

RECORDINGS FROM OMMATIDIA show trains of nerve impulses evoked by light. The upper recording from a nerve bundle (*a*) was made by one of the authors, H. K. Hartline, some 25 years ago. Shift of base line underlying nerve spikes is the generator potential. Lower recording, made with microelectrode (*b*), shows generator potential more clearly.

NERVE IMPULSES TRIGGERED BY LIGHT are directly related to intensity of steady light falling on the *Limulus* eye. Recordings were made from the optic nerve fiber arising from one ommatidium. At high light intensity (*top*) the nerve fires about 30 times per second. As intensity is reduced by factors of 10, firing is reduced in uniform steps, falling to a low of two or three impulses per second.

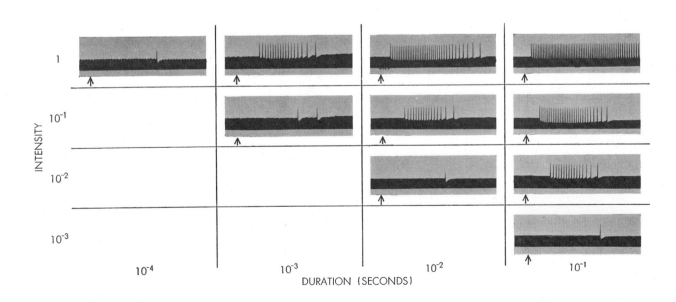

DURATION AND INTENSITY OF LIGHT have equivalent effect on the *Limulus* eye. Evidently the receptor responds to the total amount of energy received in a brief flash (*arrows*) regardless of how the energy is "packaged" in duration and intensity. Thus a brief intense flash (*top left*) evokes about the same response as a flash a 1,000th as bright lasting 1,000 times longer (*bottom right*).

11-*cis* vitamin A aldehyde to the *trans* configuration. This photochemical reaction is the first step that leads, through a chain of chemical and physical events as yet unknown, to the initiation of the generator potential of the receptor cell and finally to the discharge of impulses in the optic nerve. This is the only case in which the specific molecular mechanism is known whereby a receptor cell detects environmental conditions.

Supporting evidence that rhodopsin governs the response to a light stimulus can be found by comparing the absorption spectrum of *Limulus* rhodopsin with the sensitivity of the *Limulus* eye at various wavelengths. In 1935 Clarence H. Graham and Hartline measured the intensity of flashes at several wavelengths required to produce a fixed number of impulses in the *Limulus* optic nerve. When a sensitivity curve obtained from this experiment is superimposed on the absorption curve found by Hubbard and Wald for *Limulus* rhodopsin, the two match almost perfectly. At a wavelength of about 520 angstrom units, where rhodopsin absorbs light most strongly, the *Limulus* eye generates the highest number of impulses for a given quantity of light energy received. It turns out that the wavelength sensitivity of the *Limulus* eye is close to that of the human eye in dim light when rod vision dominates.

Many other familiar sensory experiences are manifestations of the properties of individual sense cells. Perhaps the most elementary experience is our ability to perceive when a stimulus has been increased in intensity. Under such circumstances we can be sure that the sensory fibers conveying information to the brain are firing more rapidly as the stimulus is increased. We are also familiar with the experience of sensory adaptation; for example, a strong odor usually seems to decrease in intensity after a time, although objective measurements would show that its intensity has remained constant.

We know from photography that shutter speed and lens opening can be interchanged to produce a constant exposure, which is the same as saying that intensity and duration of illumination can be interchanged (within limits) to produce a constant photochemical effect. The same equivalence holds for the human eye exposed to short flashes of light, and the equivalence can be demonstrated in the photoreceptor of *Limulus*. About the same number of nerve impulses are produced by exposing the ommatidium to a weak light for a 10th of a second as by exposing it to light 10 times as bright for a 100th of a second [*see bottom illustration on page 263*].

We also know from watching motion pictures or television that a light flickering at a high rate appears not to be flickering at all. A neural basis for this phenomenon can be seen in the generator potentials and nerve impulses recorded when a *Limulus* ommatidium is exposed to a light flickering at various rates [*see illustration at left*]. Flicker is detectable as fluctuations in the generator potential, which in turn gives rise to bursts of impulses. As the repetition rate increases, the rate of discharge becomes steadier and finally is indistinguishable from a response to continuous illumination. As can be seen from the records, this "flicker fusion" is directly attributable to the generator potential, which becomes smooth at the highest repetition rates.

The experiments described so far were carried out on single cells or single sensory units. In the eye, ear and other organs, however, receptor cells are grouped close together and usually act in concert. In fact, modern studies show that receptor cells of complex sense organs seldom act independently. In such organs the receptor cells are interconnected neurally and as a result of these connections new functional properties arise.

Although the compound eye of *Limu-*

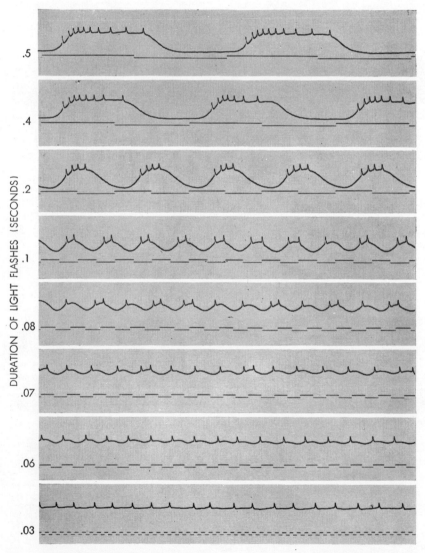

EFFECT OF FLICKERING LIGHT on the *Limulus* ommatidium provides a basis for explaining "flicker fusion": the inability to perceive a rapid flicker. The recordings show the response of the ommatidium to a light flickering at various rates; when the horizontal line is raised, the light is on. At low flicker rates the generator potential, indicated by a rise in base potential, rises and falls. As flicker rate increases, the generator potential no longer falls between flashes, and spacing between nerve impulses becomes more uniform.

lus is much less complex than the eyes of vertebrates, it still shows clearly the effects of neural interaction. In *Limulus* the activity of each photoreceptor unit is affected to some degree by the activity of adjacent ommatidia. The frequency of discharge of impulses in an optic nerve fiber from a particular ommatidium is decreased—that is, inhibited—when light falls on its neighbors. Since each ommatidium is a neighbor of its neighbors, mutual inhibition takes place. This inhibition is brought about by a branching array of nerve axons that make synaptic contact with each other in a feltwork of fine fibers behind the ommatidia. The inhibition probably results from a decrease in the magnitude of the generator potential at the site of origin of the nerve impulses, as a consequence of which the rate of firing is slowed down.

When two adjacent ommatidia are illuminated at the same time, each discharges fewer impulses than when it receives the same amount of light by itself [*see illustrations on this page*]. The magnitude of the inhibition exerted on each ommatidium (in the steady state) depends only on the frequency of the response of the other. The more widely separated the ommatidia, the smaller the mutual inhibitory effect. When several ommatidia are illuminated at the same time, the inhibition of each is given by the sum of the inhibitory effects from all others.

Inhibitory interaction can produce important visual effects. The more intensely illuminated retinal regions exert a stronger inhibition on the less intensely illuminated ones than the latter do on the former. As a result differences in neural activity from differently lighted retinal regions are exaggerated. In this way contrast is heightened and certain significant features of the retinal image tend to be accentuated at the expense of fidelity of representation.

This has been shown by illuminating the *Limulus* compound eye with a "step" pattern: a bright rectangle next to a dimmer one [*see illustration on page 266*]. The eye was masked so that only one ommatidium "observed" the pattern, which was moved to various positions on the retinal mosaic. At each position the steady-state frequency of discharge was measured. The result was a faithful reproduction (in terms of frequency of impulses) of the form of the pattern. Then the eye was unmasked so that all the ommatidia observed the pattern, and a recording was again made from the single ommatidium. This time

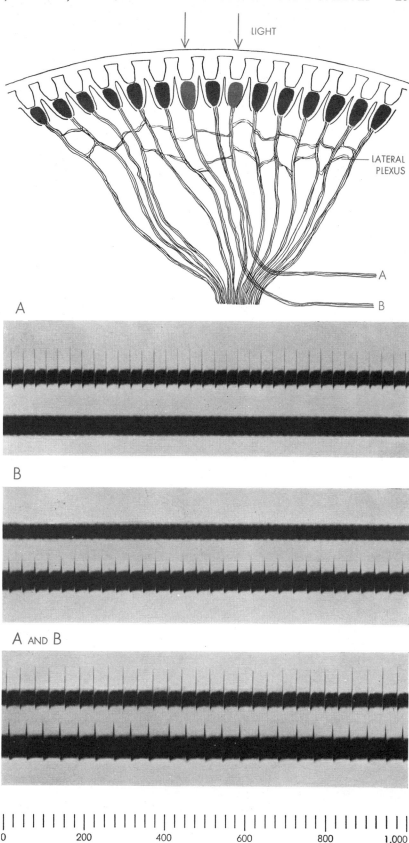

MUTUAL INHIBITION results when two neighboring ommatidia are illuminated at the same time (*top*). The inhibition is exerted by cross connections among nerve fibers. When ommatidia attached to fiber *A* and fiber *B* were illuminated separately, 34 and 30 impulses were recorded respectively in one second. Illuminated together, they fired less often.

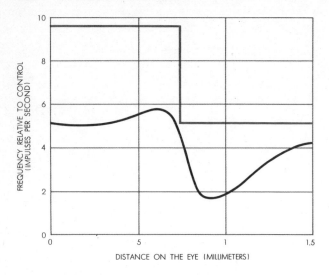 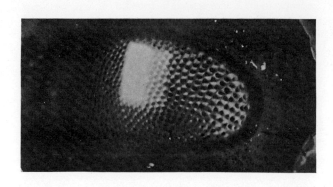

CONTRAST HEIGHTENING AT CONTOURS is demonstrated by letting "step" pattern of light, a bright area next to a darker one, fall on *Limulus* compound eye (*right*). If eye is masked so light strikes only one ommatidium, a recording of its output forms a simple step-shaped curve (*left*) as the pattern moved across the eye. If the eye is unmasked, the output of the single ommatidium is inhibited in varying degrees by the light striking its neighbors. The net effect (*lower curve*) is to heighten the contrast at light-dark boundaries.

the frequency increased on the bright side of the step and decreased near the dim side. This is expected because near the bright side of the step the neighboring ommatidia illuminated by the dim part of the step pattern have a low frequency of firing and therefore do not exert much inhibition. Consequently the frequency of discharge of the receptors on the bright side of the step is higher than its equally illuminated but more distant neighbors. Similar reasoning explains the decrease in frequency on the dim side of the step. The net effect of this pattern of response is to enhance contours, an effect we can easily demonstrate in our own vision by looking at a step pattern consisting of a series of uniform gray bands graded from white to black. Artists are quite familiar with the existence of "border contrast" and may even heighten it in their paintings. And as we all know, significant information is conveyed by contours alone, as is demonstrated by cartoons and other line drawings. Georg von Békésy of Harvard University has suggested that a similar reciprocal inhibition in the auditory system would lead to a sharpening of the sense of pitch.

There is also evidence that in many sense organs the response can be modified by neural influences exerted back onto them by higher centers of the nervous system. Thus the sensitivity of the vertebrate stretch receptor or muscle spindle is established by variations in the length of the spindle fibers, and this length is dependent both on the output of the receptor and on its interaction with higher centers. The sensitivity of the vertebrate olfactory receptors can also be altered, in all probability, by the flow of impulses from above. Similar influences, not yet well understood, also seem to be at work in the retina of the eye.

It is evident, then, that the responses of complex sense organs are determined by the fundamental properties of the individual receptor cells, by the influences they exert on one another and by control exerted on them by other organs. In this way the activity of the receptor cells is integrated into complex patterns of nervous activity that enable organisms to survive in a world of endless variety and change.

HOW CELLS MOVE

by TERU HAYASHI   September 1961

Life is movement. This definition, though incomplete, is perhaps as good as any. One is tempted to add that the movement is "constructive," or even "purposeful," but such qualifications raise definition problems of their own. So let us just say that where there is life there is movement. By this criterion we need not hesitate to describe cells as living and viruses as nonliving. Until it meets a living cell a virus is as inert as any stone.

Activity is a manifestation of the energy of living things. Generations of biologists have sought to answer the question: How does an organism move? Early workers understandably concentrated their attention on the activity of muscle, a word from the Latin for "little mouse." Later, when it was recognized that organisms are composed of cells, the search for the secret of motion moved down to the cellular level. Today we try to reach down to the level of the molecule, still seeking the final answer, which continues to elude us.

In large, complex organisms such as ourselves, the muscle cells are the cells most highly specialized for producing movement. Striated muscles activate our limbs, cardiac muscles make the heart beat, and smooth muscles push food along its digestive course. Less familiar are the movements exhibited by other cells within our bodies and by many one-celled organisms. The sperm cell, for example, is powered by a single whiplike flagellum. The paramecium and a number of other one-celled organisms are propelled through their liquid environment by fine, hairlike extensions called cilia. If the cell is stationary, as are the ciliated epithelial cells of the trachea, the cilia propel a liquid film past the cell, thereby keeping the surface of the cell washed clean. At a deeper level there is a constant turbulence within every living cell—a churning of the cell contents termed protoplasmic streaming. When this streaming results in the locomotion of the entire cell, the cell exhibits amoeboid movement. In nonmoving cells protoplasmic streaming carries a constant supply of molecular building materials to sites where large molecules are synthesized, carries these large molecules to other sites and in general provides the cell with an internal transportation system. The beautifully precise turning and positioning of the chromosomes in cell division are also a manifestation of movement within the cell [see "How Cells Divide" on page 199].

The simplest form of protoplasmic streaming occurs in the plant cell, which has a rigid cell wall with the protoplasm in thin layers just inside the wall. The innermost part of the cell is filled with a watery solution in a space called a vacuole. Close examination of the huge rhizoid (rootlike) cell in the water plant *Nitella* reveals that the thin protoplasmic layer is composed of a cortical gel layer, immediately inside the cell wall, and the more fluid endoplasm, or interior protoplasm. The endoplasm is the layer in motion; it circulates around the sides of the cell in a uniform direction, a movement called cyclosis [*see top illustration on page 270*].

A more complex type of streaming is found in the locomoting amoeba, in which the endoplasm flows forward in the direction the cell is moving. At the advancing end the endoplasmic material moves to the sides of the cell in a "fountain" flow, where it becomes transformed into the stiff, nonmoving material of the cortical gel. At the rear end of the cell the opposite series of events is taking place; the gel material is transformed into the flowing endoplasm that provides the forward-moving stream. In its simplest form, then, the amoeba moves by means of a flow of endoplasm through a tunnel of cortical gel, the tunnel wall being built up continuously at the front end of the cell from the flowing material and being torn down at the rear end to supply the material for the flow.

A still more complex movement takes place in the slime molds, such as *Plasmodium*, which are large masses of protoplasm containing many nuclei not separated from each other by cell membranes. The streaming of the protoplasm is similar to that in the amoebae but is complex because many streams exist simultaneously, moving in different directions.

Exquisite techniques for studying protoplasmic streaming have been devised by such workers as Robert D. Allen of Princeton University, R. J. Goldacre of the Chester Beatty Research Institute of the Royal Cancer Hospital in London and Noburô Kamiya of Osaka University in Japan, each of whom advocates a different hypothesis to explain protoplasmic flow. All agree that endoplasm flows passively, that there is an outside motive force urging it along. Where they disagree is on the nature of the force. Let us see how Kamiya and his associates have tried to settle the matter.

If a small mass of *Plasmodium* is divided into two parts but with a very thin connecting strand remaining, the protoplasm will flow from one side to the other for a while, then reverse and flow in the opposite direction, in an oscillatory motion. If the connecting strand is fine enough, the flow within it at any moment is simple and unidirectional. Taking advantage of this, Kamiya and his students constructed a double cham-

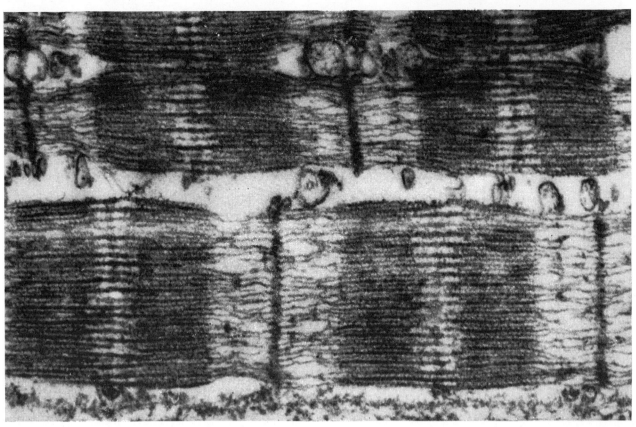

MOVEMENT IN STRIATED-MUSCLE CELL results from action of thick and thin filaments, seen here in an electron micrograph by H. E. Huxley of University College London. Dense A bands, bisected by H zones, mark overlap of thick and thin filaments. Light I bands, bisected by Z lines, are regions occupied by thin filaments alone. This is a very small part of one fibril of a rabbit muscle cell.

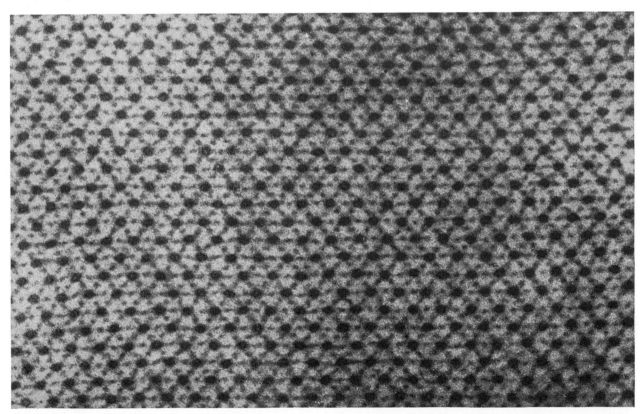

END VIEW OF FILAMENTS from an insect flight muscle shows regular hexagonal array of thick filaments (*larger spots*) and thin filaments (*smaller spots*). The electron micrograph, made by Jean Hanson of King's College in London and Huxley, enlarges the filaments approximately 250,000 diameters. This cross section shows only a part of one of the many fibrils that make up one fiber.

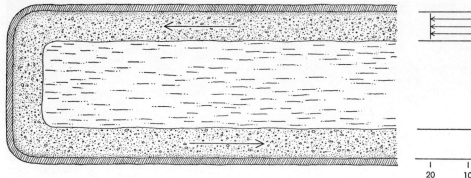
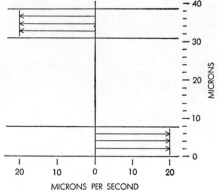

PROTOPLASMIC STREAMING, or cyclosis, in the rootlike cell of the water plant *Nitella* is diagramed. The outside wall of the cell is rigid. Just within is the very thin cortical gel layer. Next is the fluid endoplasm, which streams circularly (*arrows*) at 20 microns per second. The inner part of the cell is the vacuole, filled with a watery solution. The long cell is only 42 microns thick.

ber with a small opening in the wall between the chambers [*see illustration below*]. The two protoplasmic blobs were placed one in each chamber and connected by a single thin strand through the opening in the middle wall, so that the mold would now flow from one chamber to the other for a time, then back again. To the two chambers Kamiya attached devices to measure the rate of flow in the strand, the pressure causing the flow and the rate of oxygen consumption. Provision was also made for introducing chemical agents into the chambers during the measurements.

Here are a few of the results drawn from many precise experiments. First, the flow through the strand has the same characteristics whether it is the natural flow or a flow enforced with applied pressure. This indicates that the motion of the endoplasm is indeed a passive one, caused by local changes in pressure. Second, this motive force in the form of pressure differences has been measured and found to be about two pounds per square inch, a considerable force notwithstanding the small dimensions of the cell. Kamiya has also found that the energy for this movement comes from fermentation—that is, by the nonoxidative breakdown of glucose—rather than from the oxidative respiratory process that supplies energy for other functions.

Reviewing the information, Kamiya hypothesizes that the motive force causing the pressure change can be either a contraction of the cortical gel layer on one side of the *Plasmodium* or a shearing action that propels the endoplasm sideways across the surface of the cortical gel layer. Kamiya points out that the latter mechanism would also explain the cyclosis in *Nitella*.

Contraction of the cortical gel has also been invoked by those studying the amoeba; there is a century-old hypothesis that contraction of the cortical gel in the rear end of the advancing amoeba pushes the endoplasm forward. Goldacre favors this view on the basis of many ingenious experiments. Allen, on the other hand, finds his own observations incompatible with that hypothesis and maintains instead that a contraction of the cortical gel in the region of its formation, which he calls the "fountain zone," pulls the endoplasmic material forward.

Thus three mechanisms have been put forward to explain protoplasmic streaming in general and amoeboid movement in particular: (1) gel-endoplasm shearing or sliding, (2) rear-gel contraction and (3) front-gel contraction [*see illustrations on page 272*]. The actual parts of the cell engaged in one or more of these actions must be of molecular dimensions, because the best and most recent electron micrographs of the cells involved do not reveal any mechanical structures in the cortical gel region that could help us to decide among these choices.

Whereas visible parts are markedly lacking in streaming protoplasm, they exist in abundance in cilia and flagella. Vastly improved techniques in electron

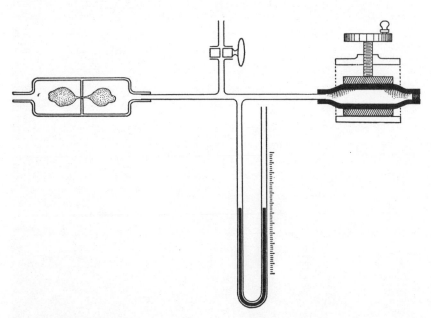

STUDIES OF STREAMING in the slime mold *Plasmodium* are carried out in this apparatus by Noburô Kamiya of Osaka University. The slime mold is in two compartments (*left*) and has a thin connecting strand. Rubber bulb with screw attached (*right*) and mercury manometer (*center*) control and measure pressure of the flow in streaming.

microscopy have revealed a wealth of exquisite fine structure in these "diverse incredibly thin feet," as Anton van Leeuwenhoek described them in 1676. In particular we have learned that these two types of motile "hair"—cilia and flagella—share the same basic structure. Barely visible under the ordinary light microscope, these fine hairs are built up of two center filaments surrounded by a ring of nine outer filaments in what is called a "9 + 2" arrangement [see illustration on page 275]. The nine outer filaments appear to be double structures, whereas the two central ones appear to be single. The widespread occurrence of this arrangement demands that any explanation of ciliary movement must include a role for these filaments.

The classic study of ciliary movement was made by Sir James Gray of the University of Cambridge with motion-picture micrography. The most common type of ciliary beat consists of an effective stroke and a recovery stroke [see top illustration on page 273]. In the effective stroke the cilium sweeps rapidly as a stiff, slightly curved rod, the only region of true bending being near its base. We can easily see that such a stroke—which resembles the power stroke made by a swimmer—will efficiently move the cell through its watery medium or move the medium past the cell. In the recovery stroke the cilium starts bending near the base and the bending proceeds as a wave toward the tip. The direction of this bending carries the cilium back to its original position.

J. R. G. Bradfield of the Cavendish Laboratory at the University of Cambridge has advanced an ingenious, if speculative, mechanism to account for the two kinds of stroke. He proposes that five of the outer filaments on one side of the cilium contract simultaneously and throughout their entire length to produce the effective stroke. The other four filaments are idle during this stroke but then contract slowly, beginning at the base, to produce the recovery stroke. The two central filaments, Bradfield believes, may act as telegraph lines to carry the message for contraction rapidly up the cilium so that the total contraction required for the effective stroke can be realized. Therefore even the apparently simple beating of a single cilium would seem to involve the complexity of message transmission. And when the cilia of many adjacent cells beat together rhythmically, as they do in the trachea and elsewhere, the co-ordinating signals must be more complicated still.

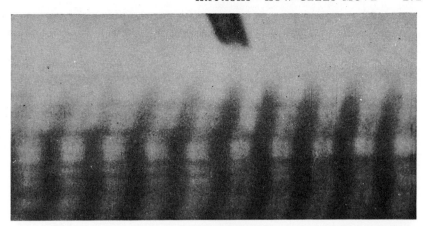
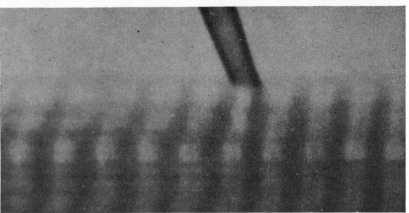
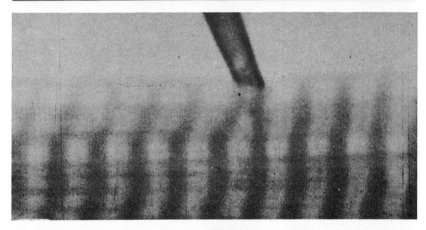
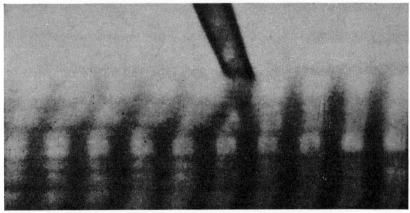

MUSCLE-FIBRIL CONTRACTION is demonstrated in sequence from a motion picture by A. F. Huxley. At top a microelectrode approaches fibril. It touches and electric current is turned on. Contraction can be seen as two dark A bands come together, pinching the light I band. The last photograph shows maximum contraction. Fibril is from frog muscle.

272 SPECIAL ACTIVITIES

The undulant motion of some flagella, which may be as much as 50 times longer than cilia, is similar to that of cilia but more involved. Waves of bending arise at the attached end and move toward the tip, exerting a pushing force on the medium and driving the cell forward. In many cases the progressing cell rotates, causing the undulating flagellum to assume the shape of a coil. This movement would seem to require localized co-ordinated contractions, presumably of the internal filaments.

Although there is yet no direct evidence for contraction in cilia and flagella, no alternative mechanism seems appealing. (One old proposal visualized the cilium as a tiny hose that lashes in reaction to water forced through it by the cell.) The chief support for the contraction hypothesis comes from the examination of muscle cells, which provide much more favorable objects for study.

In a splendid demonstration of the power of the electron microscope, H. E. Huxley of University College London and Jean Hanson of King's College in London have elucidated for us the ultrafine structure of muscle cells in the rabbit. The muscle cell, usually termed muscle fiber, is a long, thin cylinder encased in a thin membrane. This cylinder contains fine threads called myofibrils that run lengthwise throughout the full length of the cylinder. The small spaces between the myofibrils are apparently filled with that watery solution of various substances which is the liquid portion of protoplasm. This fluid can probably penetrate the fibrils easily, since the individual fibrils do not have membranes surrounding them. The fibrils themselves are bundles of even finer filaments of two kinds, a thick and a thin variety, packed in a precise geometric array [see illustrations on page 269]. The entire fiber has a diameter of only 50 microns, a thread barely visible to the naked eye; the fibrils are one to two microns thick, and the thick filaments are only 100 angstrom units in diameter, or perhaps large enough to hold three protein molecules side by side.

The action of muscle involves many things, including the message to action brought by the motor nerve, the spreading of this message over the surface of the muscle fiber, the relaying of this message to the actual contractile parts of the cell, the contracting action itself and the energy sources for the action. The essential action, however, is simply shortening or contraction, and so we shall be concerned simply with the question: How does muscle shorten?

Any proposed mechanism for action must agree with certain well-established observations, due in large part to the work of A. V. Hill and his students at the University of Cambridge. First, when muscle shortens, it gives off heat, and the amount of heat is directly proportional to the amount of shortening. Second, the tension that the active muscle can develop is at a maximum when the muscle reaches a certain length; if the muscle is stretched beyond that length, or allowed to contract to a

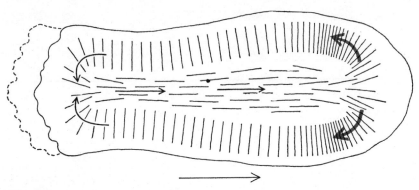

THEORIES OF AMOEBOID MOVEMENT are diagramed on this page. Robert D. Allen holds that semiliquid endoplasm streaming up center of the cell is pulled forward by contraction of cortical gel near the head (above). Colored arrows in each diagram indicate point at which force is applied. In all three diagrams endoplasm is turning to gel at head.

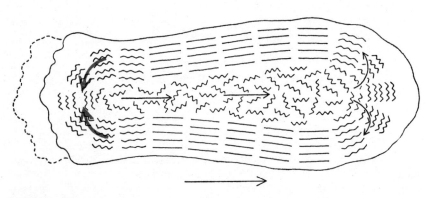

CONTRACTION AT TAIL, where gel is liquefying into endoplasm, accounts for movement, according to theory supported by R. J. Goldacre. The contraction would squeeze the endoplasm forward, in the manner of toothpaste being squeezed from a tube. Wavy lines are contracted molecules of endoplasm, which is more liquid than in Allen's theory.

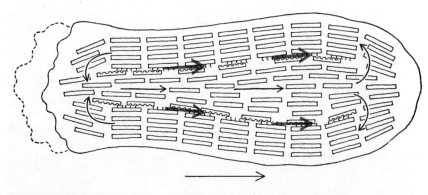

GEL-ENDOPLASM SHEARING or sliding is third theory. Chemical "ratchets" on the inner edge of the gel would push forward individual molecules of endoplasm. The flow of these molecules would carry along the endoplasm molecules nearer the center. As in top and center diagrams, endoplasm turns to gel at the head and gel turns to endoplasm at the tail.

shorter length, the tension decreases from the maximum. Finally, when the muscle is contracting, it is in an "active state" in which it strongly resists being stretched. When relaxed, the muscle is plastic and easily extensible.

From such observations it is plausible to infer that the muscle cell contains a group of contractile elements in a chemical relationship with its surrounding cell material, a picture that has been fully borne out in electron microscope studies. It is generally agreed today that the fibrils represent the contractile elements of the muscle cell.

In 1954 A. F. Huxley (no relation to the H. E. Huxley mentioned earlier) and his associates at the University of Cambridge reported the first of a series of observations of the muscle fibril during contraction, using beautifully simple and direct methods. They placed a tiny electrode against the surface of a muscle and stimulated it with carefully regulated shocks that caused only a limited local reaction, which they observed with an interference microscope. Motion pictures of the reaction showed conclusively that during contraction certain striated regions of the muscle (the so-called *I* and *A* regions) move together [see illustration on page 271].

From these and similar studies, therefore, the two Huxleys independently and simultaneously proposed that contraction takes place when the two sets of filaments, the fine and the thick, telescope into each other. They suggest that the force that makes the filaments slide past each other comes from an interaction between the thick and thin filaments—a sort of chemical, make-and-break ratchet capable of clawing one set of filaments past the other. The proposed mechanism seems quite compatible with many of the detailed facts obtained from physiological studies. In only a few muscles, however, does the electron microscope show two sets of interdigitating and geometrically perfect filaments; they have been found only in the rabbit, the frog and in the flight muscles of a few insects. No indication of such an organization appears, for example, in the best electron micrographs of smooth muscle of the viscera. Similarly, no filament structure is found in amoebae, and although filaments are found in cilia, they are not arranged in an interdigitating manner.

To mention the movement of amoebae and cilia in the same breath with muscle action may seem rather farfetched. Yet if we recall the various hypotheses that have been presented, we note that movement is ascribed either to contraction or

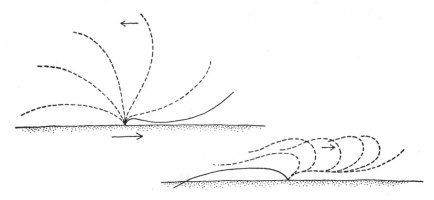

MOTION OF A CILIUM on an organism moving toward the right involves a sweeping effective stroke (*left*) followed by a graceful recovery stroke (*right*). Cilia are quite short.

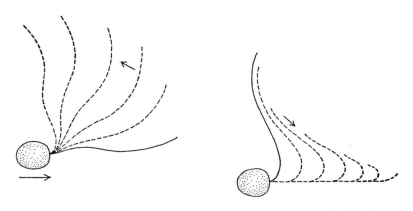

MOTION OF A FLAGELLUM propels the organism *Monas* to the right. Effective stroke is at left, recovery at right. Flagella are longer than cilia and can move in several ways.

to a sliding of one thing past another in all cases, so that perhaps it is not unreasonable to look for a common basis for all biological movement. To do this we must move from the structural level of the cell toward the molecular level.

A small boy, confronted by a ticking watch, is driven by curiosity to take it apart to see how it works; the physiologist has similarly tried to take apart the complex machine we know as the muscle cell. There are different degrees of taking things apart, however. We may take something apart a little or take it apart completely, and perhaps attempt to put the completely dissociated structure back together again, even if only partly. Physiologists refer to partly or wholly disintegrated muscle as a simplified system, and the study of such systems has been extremely fruitful.

The original simplified system is the one devised around 1949 by Albert Szent-Györgyi (now at the Institute for Muscle Research in Woods Hole, Mass.), who worked with the same rabbit muscle later studied by the Huxleys. He tied a freshly dissected muscle cell to a frame at its natural length and then soaked it in a cold solution of glycerine overnight. The muscle cell is killed by this treatment, and many of its components are leached out. Its structural integrity remains, however, and it can be preserved in a freezer for months if kept in the glycerine solution. If this unmistakably dead muscle cell is warmed to room temperature in a dilute salt solution and exposed to the action of adenosine triphosphate (ATP), a remarkable thing happens: the dead cell contracts. ATP, of course, is the substance that provides the energy for many cellular functions [see "How Cells Transform Energy" on page 85].

If muscle cells are primed to move by ATP, why not other cells that are capable of motion? This thought occurred to H. Hoffmann-Berling at the Max Planck Institute for Medical Research in Heidelberg. He glycerinated sperm cells and young connective-tissue cells (fibroblasts), which exhibit a form of amoeboid movement. When ATP was applied to these dead cells, each moved in its characteristic fashion: the fibroblast cells contracted and the sperm tails lashed

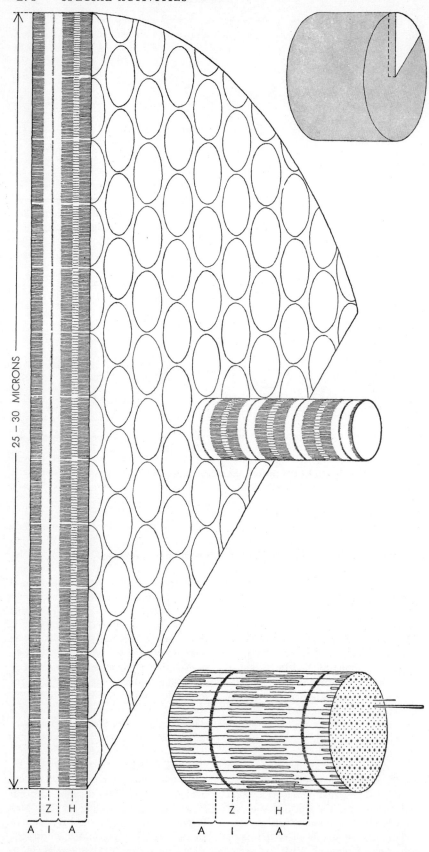

SECTION OF STRIATED-MUSCLE FIBER, highly schematic, shows how fiber is related to fibrils and to the filaments that cause the striations. The large, pie-shaped segment has been cut from a single fiber, as shown at top right. One such fiber is a long thread barely visible to the naked eye. The circles in the wedge are the numerous fibrils that pack a fiber. One of them projects. At bottom a single fibril is enlarged to show how interdigitating filaments make up the striations. One thick and one thin filament project from the fibril.

vigorously. Similar studies have since been made on a variety of cilia and flagella, cancer cells and tissue-culture cells; many are activated by ATP. These dramatic results furnish a strong argument for the idea that cell movement in general has a common molecular basis.

When the muscle cell is taken apart completely, two important and interesting proteins are found: actin and myosin. These two substances make up the bulk of the structural material of the muscle cell. In the test tube they combine to form actomyosin, and they were originally studied in this form. In 1939 V. A. Engelhardt and M. N. Ljubimova of the Institute of Biochemistry of the Academy of Sciences in Moscow found that actomyosin is an enzyme capable of releasing the chemical energy of ATP. Soon afterward Szent-Györgyi reported that a reconstituted gel of actomyosin will contract suddenly when ATP is applied. It has since been shown by H. H. Weber at the Max Planck Institute for Medical Research and in our laboratory at Columbia University that artificial fibers formed of actomyosin can contract, perform work and develop tension with much the same characteristics as living muscle. The conclusion seems inescapable that actin and myosin form the molecular basis of muscle contraction.

Elaborating on this view, H. E. Huxley has adduced evidence to show that the myosin is located in the thick filaments of the fibril, whereas actin seems to be localized in the thin filaments. He visualizes that the interaction of these proteins in the two different filaments provides the ratchet mechanism for pulling the filaments past each other.

If there is a common molecular mechanism for cell movement, then contractile proteins similar to actomyosin should be present in other types of cell. The confirmation of this reasoning has not been easy; whereas muscle cells are specialized for movement and therefore contain large amounts of actomyosin, movement is just one of many functions carried out by the amoeba and the slime mold, and they would not be expected to contain much contractile protein. In 1952, however, Ariel G. Loewy, then at the University of Pennsylvania, succeeded in isolating from slime mold a contractile protein that has properties strikingly similar to those of actomyosin. Thus it would seem that the motive force for protoplasmic streaming is provided by the action of contractile protein. In support of this view, Kamiya's group has shown that ATP extracted from *Plasmodium* will react with rabbit actomyosin fully as well as with the ATP obtained from rabbit muscle.

What of ciliary or flagellar movement? In 1958 Engelhardt and a co-worker reported finding an actomyosin-like protein in sperm cells, which they have named spermosin. So far this is the only isolation of a contractile protein associated with flagellar movement. It has been shown, however, that flagella broken off from various types of cell contain enzymes capable of releasing the energy of ATP, and Leonard Nelson of Emory University has ingeniously combined chemistry and electron microscopy to show that the splitting of ATP is restricted to the outer nine filaments of the 9 + 2 arrangement.

We see, therefore, that the intensive study of simplified systems leads us to two general and important conclusions. The first of these is that the basis of muscle movement is a combined protein called actomyosin, which is capable of producing a contraction when provided with energy by ATP. The second is that this concept of a molecular basis for contraction can be extended to other types of cellular movement.

It may be recalled that the hypotheses advanced to account for the various types of movement inside cells involve either a contraction or a sliding of one substance past another. We can now appreciate that the difference between these alternatives may be more apparent than real, and that contraction may involve a sliding. For those cells which do not possess oriented filamentous structures, the sliding-contraction may be of molecular dimensions. Such an idea is, of course, highly speculative, but the widespread attention being given to the actin and myosin molecules may demonstrate the reality of the sliding-contraction mechanism or may lead to a totally new concept.

There are always hazards in assuming that substances observed to behave in a certain way in a test tube behave the same way inside a living cell. And sometimes paradoxes arise. To illustrate, a number of investigators have posed the following question: If actomyosin is the machinery of contraction and ATP the fuel for this contraction, the contracting muscle should show a depletion of ATP; is this true?

To answer the question it is possible to carry out an experiment with identical muscles, one from each hind leg of a frog. One muscle serves as a control; the other is stimulated to contract. Both are then frozen quickly and analyzed for their chemical content; the technique has been refined so that the chemical change caused by a single twitch of the muscle can be detected. The results are discomforting because they show that very little ATP disappears, certainly not enough to account for the energy exhibited by the muscle. However, another energy-rich biochemical compound—creatine phosphate—does disappear.

The facts seem irreconcilable. Actomyosin contracts with ATP and not with creatine phosphate. On the other hand, it is the creatine phosphate that disappears in active muscle and, as Francis D. Carlson of Johns Hopkins University has recently shown, it disappears in amounts that are directly proportional to the number of contractions the muscle performs.

Yet all may not be lost. Carlson has drawn up a plausible scheme to reconcile these facts. Briefly, he proposes that the energy from creatine phosphate is funneled through a "compartmentalized" ATP, which in turn feeds the energy into the actomyosin. This ATP, then, works like the middle man in a bucket brigade, receiving a bucket of energy from creatine phosphate, passing it to the muscle and immediately obtaining another load of energy. The energy-laden ATP does not get used up because it is constantly re-formed.

It turns out, indeed, that a candidate for the role of compartmentalized ATP can be found in muscle. The candidate is the relatively little-studied protein actin, each molecule of which has an ATP molecule bound to it. A number of investigators are now looking into the

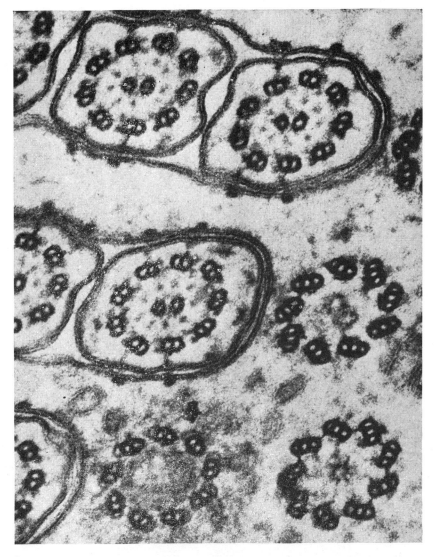

"9 + 2" ARRANGEMENT of filaments in flagella is clear in this electron micrograph by I. R. Gibbons of Harvard University and A. V. Grimstone of the University of Cambridge. Sections with membranes around them are flagella of the organism *Pseudotrichonympha*; other sections are the bases of flagella. Each of the filaments in the outer ring is double.

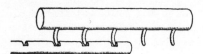 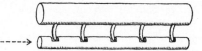 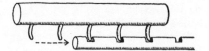

POSSIBLE RATCHET ACTION in striated muscle is diagramed. Thick and thin filaments of relaxed muscle are at left. As muscle contracts it reaches maximum tension (*center*); further contraction lowers tension because fewer ratchets are engaged (*right*).

matter and we may learn before long whether or not actin meets the requirements of Carlson's hypothesis.

Assuming that the story runs smoothly to its denouement and we are finally able to describe in detail how actin, myosin and ATP produce contractile movements in cells, is this the end of the problem? It would be pleasant to think so. But scientific problems have a way of eluding solution. There is a sense in which "the more we find out, the less we know," which is merely to say that the resolution of one problem usually opens up a number of others. Once we understand biological movement on the molecular level we may wish we could understand it on the submolecular, or atomic, level. Such is the charm and fascination of scientific inquiry.

CROSS BRIDGES connect thick and thin filaments of striated rabbit muscle, as seen in this electron micrograph by H. E. Huxley. This is part of the *A* band, enlarged some 900,000 diameters. The bridges may be the ratchets in a system that moves filaments.

THE CONTRACTION OF MUSCLE

by H. E. HUXLEY November 1958

A basic characteristic of all animals is their ability to move in a purposeful fashion. Animals move by contracting their muscles (or some primitive version of them), so muscle contraction is one of the key processes of animal life. Muscle contraction has been intensively studied by a host of investigators, and their labors have yielded much valuable information. We still, however, cannot answer the fundamental question: How does the molecular machinery of muscle convert the chemical energy stored by metabolism into mechanical work? Recent studies, notably those utilizing the great magnifications of the electron microscope, have nonetheless enabled us to begin to relate the behavior of muscle to events at the molecular level. At the very least we are now in a position to ask the right sort of question about the detailed molecular processes which remain unknown.

Muscles are usually classified as "striated" or "smooth," depending on how they look under the ordinary light microscope. The classification has a good deal of functional significance. The muscles which vertebrates such as mice or men use to move their bodies or limbs—muscles which act quickly and under voluntary control—are crossed by microscopic striations. The muscles of the gut or uterus or capillaries—muscles which act slowly and involuntarily—have no striations; they are "smooth." In this article I shall discuss only striated muscles, because our knowledge of them is in a much more advanced state. I shall be surprised, however, if nothing I say is relevant to smooth muscles.

Striated muscles are made up of muscle fibers, each of which has a diameter of between 10 and 100 microns (a micron is a thousandth of a millimeter). The fibers may run the whole length of the muscle and join with the tendons at its ends. About 20 per cent of the weight of a muscle fiber is represented by protein; the rest is water, plus a small amount of salts and of substances utilized in metabolism. Around each fiber is an electrically polarized membrane, the inside of which is about a 10th of a volt negative with respect to the outside.

If the membrane is temporarily depolarized, the muscle fiber contracts; it is by this means that the activity of muscles is controlled by the nervous system. An impulse traveling down a motor nerve is transmitted to the muscle membrane at the motor "end-plate"; then a wave of depolarization (the "action potential") sweeps down the muscle fiber and in some unknown way causes a single twitch. Even when a frog muscle is cooled to the freezing point of water, the depolarization of the muscle membrane throws the whole fiber into action within 40 thousandths of a second. When nerve impulses arrive on the motor nerve in rapid succession, the twitches run together and the muscle maintains its contraction as long as the stimulation continues (or the muscle becomes exhausted). When the nerve stimulation stops, the muscle automatically relaxes.

The Energy Budget of Muscle

Striated muscles can shorten at speeds up to 10 times their length in a second, though of course the amount of shortening is restricted by the way in which the animal is put together. Such muscles can exert a tension of about three kilograms for each square centimeter of their cross section—some 42 pounds per square inch. They exert maximum tension when held at constant length, so that the speed of shortening is zero. Even though a muscle in this state does no external work, it needs energy to maintain its contraction; and since the energy can do no work, it must be dissipated as heat. This so-called "maintenance heat" slightly warms the muscle.

When the muscle shortens, it exerts less tension; the tension decreases as the speed of shortening increases. One might suspect that the decrease of tension is due to the internal viscosity or friction in the muscle, but it is not. If it were, a muscle shortening rapidly would liberate more heat than one shortening slowly over the same distance, and this effect is not observed.

The energy budget of muscle has been investigated in great detail, particularly by A. V. Hill of England and his colleagues. Studies of this kind have shown that a shortening muscle does liberate extra heat, but in proportion to the *distance* of shortening rather than to the speed. Curiously this "shortening heat" is independent of the load on the muscle: a muscle produces no more—and no less—shortening heat when it lifts a large load than when it lifts a small one through the same distance.

But a muscle lifting a large load ob-

FILAMENTS in an insect flight-muscle are seen from the end in the electron micrograph on the opposite page. Thick filaments (*larger spots*) and thin filaments (*smaller spots*) lie beside one another in a remarkably regular hexagonal array. Some of the thick filaments appear to be hollow. This electron micrograph, which enlarges the filaments some 400,000 diameters, was made by Jean Hanson of the Medical Research Council Unit at Kings College and the author.

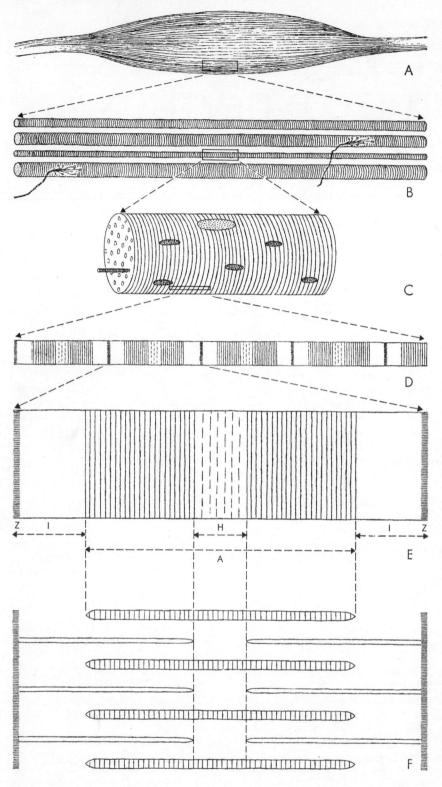

viously does more work than a muscle lifting a small load, so if the shortening heat remains constant, the total energy (heat plus work) expended by the contracting muscle must increase with the load. The chemical reactions which provide the energy for contraction must therefore be controlled not only by the change in the length of the muscle, but also by the tension placed on the muscle during the change. This is a remarkable property, of great importance to the efficiency of muscle, and new information about the structure of muscle has begun to explain it.

From the chemical point of view, the contractile structure of muscle consists almost entirely of protein. Perhaps 90 per cent of this substance is represented by the three proteins myosin, actin and tropomyosin. Myosin is especially abundant: about half the dry weight of the contractile part of the muscle consists of myosin. This is particularly significant because myosin is also the enzyme which can catalyze the removal of a phosphate group from adenosine triphosphate (ATP). And this energy-liberating reaction is known to be closely associated with the event of contraction, if not actually part of it.

Myosin and actin can be separately extracted from muscle and purified. When these proteins are in solution together, they combine to form a complex known as actomyosin. Some years ago Albert Szent-Györgyi, the noted Hungarian biochemist who now lives in the U. S., made the striking discovery that if actomyosin is precipitated and artificial fibers are prepared from it, the fibers will contract when they are immersed in a solution of ATP! It seems that in the interaction of myosin, actin and ATP we have all the essentials of a contractile system. This view is borne out by experiments on muscles which have been placed in a solution of 50 per cent glycerol and 50 per cent water, and soaked for a time in a deep-freeze. After this procedure, and some further washing, practically everything can be removed from the muscle except myosin, actin and tropomyosin; and this residual structure will still contract when it is supplied with ATP.

The Structure of the Fiber

The most straightforward way to try to find out how the muscle machine works is to study its structure in as much detail as possible, using all the techniques now at our disposal. This has proved to be a fruitful approach, and I

STRIATED MUSCLE IS DISSECTED in these schematic drawings. A muscle (A) is made up of muscle fibers (B) which appear striated in the light microscope. The small branching structures at the surface of the fibers are the "end-plates" of motor nerves, which signal the fibers to contract. A single muscle fiber (C) is made up of myofibrils, beside which lie cell nuclei and mitochondria. In a single myofibril (D) the striations are resolved into a repeating pattern of light and dark bands. A single unit of this pattern (E) consists of a "Z-line," then an "I-band," then an "A-band" which is interrupted by an "H-zone," then the next I-band and finally the next Z-line. Electron micrographs (*see opposite page*) have shown that the repeating band pattern is due to the overlapping of thick and thin filaments (F).

STRIATED MUSCLE from a rabbit is enlarged 24,000 diameters in this electron micrograph. Each of the diagonal ribbons is a thin section of a muscle fibril. Clearly visible are the dense A-bands, bisected by H-zones; and the lighter I-bands, bisected by Z-lines.

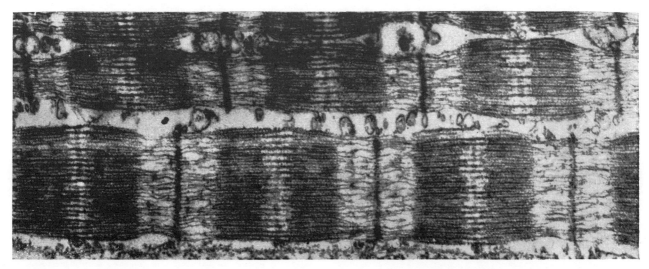

EXTREMELY THIN SECTION of a striated muscle is shown at much greater magnification. The section is so thin that in some places it contains only one layer of filaments. The way in which overlapping thick and thin filaments give rise to the band pattern can be clearly seen. Although the magnification of this electron micrograph is much larger than that of the micrograph at top of page, distance between the narrow Z-lines is less. This is because the section was longitudinally compressed by the slicing process.

shall briefly describe its results. Much of the work I shall discuss I have done in collaboration with Jean Hanson of the Medical Research Council Unit at King's College in London.

The contractile structure of a muscle fiber is made up of long, thin elements which we call myofibrils. A myofibril is about a micron in diameter, and is cross-striated like the fiber of which it is a part. Indeed, the striations of the fiber are due to the striations of the myofibril, which are in register in adjacent myofibrils. The striations arise from a repeating variation in the density, *i.e.*, the concentration of protein along the myofibrils.

The pattern of the striations can be seen clearly in isolated myofibrils, which are obtained by whipping muscle in a Waring blendor. Under a powerful light microscope there is a regular alternation of dense bands (called A-bands) and lighter bands (called I-bands). The central region of the A-band is often less dense than the rest of the band, and is known as the H-zone. When a

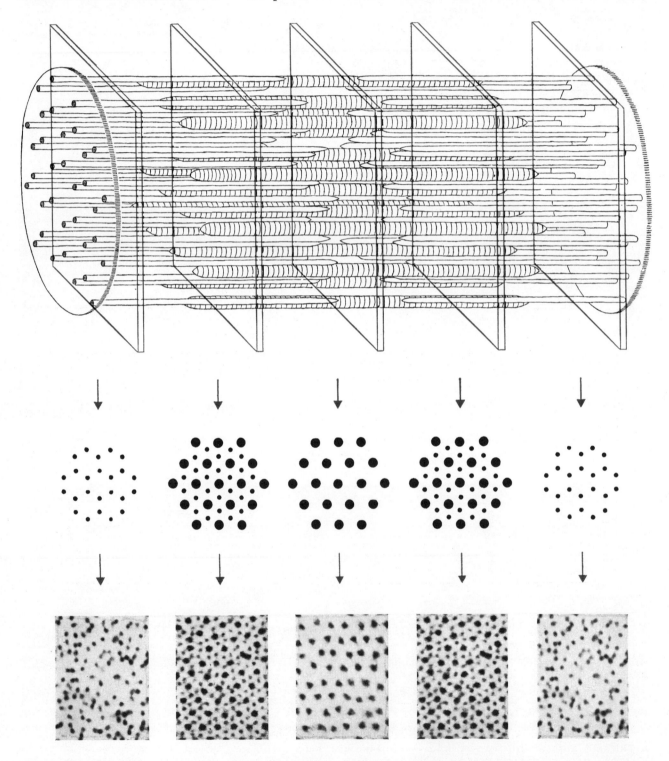

TRANSVERSE SECTIONS through a three-dimensional array of filaments in vertebrate striated muscle (*top*) show how the thick and thin filaments are arranged in a hexagonal pattern (*middle*). At bottom are electron micrographs of the corresponding sections.

striated muscle from a vertebrate is near its full relaxed length, the length of one of its A-bands is commonly about 1.5 microns, and the length of one of its I-bands about .8 micron. The I-band is bisected by a dense narrow line, the Z-membrane or Z-line. From one Z-line to the next the repeating unit of the myofibril structure is thus: Z-line, I-band, A-band (interrupted by the H-zone), I-band and Z-line.

When myofibrils are examined in the electron microscope, a whole new world of structure comes into view. It can be seen that the myofibril is made up of still smaller filaments, each of which is 50 or 100 angstrom units in diameter (an angstrom unit is a 10,000th of a micron). These filaments were observed in the earliest electron micrographs of muscle, made by Cecil E. Hall, Marie A. Jakus and Francis O. Schmitt of the Massachusetts Institute of Technology, and by M. F. Draper and Alan J. Hodge of Australia. And now thanks to recent advances in the technique of preparing specimens for the electron microscope,

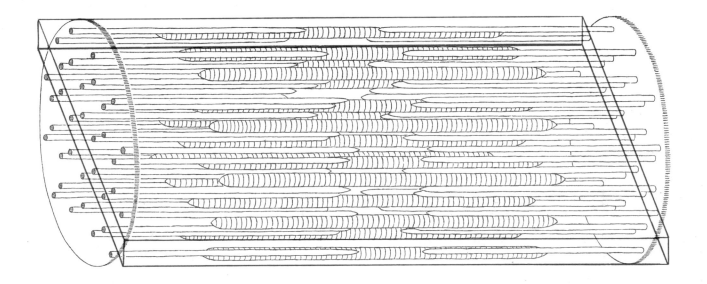

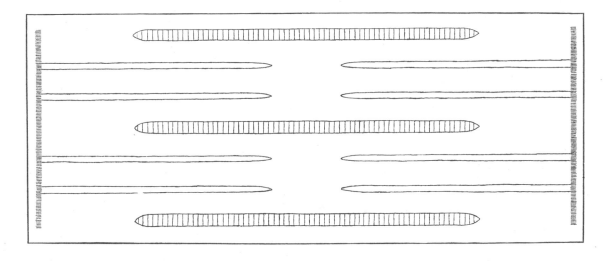

LONGITUDINAL SECTION through the same array shows how two thin filaments lie between two thick ones. This pattern is a consequence of the fact that one thin filament is centered among three thick ones. At bottom is a micrograph of the corresponding section.

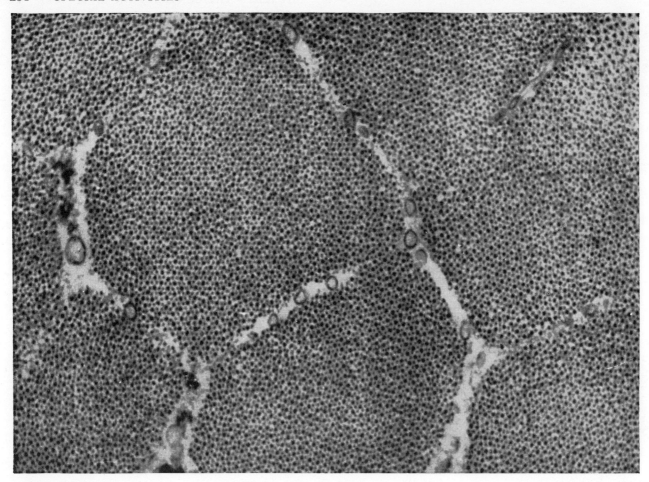

SEVERAL FIBRILS in a vertebrate striated muscle are seen from the end in an electron micrograph which enlarges them 90,000 diameters. Within each fibril is the hexagonal array of its filaments. This pattern, in which one thin filament lies symmetrically among three thick ones, differs from the pattern in the insect muscle on page 278, in which one thin filament lies between two thick ones.

it is possible to examine the arrangement of the filaments in considerable detail.

For this purpose a piece of muscle is first "fixed," that is, treated with a chemical which preserves its detailed structure during subsequent manipulations. Then the muscle is "stained" with a compound of a heavy metal, which increases its ability to deflect electrons and thus enhances its contrast in the electron microscope. Next it is placed in a solution of plastic which penetrates its entire structure. After the plastic is made to solidify, the block of embedded tissue can be sliced into sections 100 or 200 angstrom units thick by means of a microtome which employs a piece of broken glass as a knife. When we look at these very thin sections in the electron microscope, we can see immediately that muscle is constructed in an extraordinarily regular and specific manner.

A myofibril is made up of two kinds of filament, one of which is twice as thick as the other. In the psoas muscle from the back of a rabbit the thicker filaments are about 100 angstroms in diameter and 1.5 microns long; the thinner filaments are about 50 angstroms in diameter and two microns long. Each filament is arrayed in register with other filaments of the same kind, and the two arrays overlap for part of their length. It is this overlapping which gives rise to the cross-bands of the myofibril: the dense A-band consists of overlapping thick and thin filaments; the lighter I-band, of thin filaments alone; the H-zone, of thick filaments alone. Halfway along their length the thin filaments pass through a narrow zone of dense material; this comprises the Z-line. Where the two kinds of filament overlap, they lie together in a remarkably regular hexagonal array. In many vertebrate muscles the filaments are arranged so that each thin filament lies symmetrically among three thick ones; in some insect flight-muscles each thin filament lies midway between two thick ones.

The two kinds of filament are linked together by an intricate system of cross-bridges which, as we shall see, probably play an important role in muscle contraction. The bridges seem to project outward from a thick filament at a fairly regular interval of 60 or 70 angstroms, and each bridge is 60 degrees around the axis of the filament with respect to the adjacent bridge. Thus the bridges form a helical pattern which repeats every six bridges, or about every 400 angstroms along the filament. This pattern joins the thick filament to each one of its six adjacent thin filaments once every 400 angstroms.

The arrangement of the filaments and their cross-bridges, as seen in the electron microscope, is so extraordinarily well ordered that one may wonder whether the fixing and staining procedures have somehow improved on nature. Fortunately this regularity is also apparent when we examine muscle by another method: X-ray diffraction. Muscle which has not been stained and fixed deflects X-rays in a regular pattern, indicating that the internal structure of muscle is also regular. The details of the diffraction pattern are in accord with the structural features observed in the electron microscope. Indeed, many of these features were originally predicted on the basis of X-ray diffraction patterns alone.

The Sliding-Filament Model

As soon as the meaning of the band pattern of striated muscle became apparent, it was obvious that changes in the pattern during contraction should give us new insight into the molecular nature of the process. Such changes can be unambiguously observed in modern light microscopes, notably the phase-contrast microscope and the interference microscope. They can be studied in living muscle fibers (as they were by A. F. Huxley and R. Niedergerke at the University of Cambridge) or in isolated myofibrils contracting in a solution of ATP (as they were by Jean Hanson and myself at M.I.T.). We all came to the same conclusions.

It has been found that over a wide range of muscle lengths, during both contraction and stretching, the length of the A-bands remains constant. The length of the I-bands, on the other hand, changes in accord with the length of the muscle. Now the length of the A-band is equal to the length of the thick filaments, so we can assume that the length of these filaments is also constant. But the length of the H-zone—the lighter region in the middle of the A-band—increases and decreases with the length of the I-band, so that the distance from the end of one H-zone through the Z-line to the beginning of the next H-zone remains approximately the same. This distance is equal to the length of the thin filaments, so they too do not alter their length by any large amount.

The only conclusion one can draw from these observations is that, when the muscle changes length, the two sets of filaments slide past each other. Of course when the muscle shortens enough, the ends of filaments will meet; this happens first with the thin filaments, and then with the thick [*see illustration at right*]. Under such conditions, in fact, new bands are observed which suggest that the ends of the filaments crumple or overlap. But these effects seem to occur as a *result* of the shortening process, and not as causes of contraction.

It has often been suggested that the contraction of muscle results from the extensive folding or coiling of the filaments. The new observations compel us to discard this idea. Instead we are obliged to look for processes which could cause the filaments to slide past one another. Although this search is only beginning, it is already apparent that

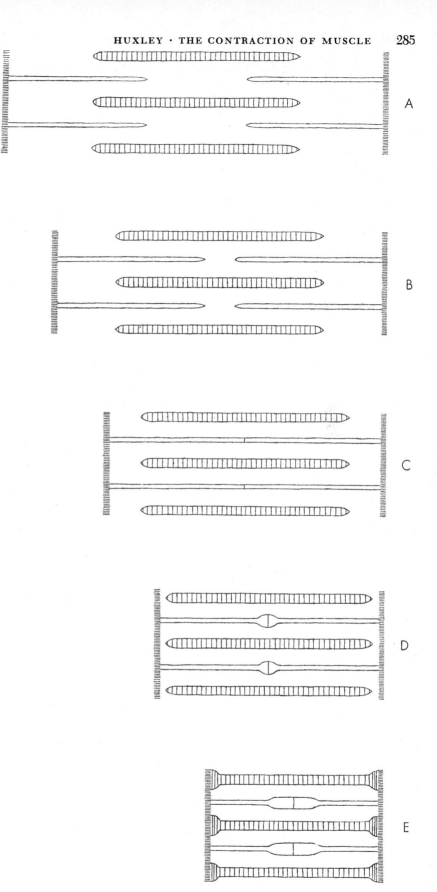

CHANGE IN LENGTH of the muscle changes the arrangement of the filaments. In A the muscle is stretched; in B it is at its resting length; in C, D and E it is contracted. In C the thin filaments meet; in D and E they crumple up. In E the thick filaments also meet adjacent thick filaments (*not shown*) and crumple. The crumpling gives rise to new band patterns.

286   SPECIAL ACTIVITIES

the sliding concept places us in a much more favorable position with respect to what we might call the intermediate levels of explanation: the description of the behavior of muscle in terms of molecular changes whose detailed nature we do not know, but whose consequences we can now compute.

There is more to be said about such matters, but first let us return to the chemical structure of muscle. If a muscle is treated with an appropriate salt solution, and then examined under the light microscope, it is observed that the A-bands are no longer present. It is also known that such a salt solution will remove myosin from muscle. This demonstrates that the thick filaments of the A-band are composed of myosin, a conclusion which has been quantitatively confirmed by comparing measurements made by chemical methods with those made by the interference microscope. Moreover, when myofibrils which have been treated with salt solution are examined in the electron microscope, they lack the thick filaments. The "ghost" myofibril that remains consists of segments of material which correspond to the arrays of thin filaments in the I-bands. If the myofibril is treated so as to extract its actin, a large part of the material in these segments is removed. This indicates that the thin filaments of the I-band are composed of actin and (probably) tropomyosin.

Thus the two main structural proteins of muscle are separated in the two kinds of filaments. As noted earlier, actin and myosin can be made to contract in a solution of ATP, but only when they are combined. We therefore conclude that the physical expression of the combination of actin and myosin is to be found in the bridges between the two kinds of filaments. It should also be said that the thick and thin filaments are too far apart for any plausible "action at a distance,"

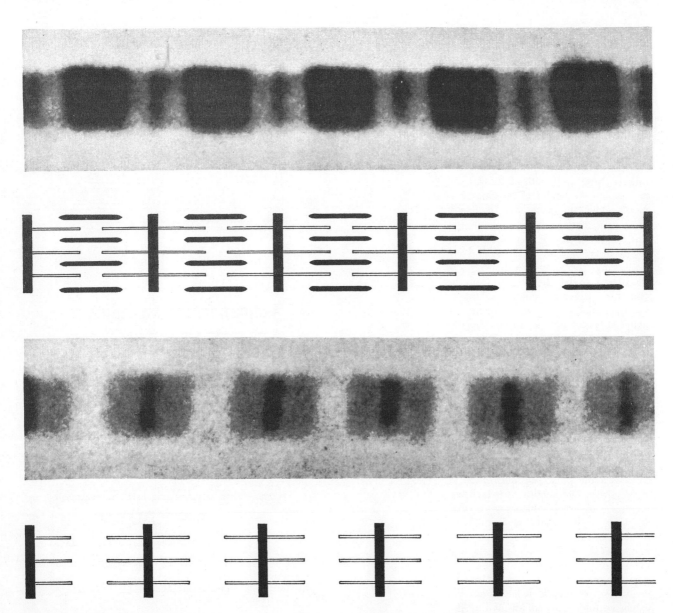

DIFFERENT CHEMICAL COMPOSITION of the thick and thin filaments is demonstrated. At top is a myofibril photographed in the phase-contrast light microscope. The wide dark regions are A-bands; between them are I-bands bisected by Z-lines. Second from top is a simplified schematic drawing of how the thick and thin filaments give rise to this pattern. Third from top is a photomicrograph of a myofibril from which the protein myosin has been chemically removed. The A-bands have disappeared, leaving only the I-bands and Z-lines. At bottom is a drawing which shows how this pattern is explained on the assumption that the thick filaments have been removed. Thus it appears that the thick filaments are composed of myosin, and the thin filaments of other material.

so it would seem likely that the sliding movement is mediated by the bridges.

The Cross-Bridges

The bridges seem to form a permanent part of the myosin filaments; presumably they are those parts of myosin molecules which are directly involved in the combination with actin. In fact, when we calculate the number of myosin molecules in a given volume of muscle, we find that it is surprisingly close to the number of bridges in the same volume. This suggests that each bridge is part of a single myosin molecule.

How could the bridges cause contraction? One can imagine that they are able to oscillate back and forth, and to hook up with specific sites on the actin filament. Then they could pull the filament a short distance (say 100 angstroms) and return to their original configuration, ready for another pull. One would expect that each time a bridge went through such a cycle, a phosphate group would be split from a single molecule of ATP; this reaction would provide the energy for the cycle.

To account for the rate of shortening and of energy liberation in the psoas muscle of a rabbit, each bridge would have to go through 50 to 100 cycles of operation a second. This figure is compatible with the rate at which myosin catalyzes the removal of phosphate groups from ATP. When the muscle has relaxed, we suppose that the removal of phosphate groups from ATP has stopped, and that the myosin bridges can no longer combine with the actin filaments; the muscle can then return to its uncontracted length. Indeed, there is evidence from various experiments that ATP from which phosphate has *not* been split can break the combination of actin and myosin. The reverse effect—the formation of permanent links between the actin and myosin filaments in the total absence of ATP—would explain the rigidity of muscles in *rigor mortis:* when the muscles' supply of ATP has been used up, they "seize" like a piston which has been deprived of lubrication.

The system I have described is sharply distinguished from most other suggested muscle mechanisms by one significant feature: a ratchet device in the linkage between the detailed molecular changes and the contraction of the muscle. This makes it possible for a movement at the molecular level to reverse direction without reversing the contraction. Thus during each contraction the molecular events responsible for the contraction can occur repeatedly at each active site in the muscle. As a result the muscle can do much more work during a single contraction than it could if only one event could occur at each active site.

Earlier in this article I mentioned that the tension exerted by a muscle falls off as its speed of shortening increases. This phenomenon can now be explained quite simply if we assume that the process by which a cross-bridge is attached to an active site on the actin filament occurs at a definite rate. There is only a certain period of time available for a bridge to become attached to an actin site moving past it, and the time decreases as the speed of shortening in-

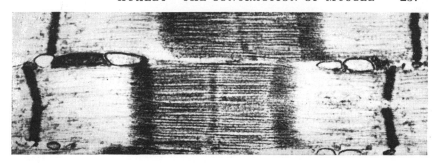

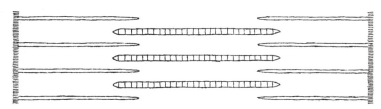

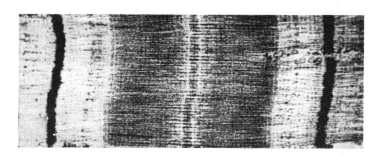

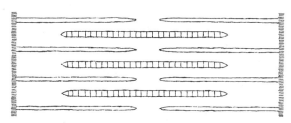

STRETCHING of muscle changes its band pattern. At top is an electron micrograph showing two myofibrils in a stretched muscle. Second is a drawing of the position of their filaments. The thick and thin filaments overlap only at their ends. Third is a micrograph of a myofibril at its resting length. At bottom is a drawing showing the position of the filaments.

creases. Thus during shortening not all the bridges are attached at a given moment; the number of ineffective bridges increases with increasing speed of shortening, and the tension consequently decreases. A. F. Huxley has worked out a detailed scheme of this general nature, and has shown that it can account for many features of contraction.

It was also indicated earlier that the total energy (heat plus work) developed by a muscle contracting over a given distance increases with the tension or load placed on the muscle. This can be explained by our mechanism if the chemical reaction which delivers the energy—say the removal of phosphate groups

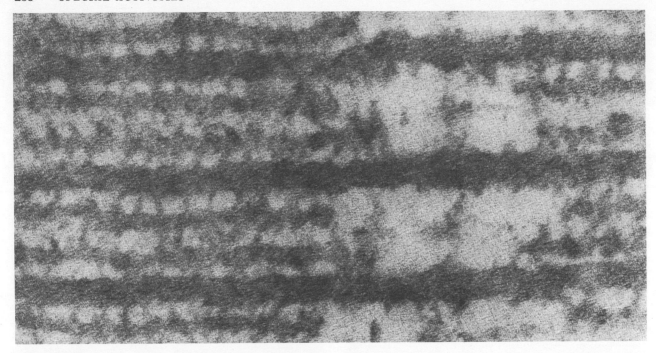

CROSS-BRIDGES between thick and thin filaments may be seen in this electron micrograph of the central region of an A-band. The micrograph enlarges the filaments 600,000 diameters. Three thick filaments are seen; between each pair are two thin filaments.

from ATP—proceeds slowly at bridges which are not attached to an actin filament, and rapidly at bridges which are attached. Since the number of bridges attached at any moment is determined by the load on the muscle, the amount of energy released in a given distance of shortening is automatically varied according to the amount of external work done. This assumption of a difference in the reaction rate at unattached bridges and at attached bridges is plausible: when myosin is placed in a solution approximating the environment of muscle, it splits ATP rather slowly; when the myosin is allowed to combine with actin, the splitting is greatly accelerated.

There are other reasons, with which I shall not burden the general reader, for believing that the sliding-filament model of muscle accords rather well with our chemical and physiological knowledge of striated muscle. The model provides a frame of reference in which we can relate to one another many different kinds of information: about muscle itself, about artificial contractile systems and about muscle proteins. The situation is promising and stimulating, and we seem to be on the right track, but we are still far from being able to describe the contraction of muscle in detailed molecular terms—perhaps farther than we think!

There remains the most fundamental question of all: Exactly how does a chemical reaction provide the motive force for the molecular movements of contraction? We have made little progress toward answering the question; indeed, the recent studies have made the problem more difficult by seeming to require that a movement of 100 angstroms

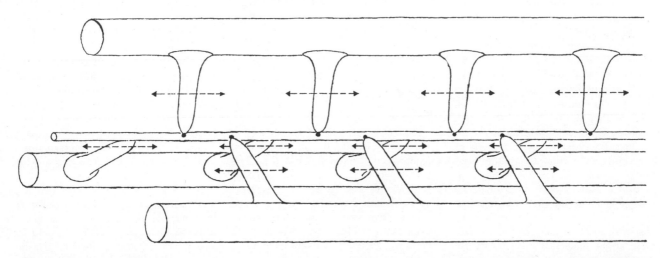

ARRANGEMENT OF CROSS-BRIDGES suggests that they enable the thick filaments to pull the thin filaments by a kind of ratchet action. In this schematic drawing one thin filament lies among three thick ones. Each bridge is a part of a thick filament, but it is able to hook onto a thin filament at an active site (dot). Presumably the bridges are able to bend back and forth (arrows). A single bridge might thus hook onto an active site, pull the thin filament a short distance, then release it and hook onto the next active site.

in part of the muscle structure be the consequence of a single chemical event. But it may be that the sliding process is effected by a more subtle mechanism than the one described here; perhaps a caterpillar-like action, in which one kind of filament crawls past the other by small repetitive changes of length, will be closer to the truth.

Two things are certain. The problem of muscular contraction will not be solved independently of other modern biological problems—those of the structure of proteins, of the action of enzymes, and of energy transfer in biological systems. And muscle itself provides as promising a system for attacking these problems as any we know.

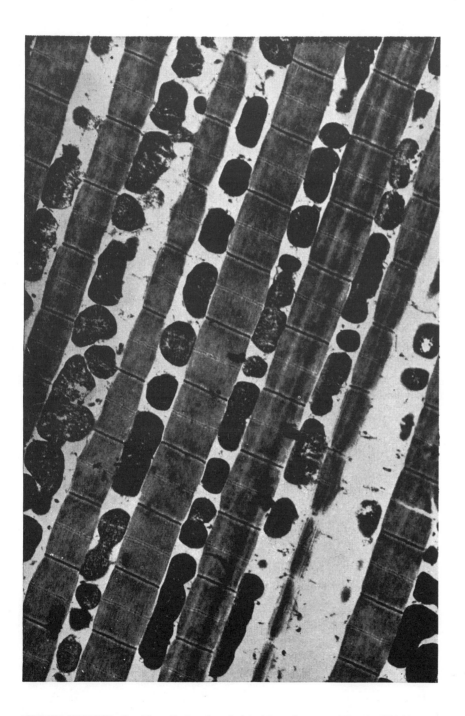

FLIGHT MUSCLE of a blow fly has broad A-bands and narrow I-bands. This is consistent with the sliding-filament hypothesis because the flight muscle of the blow fly contracts only a few per cent of its length (though it must do so several hundred times a second). The dense bodies between the myofibrils are mitochondria, particles in which foodstuff is oxidized to provide the energy for contraction.

Biographical Notes and Bibliographies

SECTION I. LEVELS OF COMPLEXITY

The Living Cell

JEAN BRACHET

The Author

JEAN BRACHET is professor of general biology at the Free University of Brussels, of which his father, the anatomist and embryologist Albert Brachet, had been the rector. As a medical student at the Free University, Brachet did research under the direction of Albert M. Dalcq, acquired an M.D. degree in 1934 and joined the faculty as an instructor in anatomy the same year. Brachet writes that he "preferred the satisfaction of doing original and independent research in biology to the responsibilities of medical practice...and decided to study the elementary forms of life, *i.e.*, cells and embryos. Such studies ultimately will lead to a better understanding of the origin of cancer and to a more logical therapeutic approach to disease." Brachet has held visiting professorships at the University of Pennsylvania, the Rockefeller Institute, the Pasteur Institute in Paris and the Indian Cancer Research Centre in Bombay.

Bibliography

THE CELL: BIOCHEMISTRY, PHYSIOLOGY, MORPHOLOGY. VOL. I: METHODS; PROBLEMS OF CELL BIOLOGY. VOL. II: CELLS AND THEIR COMPONENT PARTS. Edited by Jean Brachet and Alfred E. Mirsky. Academic Press Inc., 1959; 1961.

CELL GROWTH AND CELL FUNCTION. Torbjörn O. Caspersson. W. W. Norton & Company, Inc., 1950.
THE ULTRASTRUCTURE OF CELLS. Marcel Bessis. Sandoz Monographs, 1960.

The Smallest Living Cells

HAROLD J. MOROWITZ AND
MARK E. TOURTELLOTTE

The Authors

HAROLD J. MOROWITZ and MARK E. TOURTELLOTTE ("The Smallest Living Cells") are respectively associate professor of biophysics and research associate in biophysics at Yale University. Morowitz studied philosophy, physics and biophysics at Yale, acquiring his Ph.D. there in 1951. He spent two years at the National Bureau of Standards and another two years at the National Heart Institute before joining the faculty of Yale in 1955. Tourtellotte was graduated from Dartmouth College in 1950 and took an M.S. degree at the University of Connecticut in 1953. He remained at Connecticut as a bacteriologist in the department of animal pathology, working on the physiology and serology of the pleuropneumonia-like organisms discussed in the present article. He received his Ph.D. from Connecticut in 1959 and went to Yale on a postdoctoral fellowship from the National Institutes of Health.

Bibliography

BIOLOGY OF THE PLEUROPNEUMONIA-LIKE ORGANISIMS. *Annals of the New York Academy of Sciences*, Vol. 79, Art. 10, pages 304–758; January 15, 1960.
NUTRITION, METABOLISM, AND PATHOGENICITY OF MYCOPLASMAS. H. E. Adler and Moshé Shifrine in *Annual Review of Microbiology*, Vol. 14, pages 141–160; 1960.
THE PLEUROPNEUMONIA GROUP OF ORGANISMS: A REVIEW, TOGETHER WITH SOME NEW OBSERVATIONS. D. G. ff. Edward in *The Journal of General Microbiology*, Vol. 10, No. 1, pages 27–64; February, 1954.
PLEUROPNEUMONIA-LIKE ORGANISMS (PPLO)—MYCOPLASMATACEAE. E. Klieneberger-Nobel. Academic Press Inc., 1962.

Viruses and Genes

FRANÇOIS JACOB AND
ELIE L. WOLLMAN

The Authors

FRANÇOIS JACOB and ELIE L. WOLLMAN are respectively head of the Department of Microbial Genetics and *chef de laboratoire* in the Department of Microbial Physiology at the Pasteur Institute in Paris. Jacob's medical studies at the University of Paris were interrupted by World War II. In 1940 he escaped to England to join the Free French forces there and he later fought in both Africa and France. Jacob completed his M.D. degree after the war and went to the Pasteur Institute in 1950. In 1954 he received a D.Sc. degree from the Sorbonne. Wollman, whose parents were microbiologists at the Pasteur Institute, studied medicine and biology at the University of Paris until war intervened. During the German occupation of France, in which both of his parents were killed, Wollman finished his medical degree at the University of Lyon and served as a physician with the resistance forces and later with the French army. He became a staff member of the Pasteur Institute in 1945.

Bibliography

THE CONCEPT OF VIRUS. A. Lwoff in *The Journal of General Microbiology,* Vol. 17, No. 2, pages 239–253; October, 1957.

GENETIC CONTROL OF VIRAL FUNCTIONS. François Jacob in *The Harvey Lectures,* Series LIV, 1958–1959, pages 1–39. Academic Press Inc., 1960.

MICROBIAL GENETICS. Tenth Symposium of the Society for General Microbiology. Cambridge University Press, 1960.

PHYSIOLOGICAL ASPECTS OF BACTERIOPHAGE GENETICS. S. Brenner in *Advances in Virus Research,* Vol. 6, pages 137–158; 1959.

A SYMPOSIUM ON THE CHEMICAL BASIS OF HEREDITY. Edited by William D. McElroy. Johns Hopkins Press, 1957.

VIRUSES AS INFECTIVE GENETIC MATERIAL. S. E. Luria in *Immunity and Virus Infection,* edited by Victor A. Najjar, pages 188–195. John Wiley & Sons, Inc., 1959.

THE VIRUSES: BIOCHEMICAL, BIOLOGICAL, AND BIOPHYSICAL PROPERTIES. Edited by F. M. Burnet and W. M. Stanley. Academic Press Inc., 1956.

SECTION II. ORGANELLES

The Membrane of the Living Cell

J. DAVID ROBERTSON

The Author

J. DAVID ROBERTSON is assistant professor of neuropathology at the Harvard Medical School and associate biophysicist at McLean Hospital in Belmont, Mass. A graduate of the University of Alabama, Robertson took his M.D. at Harvard in 1945, interned at Boston City Hospital and in 1948 went to the Massachusetts Institute of Technology. After receiving a Ph.D. in biochemistry from M.I.T. in 1952 he taught for three years at the University of Kansas. From 1955 to 1960, when he went to Harvard, Robertson was research associate in anatomy at University College London.

Bibliography

THE FORMATION FROM THE SCHWANN CELL SURFACE OF MYELIN IN THE PERIPHERAL NERVES OF CHICK EMBRYOS. Betty Ben Geren in *Experimental Cell Research,* Vol. 7, pages 558–562; 1954.

HOW THINGS GET INTO CELLS. Heinz Holter in *Scientific American,* Vol. 205, No. 3, pages 167–180; September, 1961.

THE MOLECULAR STRUCTURE AND CONTACT RELATIONSHIPS OF CELL MEMBRANES. J. D. Robertson in *Progress in Biophysics and Biophysical Chemistry,* Vol. 10, pages 344–418; 1960.

TEXTBOOK OF GENERAL PHYSIOLOGY. Hugh Davson. Little, Brown & Co., 1959.

THE ULTRASTRUCTURE OF ADULT VERTEBRATE PERIPHERAL MYELINATED NERVE FIBERS IN RELATION TO MYELINOGENESIS. J. D. Robertson in *Journal of Biophysical and Biochemical Cytology,* Vol. 1, No. 4, pages 271–278; 1955.

THE ULTRASTRUCTURE OF CELL MEMBRANES AND THEIR DERIVATIVES. J. D. Robertson in *Biochemical Society Symposia,* No. 16, pages 3–43; 1959.

The Mitochondrion

DAVID E. GREEN

The Author

DAVID E. GREEN is professor of enzyme chemistry and codirector of the Institute for Enzyme Research at the University of Wisconsin. A graduate of New York University, Green acquired a Ph.D. in biochemistry from the University of Cambridge in 1934. He did research at Cambridge from 1934 until 1940, when he returned to this country to continue his work at the Harvard Medical School. In 1941 he joined the staff of the College of Physicians and Surgeons of Columbia University, where he later became head of the enzyme laboratory. He joined the Wisconsin faculty in 1948.

Bibliography

ENERGY-LINKED FUNCTIONS OF MITOCHONDRIA, edited by Britton Chance. Academic Press, 1963.

HORIZONS IN BIOCHEMISTRY: ALBERT SZENT-GYORGYI DEDICATORY VOLUME, edited by Michael Kasha and Bernard Pullman. Academic Press, 1962.

THE MITOCHONDRION AND BIOCHEMICAL MACHINES. D. E. Green and Y. Hatefi in *Science,* Vol. 133, No. 3445, pages 13–19; January, 1961.

QUINONES IN ELECTRON TRANSPORT, edited by G. E. W. Wolstenholme and Cecilia M. O'Connor. Little, Brown and Company, 1960.

STRUCTURE AND FUNCTION OF SUBCELLULAR PARTICLES. David E. Green in *Proceedings of the Fifth International Congress of Biochemistry: Vol. IX.* Pergamon Press, 1963.

SYMPOSIUM ON OXIDATIVE PHOSPHORYLATION. *Proceedings of the Federation of American Societies for Experimental Biology,* Vol. 22, No. 4, Parts I and II, 1963.

The Lysosome

CHRISTIAN DE DUVE

The Author

CHRISTIAN DE DUVE is professor and head of the department of physiological chemistry at the Catholic University of Louvain in Belgium. He is also professor at the Rockefeller Institute, where he recently founded a new laboratory. De Duve received an M.D. from Louvain in 1941 and an advanced degree in chemistry in 1946. For the next 18 months, in the laboratory of the Nobel laureate Hugo Theorell at the Nobel Institute in Stockholm, de Duve did research on the muscle protein myoglobin. He worked for six months with another Nobel prize winner, Carl F. Cori, at Washington University in St. Louis, on a pancreatic hormone. De Duve was appointed to his present post at Louvain in 1947; his association with the Rockefeller Institute began in January, 1962.

Bibliography

LYSOSOMES: A NEW GROUP OF CYTOPLASMIC PARTICLES. C. de Duve in *Subcellular Particles*. Edited by Teru Hayashi. The Ronald Press Company, 1959.

LYSOSOMES. Alex B. Novikoff in *The Cell, Vol. II: Biochemistry, Physiology and Morphology*. Edited by Jean Brachet and A. E. Mirsky. Academic Press, 1961.

Cilia

PETER SATIR

The Author

PETER SATIR is a Graduate Fellow of the Rockefeller Institute. Born and raised in New York, Satir attended Columbia University, where he received his B.A. in 1956. He then spent the summer working with the Columbia physiologist Teru Hayashi at the Marine Biological Laboratory in Woods Hole, Mass. During that summer he was invited to continue his research at the Rockefeller Institute. In 1958 Satir went to the Carlsberg Foundation Biological Institute in Copenhagen to do research with Erik Zeuthen on problems of cell growth and cell division. At present Satir is working with Keith R. Porter of the Rockefeller Institute.

Bibliography

THE SUBMICROSCOPIC MORPHOLOGY OF PROTOPLASM. Keith R. Porter in *The Harvey Lectures*, Series 51, pages 175–228; Academic Press, Inc., 1957.

SECTION III. ENERGETICS

How Cells Transform Energy

ALBERT L. LEHNINGER

The Author

ALBERT L. LEHNINGER has since 1952 been DeLamar Professor of Physiological Chemistry and director of the department of physiological chemistry at the Johns Hopkins School of Medicine. He received his B.A. from Wesleyan University in 1939 and his M.S. and Ph.D. from the University of Wisconsin respectively in 1940 and 1942. After teaching at Wisconsin until 1945, Lehninger joined the faculty of the University of Chicago. In 1951 he went to the University of Frankfurt as an exchange professor, and in 1951 and 1952 he was a Guggenheim fellow and Fulbright research professor at the University of Cambridge. In 1948 Lehninger discovered that the enzymes involved in the citric acid and respiratory cycles of energy transformation in the cell are located in the mitochondria.

Bibliography

ENERGY RECEPTION AND TRANSFER. Melvin Calvin in *Biophysical Science—a Study Program*, edited by J. L. Oncley et al., pages 147–156. John Wiley & Sons, Inc., 1959.

FREE ENERGY AND THE BIOSYNTHESIS OF PHOSPHATES. M. R. Atkinson and R. K. Morton in *Comparative Biochemistry*, edited by Marcel Florkin and Howard S. Mason, Vol. II, pages 1–95. Academic Press Inc., 1960.

RESPIRATORY-ENERGY TRANSFORMATION. Albert L. Lehninger in *Biophysical Science—a Study Program*, edited by J. L. Oncley et al., pages 136–146. John Wiley & Sons, Inc., 1959.

The Role of Light in Photosynthesis

DANIEL I. ARNON

The Author

DANIEL I. ARNON is professor of cell physiology at the University of California and biochemist at the University's Experiment Station. He did his undergraduate and graduate work at the University of California, where he specialized in plant physiology and biochemistry and acquired a Ph.D. in 1936. Except for service with the U. S. Air Force in World War II, a year at the University of Cambridge and a year as Fulbright research fellow at the Max Planck Institute for Cell Physiology in Dahlem, Arnon has spent his entire professional career at California. His early research in trace elements in plants led to the discovery of the role of molybdenum in green plants (with P. R. Stout) and of vanadium in green algae. His interest in the biochemical function of micronutrients in plant nutrition led him to investigate photosynthesis.

Bibliography

THE CHLOROPLASTS AS A FUNCTIONAL UNIT IN PHOTOSYNTHESIS. Daniel I. Arnon in *Handbuch der Pflanzenphysiologie*, Vol. 5, pages 773–829. Springer-Verlag, 1960.

CONVERSION OF LIGHT INTO CHEMICAL ENERGY IN PHOTOSYNTHESIS. Daniel I. Arnon in *Nature*, Vol. 184, pages 10–21; July 4, 1959.

The Path of Carbon in Photosynthesis

J. A. BASSHAM

The Author

J. A. BASSHAM is research chemist and lecturer in chemistry at the University of California, where he received his B.S. in 1945. Bassham did his doctoral research at the University of California under Melvin Calvin on the path of carbon in photosynthesis and received his Ph.D. in 1949. Since then he has been in Calvin's Bio-Organic Chemistry Group at the Lawrence Radiation Laboratory, except for a two-year tour of duty in the Navy and a year in H. A. Kreb's laboratory at the University of Oxford.

Bibliography

NEW ASPECTS OF PHOTOSYNTHESIS. J. A. Bassham in *Journal of Chemical Education*, Vol. 38, No. 3, pages 151–155; March, 1961.

THE NURTURE OF CREATIVE SCIENCE AND THE MEN WHO MAKE IT. THE PHOTOSYNTHESIS STORY: A CASE HISTORY. Melvin Calvin in *Journal of Chemical Education*, Vol. 35, No. 9, pages 428–432; September, 1958.

THE PATH OF CARBON IN PHOTOSYNTHESIS. Melvin Calvin and James A. Bassham. Prentice-Hall, Inc., 1957.

THE PATH OF CARBON IN PHOTOSYNTHESIS. XXI: THE CYCLIC REGENERATION OF CARBON DIOXIDE ACCEPTOR. J. A. Bassham, A. A. Benson, Lorel D. Kay, Anne Z. Harris, A. T. Wilson and M. Calvin in *Journal of the American Chemical Society*, Vol. 76, No. 7, pages 1760–1770; April 5, 1954.

PHOTOSYNTHESIS. J. A. Bassham in *Journal of Chemical Education, Vol. 36, No. 11, pages 548–554; November, 1959.

THE PHOTOSYNTHESIS OF CARBON COMPOUNDS. Melvin Calvin and James A. Bassham. W. A. Benjamin, Inc., 1962.

Biological Luminescence

WILLIAM D. MC ELROY AND
HOWARD H. SELIGER

The Authors

WILLIAM D. McELROY and HOWARD H. SELIGER are respectively director of the McCollum-Pratt Institute at Johns Hopkins University and research associate at the Institute. McElroy, also chairman of the department of biology at Johns Hopkins, acquired a Ph.D. in biochemistry from Princeton University in 1943 and from 1942 to 1945 was engaged in research on various war projects for the Office of Scientific Research and Development. Following a year of postdoctoral work with George W. Beadle at Stanford University, McElroy went to Johns Hopkins in 1945. At the McCollum-Pratt Institute, which he has directed since 1949, McElroy has been concerned primarily with the mechanism of light emission from chemical reactions, particularly those of biological origin. In addition to this work McElroy serves in an editorial capacity with several journals and as executive editor of Archives of Biochemistry and Biophysics. He is the author of some half-dozen books and was coauthor (with C. P. Swanson) of "Trace Elements" in the January 1953 issue of SCIENTIFIC AMERICAN. Seliger, whose chief research interests are energy transfer in bioluminescent process and the physics of light-producing chemical processes, was originally trained as a nuclear physicist and received his Ph.D. from the University of Maryland in 1954. Before taking his present job in 1958 he had been supervisory physicist of the Radioactivity Section of the National Bureau of Standards.

Bibliography

BIOLUMINESCENCE. E. Newton Harvey. Academic Press, Inc., 1952.

A SYMPOSIUM ON LIGHT AND LIFE. Edited by William D. McElroy and Bentley Glass. Johns Hopkins Press, 1961.

SECTION IV. SYNTHESIS

How Cells Make Molecules

VINCENT G. ALLFREY AND
ALFRED E. MIRSKY

The Authors

VINCENT G. ALLFREY and ALFRED E. MIRSKY are respectively associate professor of molecular biology and Member of the Rockefeller Institute. Allfrey took a B.S. in chemistry at the College of the City of New York while working as a laboratory technician with Mirsky at the Rockefeller Institute. Following service with the Army Medical Corps in 1944 and 1945, Allfrey acquired an M.S. at Columbia University in 1948 and a Ph.D. there in 1949. Since joining the Rockefeller Institute in 1949 he has worked on the problems of isolating cell nuclei and on the genetic control of protein synthesis. Mirsky, who is editor of the Journal of General Physiology, received a B.A. from Harvard University in 1922 and a Ph.D. in physiology from the University of Cambridge in 1926. He went to the Rockefeller Institute in 1927.

Bibliography

ENZYMATIC SYNTHESIS OF DEOXYRIBONUCLEIC ACID. Arthur Kornberg in The Harvey Lectures, 1957–1958, Series 53, pages 83–112; 1959.

HISTORICAL AND CURRENT ASPECTS OF THE PROBLEM OF PROTEIN SYNTHESIS. Paul C. Zamecnik in The Harvey Lectures, 1958–1959, Series 54, pages 256-281; 1960.

THE INTERPHASE NUCLEUS. Alfred E. Mirsky and Syozo Osawa in The Cell, Vol. II, pages 677–770. Academic Press Inc., 1961.

THE ISOLATION OF SUBCELLULAR COMPONENTS. Vincent C. Allfrey in The Cell, Vol. I, pages 193–290. Academic Press Inc., 1959.

The Genetic Code: II

MARSHALL W. NIRENBERG

The Author

MARSHALL W. NIRENBERG is head of the Section of Biochemical Genetics at the National Heart Institute, one of the nine National Institutes of Health. Nirenberg took a B.S. at the University of Florida in 1948. After receiving an M.S. in biology from the University of Florida in 1952, Nirenberg went to the department of biological chemistry at the University of Michigan, where he acquired a Ph.D. in 1957. A two-year postdoctoral fellowship from the American Cancer Society brought him to the National Institute of Arthritis and Metabolic Diseases later the same year, where he remained until he took his present post in June, 1962.

Bibliography

THE DEPENDENCE OF CELL-FREE PROTEIN SYNTHESIS IN E. COLI UPON NATURALLY OCCURRING OR SYNTHETIC POLYRIBONUCLEOTIDES. Marshall W. Nirenberg and J. Heinrich Matthaei in Proceedings of the National Academy of Sciences of the U.S.A., Vol. 47, No. 10, pages 1588–1602; October, 1961.

A PHYSICAL BASIS FOR DEGENERACY IN THE AMINO ACID CODE. Bernard Weisblum, Seymour Benzer and Robert W. Holley in Proceedings of the National Academy of Sciences of the U.S.A., Vol. 48, No. 8, pages 1449-1453; August, 1962.

POLYRIBONUCLEOTIDE - DIRECTED PROTEIN SYNTHESIS USING AN E. COLI CELL-FREE SYSTEM. M. S. Bretscher and M. Grunberg-Manago in Nature, Vol. 195, No. 4838, pages 283–284; July 21, 1962.

QUALITATIVE SURVEY OF RNA CODEWORDS. Oliver W. Jones, Jr., and Marshall W. Nirenberg in Proceedings of the National Academy of Sciences of the U.S.A., Vol. 48, No. 12, pages 2115–2123; December, 1962.

SYNTHETIC POLYNUCLEOTIDES AND THE AMINO ACID CODE, IV. J. F. Speyer, P. Lengyel, C. Basilio and S. Ochoa in Proceedings of the National Academy of Sciences of the U.S.A., Vol. 48, No. 3, pages 441–448; March, 1962.

Polyribosomes

ALEXANDER RICH

The Author

ALEXANDER RICH is professor of biophysics at the Massachusetts Institute of Technology. A graduate of Harvard College, Rich obtained an M.D. from the Harvard Medical School in 1949. He did research in chemistry at the California Institute of Technology from 1949 to 1954, when he became chief of the section on physical chemistry at the National Institute of Health in Bethesda, Md. Rich was visiting scientist at the Cavendish Laboratory in Cambridge, England, during 1955. He joined the M.I.T faculty in 1958.

Bibliography

ELECTRON MICROSCOPE STUDIES OF RIBOSOMAL CLUSTERS SYNTHESIZING HEMOGLOBIN. Jonathan R. Warner, Alexander Rich and Cecil E. Hall in *Science*, Vol. 138, No. 3548, pages 1399-1403; December 28, 1962.

FUNCTION OF AGGREGATED RETICULOCYTE RIBOSOMES IN PROTEIN SYNTHESIS. Alfred Gierer in *Journal of Molecular Biology*, Vol. 6, No. 2, pages 148-157; February, 1963.

MECHANISM OF POLYRIBOSOME ACTION DURING PROTEIN SYNTHESIS. Howard M. Goodman and Alexander Rich in *Nature*, Vol. 199, No. 4891, pages 318-322; July 27, 1963.

A MULTIPLE RIBOSOMAL STRUCTURE IN PROTEIN SYNTHESIS. Jonathan R. Warner, Paul M. Knopf and Alexander Rich in *Proceedings of the National Academy of Sciences of the U.S.A.*, Vol. 49, No. 1, pages 122-129; January, 1963.

The Three-Dimensional Structure of a Protein Molecule

JOHN C. KENDREW

The Author

JOHN C. KENDREW is deputy director of the Medical Research Council Unit for Molecular Biology in the Cavendish Laboratory at the University of Cambridge. Kendrew took his B.A., M.A. and Ph.D. degrees at Cambridge, receiving the last in 1949. A Fellow of the Royal Society and editor-in-chief of the *Journal of Molecular Biology*, Kendrew has been engaged in research on the structure of proteins since 1946.

Bibliography

BIOPHYSICAL CHEMISTRY. Edited by John T. Edsall and Jeffries Wyman. Academic Press Inc., 1958.

THE CRYSTALLINE STATE. Edited by Sir W. H. Bragg and W. L. Bragg. G. Bell and Sons Ltd., 1933.

THE MOLECULAR BASIS OF EVOLUTION. Christian B. Anfinsen. John Wiley & Sons, Inc., 1959.

X-RAY ANALYSIS AND PROTEIN STRUCTURE. F. H. C. Crick and J. C. Kendrew in *Advances in Protein Chemistry*, Vol. 12, pages 133-214; 1957.

Collagen

JEROME GROSS

The Author

JEROME GROSS is assistant professor of medicine at the Harvard Medical School and associate biologist at Massachusetts General Hospital. Gross studied biophysics at the Massachusetts Institute of Technology, receiving a B.S. in 1939. His interest in approaching problems of medicine from the viewpoint of physical biology took him to the New York University College of Medicine; there his original interests were further stimulated by work in the rheumatic fever clinic, by a lecture he heard in 1941 on the recently developed electron microscope and by a visit to M.I.T., where F. O. Schmitt was using the electron microscope in his research on collagen. He received his M.D. degree in 1943, served for three years in the Army Medical Corps and then returned to M.I.T. as Schmitt's assistant to do research on the molecular biology of connective tissue. Gross joined the faculty of the Harvard Medical School in 1948 and the staff of Massachusetts General Hospital in 1951.

SECTION V. DIVISION AND DIFFERENTIATION

How Cells Divide

DANIEL MAZIA

The Author

DANIEL MAZIA is professor of zoology at the University of California. He received his A. B. in 1933 and his Ph.D. in zoology in 1937 from the University of Pennsylvania, where he did graduate work under the direction of L. V. Heilbrunn. After a year as a National Research Council fellow at Princeton University, working under E. Newton Harvey, Mazia went to the University of Missouri. There he became interested in the role of nucleic acids in chromosome structure and later in the functions of cell nuclei. He joined the faculty of the University of California in 1950. Mazia began doing research on the process of mitosis in 1952, when a former classmate of his at Pennsylvania, Katsuma Dan of Tokyo Metropolitan University, visited his laboratory. The result of their collaboration was the first isolation of the mitotic apparatus from a cell.

Bibliography

MITOSIS AND THE PHYSIOLOGY OF CELL DIVISION. Daniel Mazia in *The Cell*, Vol. III, pages 77-412. Academic Press Inc., 1961.

SECOND CONFERENCE ON THE MECHANISMS OF CELL DIVISION. *Annals of the New York Academy of Sciences*, Vol. 90, Article 2, pages 345-613; October 7, 1960.

How Cells Associate

A. A. MOSCONA

The Author

A. A. MOSCONA is professor of zoology at the University of Chicago. He received a Ph.D. in zoology from the Hebrew University in Jerusalem in 1950, spent two years at the Strangeways Research Laboratory of the University of Cambridge and from 1953 to 1955 was associate professor of physiology at the University of Jerusalem. Before joining the faculty at Chicago in 1958, Moscona had held a two-year research fellowship at the Rockefeller Institute.

Bibliography

GUIDING PRINCIPLES IN CELL LOCOMOTION AND CELL AGGREGATION. Paul Weiss in *Experimental Cell Research*, Supplement 8, pages 260–281; 1961.

IMMUNOLOGICAL RECOGNITION OF SELF. F. M. Burnet in *Science*, Vol. 133, No. 3449, pages 307–311; February 3, 1961.

ROTATION-MEDIATED HISTOGENIC AGGREGATION OF DISSOCIATED CELLS. A. Moscona in *Experimental Cell Research*, Vol. 22, pages 455–475; 1961.

How Cells Specialize

MICHAIL FISCHBERG AND ANTONIE W. BLACKLER

The Authors

MICHAIL FISCHBERG and ANTONIE W. BLACKLER worked together in the Embryology Laboratory at the University of Oxford and continued their joint research at the University of Geneva, where Fischberg was appointed to the chair of zoology and directorship of the department of zoology, and Blackler to a visiting professorship. Fischberg was born in St. Petersburg in Russia and was educated in Switzerland, receiving his Ph.D. from the University of Zürich in 1946. After a year of postdoctoral research at the University of Basel, he went to the Institute of Animal Genetics in Edinburgh on a two-year research fellowship from the Swiss Academy of Medicine. Fischberg became Jenkinson Memorial Lecturer in Embryology and head of the Embryology Laboratory at Oxford in 1951. In 1961 he was a visiting professor at the Rockefeller Institute. Blackler, who joined Fischberg's laboratory in 1959, took his B.Sc. at University College London in 1953 and his Ph.D. there in 1956.

Bibliography

GAMETOGENESIS. Jean Brachet in *The Biochemistry of Development*, pages 1–44. Pergamon Press, 1960.

CHANGES IN THE NUCLEI OF DIFFERENTIATING ENDODERM CELLS AS REVEALED BY NUCLEAR TRANSPLANTATION. Robert Briggs and Thomas J. King in *Journal of Morphology*, Vol. 100, No. 2, pages 269–311; March, 1957.

NUCLEAR TRANSFER IN AMPHIBIA AND THE PROBLEM OF THE POTENTIALITIES OF THE NUCLEI OF DIFFERENTIATING TISSUES. M. Fischberg, J. B. Gurdon and T. R. Elsdale in *Experimental Cell Research*, Supplement 6, pages 161–178; 1959.

Chromosome Puffs

WOLFGANG BEERMANN AND ULRICH CLEVER

The Authors

WOLFGANG BEERMANN and ULRICH CLEVER work at the Max Planck Institute for Biology in Tübingen. They are also members of the faculty of the University of Tübingen. Beermann is a director of the Max Planck Institute in Tübingen and professor of zoology at the university. A native of Hanover, he received his doctorate from the University of Göttingen in 1952. He did research at the Max Planck Institute for Marine Biology in Wilhemshaven from 1952 to 1954, when he was appointed assistant professor at the Zoological Institute of the University of Marburg. He took up his present post in 1958. Clever is a research associate at the Max Planck Institute in Tübingen and lecturer in zoology and genetics at the university. He received his doctorate from Göttingen in 1957 and did research for a year at the Federal Research Institute for Viticulture before going to Tübingen in 1958.

Bibliography

CHROMOSOMES AND CYTODIFFERENTIATION. Joseph G. Gall in *Cytodifferentation and Macromolecular Synthesis*, edited by Michael Locke. Academic Press, 1963.

NUCLEIC ACIDS AND CELL MORPHOLOGY IN DIPTERAN SALIVARY GLANDS. Hewson Swift in *The Molecular Control of Cellular Activity*, edited by John M. Allen. McGraw-Hill Book Company, 1962.

RIESENCHROMOSOMEN. Wolfgang Beermann in *Protoplasmatologia*, Vol. VI/D. Springer-Verlag, 1962.

UNTERSUCHUNGEN AN RIESENCHROMOSOMEN ÜBER DIE WIRKUNGSWEISE DER GENE. Ulrich Clever in *Materia Medica Nordmark*, Vol. 15, No. 10, pages 438–452; July, 1962.

SECTION VI. SPECIAL ACTIVITIES

How Cells Communicate

BERNHARD KATZ

The Author

BERNHARD KATZ is professor and head of the biophysics department of University College London. Born in Leipzig, Germany, in 1911, Katz acquired an M.D. degree at the University of Leipzig in 1934. From 1935 to 1939 he worked under the direction of A. V. Hill at University College, where he received a Ph.D. in 1938. Katz served with the Royal Air Force in the Pacific from 1942 to 1945 and then returned to University College. A Fellow of the Royal Society since 1952, Katz was Herter Lecturer at Johns Hopkins University in 1958 and Dunham Lecturer at Harvard University this past year.

Bibliography

BIOPHYSICAL ASPECTS OF NEURO-MUSCULAR TRANSMISSION. J. del Castillo and B. Katz in *Progress in Biophysics and Biophysical Chemistry*, Vol. 6, pages 121–170; 1956.

IONIC MOVEMENT AND ELECTRICAL ACTIVITY IN GIANT NERVE FIBRES. A. L. Hodgkin in *Proceedings of the Royal Society*, Series B, Vol. 148, No. 930, pages 1–37; January 1, 1958.

MICROPHYSIOLOGY OF THE NEUROMUSCULAR JUNCTION, A PHYSIOLOGICAL "QUANTUM OF ACTION" AT THE MYONEURAL JUNCTION. Bernhard Katz in *Bulletin of the Johns Hopkins Hospital*, Vol. 102, No. 6, pages 275–312; June, 1958.

THE PHYSIOLOGY OF NERVE CELLS. John Carew Eccles. The Johns Hopkins Press, 1957.

How Cells Receive Stimuli

WILLIAM H. MILLER,
FLOYD RATLIFF, AND
H. K. HARTLINE

The Authors

WILLIAM H. MILLER, FLOYD RATLIFF and H. K. HARTLINE have collaborated in studies of visual receptors and neural interaction at the Rockefeller Institute since 1955. Miller, assistant professor of biophysics, is a graduate of Haverford College and received his M.D. degree from Johns Hopkins University in 1954. He first began doing research on the eyes of invertebrates in Hartline's laboratory at Johns Hopkins. After an internship at Baltimore City Hospital, Miller joined the Rockefeller Institute in 1955. Ratliff, associate professor of biophysics, took a B.A. at Colorado College in 1947 and a Ph.D. in psychology at Brown University in 1950. He went to Johns Hopkins the following year on a National Research Council fellowship to study retinal interaction with Hartline. Before going to the Rockefeller Institute in 1954, Ratliff was assistant professor of the newly established Jenkins Department. Hartline joined the Rockefeller Institute as Member and professor in 1953. A graduate of Lafayette College, he took his M.D. degree at Johns Hopkins in 1927, did graduate work in physics at the same institution and from 1931 to 1949 was a staff member of the Johnson Research Foundation at the University of Pennsylvania. He went to Johns Hopkins in 1949 as professor and chairman of the newly established Jenkins Department of Biophysics.

Bibliography

INITIATION OF IMPULSES AT RECEPTORS. J. A. B. Gray in *Handbook of Physiology*, Vol. I, Section I: *Neurophysiology*, pages 123–145. American Physiological Society, 1959.

THE NEURAL MECHANISMS OF VISION. H. K. Hartline in *The Harvey Lectures, 1941–1942*. Series 37, pages 39–68; 1942.

RECEPTORS AND SENSORY PERCEPTION. R. Granit. Yale University Press, 1955.

SENSORY COMMUNICATION. Edited by Walter A. Rosenblith. John Wiley & Sons, Inc., 1961.

How Cells Move

TERU HAYASHI

The Author

TERU HAYASHI is professor of zoology at Columbia University. Growing up in Atlantic City, N.J., during the depression, he "managed to get through the local high school and subsequently Ursinus College between odd jobs as a baker's apprentice, auctioneer's helper, aquaplane performer in water carnivals and announcer for bingo and other games of chance." His professor of physics at Ursinus, John Mauchly, a pioneer in the development of electronic computers, got Hayashi interested in an academic career. Switching from physics to biology, he studied under L. V. Heilbrunn at the University of Pennsylvania and then went to the University of Missouri, where, as a student of Daniel Mazia's, he took a Ph.D. in cell physiology in 1943. He joined the Columbia faculty later that year.

Bibliography

AMEBOID MOVEMENT. Robert D. Allen in *The Cell*, Vol. II, pages 135–216. Academic Press Inc., 1961.

CHEMISTRY OF MUSCULAR CONTRACTION. A. Szent-Györgyi. Academic Press Inc., 1951.

CILIA AND FLAGELLA. Don W. Fawcett in *The Cell*, Vol. II, pages 217–297. Academic Press Inc., 1961.

CILIARY MOVEMENT. James Gray. Cambridge University Press, 1928.

THE STRUCTURE AND FUNCTION OF MUSCLE. Vol. I: STRUCTURE. Vol. II: BIOCHEMISTRY AND PHYSIOLOGY. Edited by G. H. Bourne. Academic Press Inc., 1960.

The Contraction of Muscle

H. E. HUXLEY

The Author

H. E. HUXLEY comes from Birkenhead, Cheshire, and studied theoretical physics at the University of Cambridge, where his undergraduate career was interrupted by five years of radar research for the Royal Air Force (his work in this line was rewarded with a membership in the Order of the British Empire in the New Year's Honours List of 1947–48). Huxley returned to Cambridge to study for the Physics Tripos examinations. "I had originally intended to become a nuclear physicist," he says. "But when I finished off my undergraduate degree, it seemed a more attractive proposition to think of applying the ideas and methods of the physicist to the biological field. Whilst working in the Medical Research Council Unit in Cambridge on crystalline proteins, I was struck by the fact that although the hydrated crystals gave excellent X-ray diffraction patterns, the dry crystals gave very poor ones. I wondered whether the same thing might not be true of muscle, previously examined only in a dried state. Apparatus was assembled to get a low-angle X-ray diffraction pattern from wet, living muscle. I found immediately that this material gave an excellent pattern." As a Commonwealth Fellow at the Massachusetts Institute of Technology, Huxley continued this research with the electron microscope, and later, after returning to Cambridge, with phase-contrast and interference microscopy. In 1956 he joined the staff of University College London. He is not related to Thomas, Aldous or Julian Huxley.

Bibliography

CHEMISTRY OF MUSCULAR CONTRACTION. Second Edition, Revised and Enlarged. A. Szent-Györgyi. Academic Press, Inc., 1951.

THE DOUBLE ARRAY OF FILAMENTS IN CROSS-STRIATED MUSCLE. H. E. Huxley in *The Journal of Biophysical and Biochemical Cytology*, Vol. 3, No. 5, pages 631–646; September 25, 1957.

FACTS AND THEORIES ABOUT MUSCLE. D. R. Wilkie in *Progress in Biophysics and Biophysical Chemistry*, Vol. 4, pages 288–322; 1954.

MUSCLE STRUCTURE AND THEORIES OF CONTRACTION. H. E. Huxley in *Progress in Biophysics and Biophysical Chemistry*, Vol. 7, pages 255–312; 1957.

THE TRANSFERENCE OF THE MUSCLE ENERGY IN THE CONTRACTION CYCLE. H. H. Weber and Hildegard Portzehl in *Progress in Biophysics and Biophysical Chemistry*, Vol. 4, pages 60–107; 1954.